Nontechnical Guide to

Petroleum Geology, Exploration, Drilling, and Production

2nd Edition

Nontechnical Guide to
Petroleum Geology, Exploration, Drilling, and Production

2nd Edition

Norman J. Hyne, Ph.D.

Copyright ©2001 by
PennWell Corporation
1421 S. Sheridan Road
Tulsa, Oklahoma 74112-6600 USA

800-752-9764
+1-918-831-9421
sales@pennwell.com
www.pennwell.com
www.pennwellbooks.com

Director: Mary McGee
Production / Operations Manager: Traci Huntsman
Cover Designer / Book Designer: Joey Zielazinski

Library of Congress Cataloging-in-Publication Data

Hyne, Norman J., Ph. D.
Nontechnical Guide to Petroleum Geology, Exploration, Drilling,
and Production, 2nd edition

Includes index
ISBN 0-87814-823-X
ISBN13 918-0-87814-823-3

Printed in the United States of America.

12 13 14   13 12 11

TABLE OF CONTENTS

LIST OF FIGURES

LIST OF PLATES

preface

This book contains an enormous amount of useful information on the upstream petroleum industry. It is designed for easy reading, and the information is readily accessible. The introductory chapter should be read first. It is an excellent overview that shows how everything in petroleum geology, exploration, drilling, and production is interrelated.

Each subject is has its own chapter that is well illustrated with figures and plates. The rock and minerals and seismic examples are in color. Industry terms are defined in the text and shown in italics. All measurements are in both English and metric units. A useful index and extensive glossary are located at the back of the book, as well as an interesting list of petroleum records.

Cover photos courtesy American Petroleum Institute, Parker Drilling, and Norman Hyne. Author photo courtesy Marshall Heim.

introduction

Both crude oil and natural gas are mixtures of molecules formed by carbon and hydrogen atoms. There are many different types of crude oils and natural gases, some more valuable than others. Heavy crude oils are very thick and viscous and are difficult or impossible to produce, whereas light crude oils are very fluid and relatively easy to produce. Less valuable are sour crude oils that contain sulfur and sour natural gasses that contain hydrogen sulfide. Some natural gases burn with more heat than others, contain natural gas liquids and gasoline, and are more valuable.

In order to have a commercial deposit of gas or oil, three geological conditions must have been met. First, there must be a source rock in the subsurface of that area that generated the gas or oil at some time in the geological past. Second, there must be a separate, subsurface reservoir rock to hold the gas or oil. Third, there must be a trap on the reservoir rock to concentrate the gas or oil into commercial quantities.

The uppermost crust of the earth in oil- and gas-producing areas is composed of sedimentary rock layers. Sedimentary rocks are the source and reservoir rocks for gas and oil. These rocks are called sedimentary rocks because they are composed of sediments. Sediments are 1) particles such as sand grains that were formed by the breakdown of pre-existing rocks and transported, 2) seashells, or 3) salt that precipitated from of water. The sedimentary rocks that make up the earth's crust are millions and sometimes billions of years old. During the vast expanse of geological time, sea level has not been constant. Many times in the past, the seas have risen to cover the land and then fallen to expose the land. During these times, sediments were deposited. These sediments are relatively simple materials such as sands deposited along beaches, mud on the sea bottom, and beds of seashells. These ancient sediments, piled layer upon layer, form the sedimentary rocks that are drilled to find and produce oil and gas.

The source of gas and oil is the organic matter that is buried and preserved in the ancient sedimentary rocks. These rocks contain not only inorganic particles such as sand grains and mud, but also dead plant and animal material. The most common organic-rich sedimentary rock (the source rock for most of the gas and oil) is black shale. It was deposited as organic-rich mud on an ancient ocean bottom. In the subsurface, temperature is the most important factor in turning organic matter into oil. As the source rock is covered with more sediments and buried deeper in the earth, it becomes hotter and hotter. The minimum temperature for the formation of oil, about 150°F (65°C), occurs at a depth of about 7000 ft (2130 m) below the surface (Fig. I–1). Oil is generated from there and down to about 300°F (150°C) at about 18,000 ft (5500 m). The reactions that change organic matter into oil are complex and take a long time. If the source rock is buried deeper where the temperatures are above 300°F (150°C), the remaining organic matter will generate natural gas.

Gas and oil are relatively light in density compared to water that also occurs in the subsurface sedimentary rocks. After oil and gas have been generated, they rise due to buoyancy through fractures in the subsurface rocks. The rising gas and oil can intersect a layer of reservoir rock. A reservoir rock is a sedimentary rock that contains billions of tiny spaces called pores. A common sedimentary rock is sandstone composed of sand grains similar to the sand grains on a beach or in a river channel. Sand grains are like spheres, and there is no way the grains will fit together perfectly. There are pore spaces between the sand grains on a beach and in a sandstone rock. Limestone, another common sedimentary rock, is deposited as shell beds or reefs, and there are pores between the shells and corals. The gas and oil flow into the pores of the reservoir rock layer.

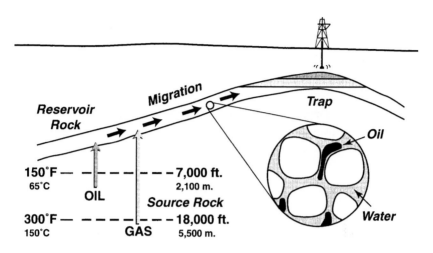

Fig. I-1 *Generation and migration of gas and oil*

Any fluid (water, gas, or oil), either on the surface or in the subsurface, will always flow along the path of least resistance, the easiest route. In the subsurface, the path of least resistance is along a reservoir rock layer. This is because most of the pore spaces interconnect, and the fluid can flow from pore to pore to pore up the angle of the rock layer toward the surface. The ease in which the fluid can flow through the rock is called permeability, and the movement of the gas and oil up the angle of the reservoir rock toward the surface is called migration. Because of migration, the gas and oil can end up a considerable distance, both vertically and horizontally, from where it was originally formed. (Fig. I-1)

As the gas and oil migrates up along the reservoir rock, it can encounter a trap. A trap is a high point in the reservoir rock where the gas or oil is stopped and concentrated. Because the pores in the reservoir rock are filled with water, the gas and oil will flow to the highest part of the reservoir rock. One type of trap is a natural arch in the reservoir rock (Fig. I–2) called a dome or anticline.

In the trap, the fluids separate according to their density. The gas is the lightest and goes to the top of the trap to form the free gas cap.

The oil goes to the middle to form the oil reservoir. The salt water, the heaviest, goes to the bottom.

To complete the trap, a caprock must overlie the reservoir rock. The caprock is a seal and doesn't allow fluids to flow through it. Without a caprock, the oil and gas would leak up to the surface of the ground. Two common sedimentary rocks that can be seals are shale and salt.

How are subsurface deposits of gas and oil located? During the early days of drilling, it was thought that there were large, flowing underground rivers and subsurface pools of oil. Early drillers, however, had some success because many subsurface traps are leaky. There are small fractures in the caprock, and some of the oil and gas leaks up and seeps onto the surface. The early drillers located their wells on the seeps.

By the early 1900s, the principles of subsurface gas and oil deposits were becoming better known. Oil companies realized that by mapping how the sedimentary rock layers crop out on the surface of the ground, the rock layers could be projected into the subsurface, and traps could be located (Fig. I–3). Geologists were hired to map rock outcrops.

Later, seismic method was developed to detect hidden traps in the sub-surface. Seismic exploration uses a source and detector (Fig. I–4). The source, such as dynamite, is located on or near the surface and gives off an impulse of

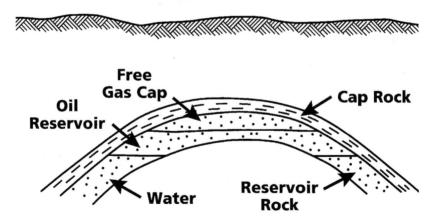

Fig. I–2 Petroleum trap

sound energy into the subsurface. The sound energy bounces off sedimentary rock layers and returns to the surface to be recorded by the detector. Sound echoes are used to make an image of the subsurface rock layers.

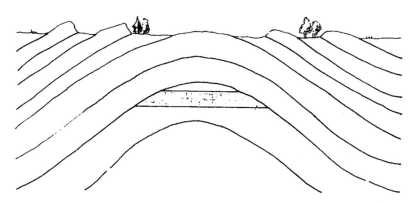

Fig. I–3 *Rock outcrops*

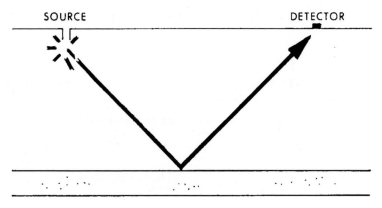

Fig. I–4 *The seismic method*

The only way to know for sure if a trap contains commercial amounts of gas and oil is to drill a well. A well drilled to find a new gas or oil field is called a wildcat well. Most wildcat wells are dry holes with no commercial amounts of gas or oil. The well is drilled using a rotary drilling rig (Fig. I–5). There can be thousands of feet of drillpipe with a bit on the end, called the drillstring, suspended in the well. By rotating the drillstring from the surface, the bit on the bottom is turned and cuts the hole.

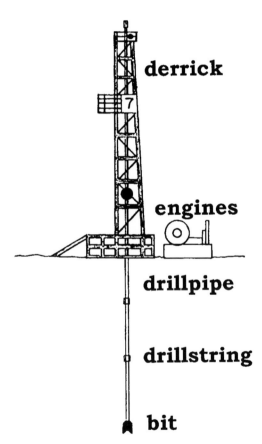

Fig. I–5 *Rotary drilling rig*

As the well is drilled deeper, more drillpipe is added. The power is supplied by diesel engines. A steel tower above the well, the derrick, is used to raise and lower equipment. The well can be drilled either almost straight down as a straight hole or out at an angle as a deviated well.

An important system on the rig is the circulating mud system. The drilling mud is pumped down the inside of the drillpipe where it jets out from nozzles in the bit and returns up the outside of the drillpipe to the surface (Fig. I–6). The drilling mud removes the rock chips made by the bit, called well cuttings, from the bottom of the hole. This prevents them from clogging up the bottom of the well. The well is always kept filled to the top with heavy drilling mud as it is being drilled. This prevents any fluids such as water, gas, and oil from flowing out of the subsurface rocks and into the well. If gas and oil flowed up onto the floor of the drilling rig, it could catch fire. Even if only water flowed out of the surrounding rock into the well, the sides of the well could cave in and the well could be lost. The drilling mud keeps the fluids back in the surrounding rocks. Offshore wells are drilled the same as on land. For offshore wildcat wells, the rig is mounted on a barge, floating platform, or ship that can be moved. Once an offshore field is located, a production platform is then installed to drill the rest of the wells and produce the gas and oil.

Because the drilling mud keeps gas and oil back in the rocks, a sub-surface deposit of gas or oil can be drilled without any indication of the gas or oil. To evaluate the well, a service company runs a wireline well log. A logging truck is driven out to the well. A long cylinder containing instruments called a sonde is unloaded from the truck and lowered down the well on a wireline (Fig. I–7). As the sonde is brought back up the well, the instruments remotely sense the electrical, sonic, and radioactive properties of the surrounding rocks and their fluids. These measurements are recorded on a long strip of paper called a well log (Fig. I–8). It is used to determine the composition of each rock layer, whether the rock layer has pores, and what fluid (water, gas, or oil) is in the pores.

Depending on the test results, the well can be plugged and abandoned as a dry hole or completed as a producer. Setting pipe is synonymous with completing a well. To set pipe, a long length of large diameter steel pipe (casing) is lowered down the hole. Wet cement is then pumped between the

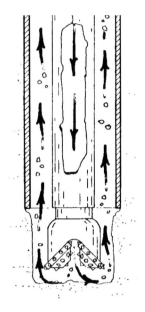

Fig. I–6 *Well-cutting removal by circulating drilling mud*

casing and the well walls and allowed to set (Fig. I–9). This stabilizes the hole. In most wells, the casing is done in stages called a *casing program* during which the well is drilled, cased, drilled deeper, cased again, drilled deeper, and cased again.

In order for the gas or oil to flow into the well, the casing is shot with explosives to form holes called perforations (Fig. I–10). A long length of narrow diameter steel pipe (tubing) is then suspended down the center of the well. The produced fluids (water, gas, and oil) are brought up the tubing string to the surface to prevent them from touching and corroding the casing that is harder to repair. The tubing is relatively easy to repair during a workover.

In a gas well, gas flows to the surface by itself. There are some oil wells, early in the development of the oilfield, in which the oil has enough pressure to flow up to the surface. Gas wells and flowing oil wells are completed with a series of fittings and valves called a Christmas tree on the surface to control the flow (Fig. I–11).

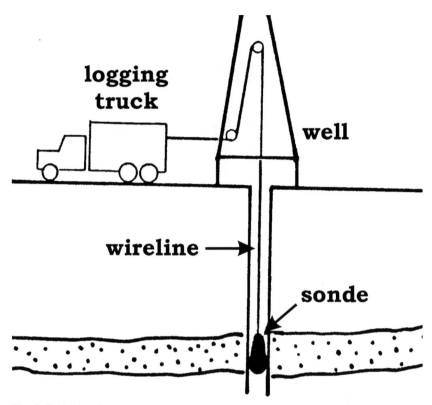

Fig. I–7 *Well logging*

Most oil wells, however, do not have enough pressure for the oil to flow to the surface and artificial lift must be used. A common artificial lift system is a sucker-rod pump (Fig. I–11). A downhole pump on the bottom of the tubing string is driven by a beam-pumping unit on the surface. The pump lifts the oil up the tubing to the surface.

On the surface, gas is prepared for delivery to a pipeline by gas-conditioning equipment that removes impurities such as water vapor and corrosive gasses. Valuable natural gas liquids are removed from the gas in a

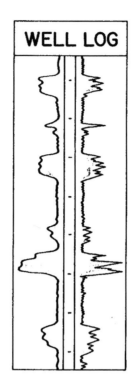

Fig. I–8 *Wireline well log*

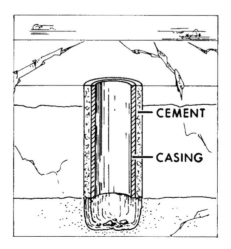

Fig. I–9 *Casing in a well*

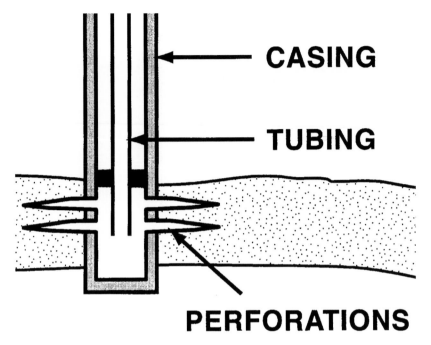

Fig. I–10 *Perforations and tubing in a well*

natural gas processing plant. For oil, a long, steel tank, called a separator, is used to separate natural gas and salt water from the oil. The oil is then stored in steel stock tanks.

Production from wells can be increased by acid and frac jobs. Acid is pumped down a well to dissolve some of the reservoir rock adjacent to the wellbore during an acid job. During a frac job, the reservoir rock is hydrauli-cally fractured with a liquid pumped under high pressure down the well. Periodically, production from the well must be interrupted for repairs or remedial work during a workover.

As fluids are produced from the subsurface reservoir, the pressure on the remaining fluids drops. The production of oil and gas from a well or a field

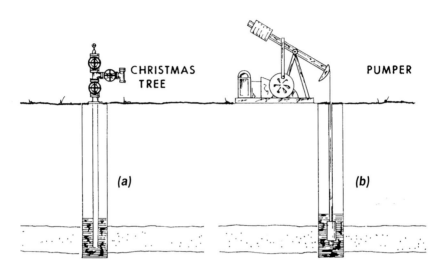

Fig. I–11 *Gas (a) and oil well (b) completions*

decreases with time on a decline curve. The shape of the decline curve and the ultimate amount of oil or gas produced depend on the reservoir drive, the natural energy that forces the oil or gas through the subsurface reservoir and into the well. Ultimate recovery of gas from a gas reservoir is often about 80% of the gas in the reservoir. Oil reservoirs, however, are far more variable and less efficient. They range from 5% to 80% recovery but average only 30% of the oil in the reservoir. This leaves 70% of the oil remaining in the pressure-depleted reservoir.

After the natural reservoir drive has been depleted in an oilfield, water-flood, and enhanced oil recovery can be attempted to produce some of the remaining oil. During a waterflood, water is pumped under pressure down injection wells into the depleted reservoir to force some of the remaining oil through the reservoir toward producing wells (Fig. I–12). Enhanced oil recovery involves pumping fluids that are not natural to the reservoir, such as carbon dioxide or steam, down injections wells to obtain more production.

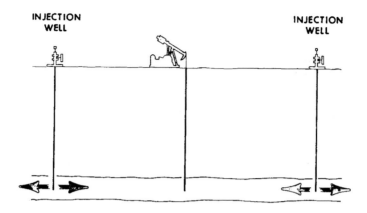

Fig. I–12 Waterflood

After the well has been depleted, it is plugged and abandoned. Cement must be poured down the well to seal the depleted reservoir and to protect any subsurface fresh water reservoirs. A steel plate is then welded to the top of the well.

one

THE NATURE OF OIL AND GAS

Petroleum

The word *petroleum* comes from the Greeks. *Petro* means rock, and *oleum* means oil. In it strictest sense, petroleum includes only crude oil. By usage, however, petroleum includes both crude oil and natural gas.

Chemical Composition

The chemical composition by weight of typical crude oil and natural gas is shown in Table 1–1. The two most important elements in both crude oil

Table 1–1
The chemical composition of typical crude oil and natural gas

	crude oil	natural gas
Carbon	84–87%	65–80%
Hydrogen	11–14%	1–25%
Sulfur	0.06–2%	0–0.2%
Nitrogen	0.1–2%	1–15%
Oxygen	0.1–2%	0%

(modified from Levorsen, 1967)

and natural gas are carbon and hydrogen. Because of this, crude oil and natural gas are called *hydrocarbons*.

The difference between crude oil and natural gas is the size of the hydrocarbon molecules. Under surface temperature and pressure, any hydrocarbon molecule that has one, two, three, or four carbon atoms occurs as a gas. Natural gas is composed of a mixture of the four short hydrocarbon molecules. Any hydrocarbon molecule with five or more carbon atoms occurs as a liquid. Crude oil is a mixture of more than 100 hydrocarbon molecules that range in size from 5 to more than 60 carbons in length. The hydrocarbon molecules in oil form straight chains, chains with side branches, and circles.

Crude oil

Hydrocarbon Molecules

Four types of hydrocarbon molecules, called the *hydrocarbon series*, occur in each crude oil. The relative percentage of each hydrocarbon series molecule varies from oil to oil, controlling the chemical and physical properties of that oil. The hydrocarbons series includes paraffins, naphthenes, aromatics, and asphaltics. Hydrocarbons that have only single bonds between carbon atoms are called *saturated*. If they contain one or more double bonds, they are *unsaturated*.

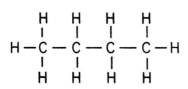

Fig. 1–1 *Paraffin molecule*

The *paraffin* or *alkane molecule* is a straight chain of carbon atoms with saturated (single) bonds between the carbon atoms (Fig. 1–1). The general formula for paraffins is C_nH_{2n+2}. They are five carbon atoms and longer in length. If the paraffin molecule is longer than 18 carbons in length, it is a wax and forms a *waxy crude oil*.

The *naphthene* or *cycloparaffin molecule* is a closed circle with saturated bonds between the carbon atoms (Fig. 1–2). The general formula for naphthenes is C_nH_{2n}. These molecules are five carbon atoms and longer in length. Oils with high naphthene content tend to have a large asphalt content that reduces the value of the oil.

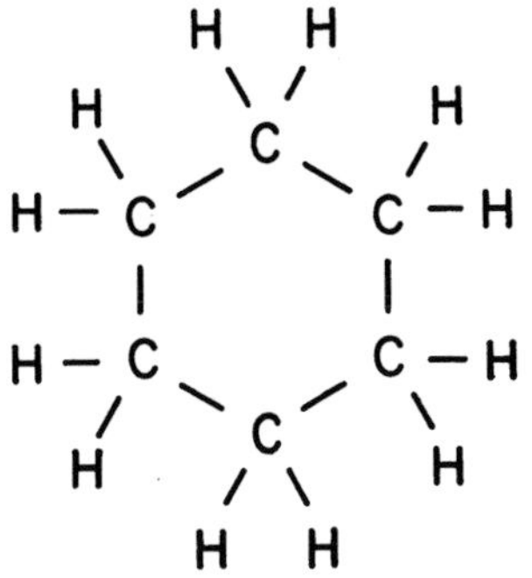

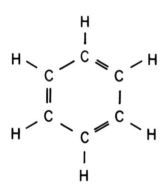

Fig. 1–2 Naphthene molecule

The *aromatic* or *benzene molecule* is a closed ring with some unsaturated (double) bonds between carbon atoms (Fig. 1–3). Their general formula is C_nH_{2n-6}. Aromatic molecules are six carbon atoms and longer in length. At the refinery, an aromatic-rich crude oil yields the highest-octane gasoline and makes a valuable feedstock for the petrochemical industry. The refiner often pays a premium for aromatic crude oil. Fresh from the well, normal crude oil has a pungent odor of gasoline. An aromatic-rich crude oil, however, has a fruity odor.

The asphaltic molecule has 40 to more than 60 carbon atoms. Asphalt is brown to black in color and is solid to semisolid under surface conditions. It has a high boiling point.

Table 1–2 shows the hydrocarbon series content of crude oil.

There are two types of crude oils at the refinery. An *asphalt-based crude oil* contains little or no paraffin wax. It is usually black.

Fig. 1–3 Aromatic molecule

When refined, it yields a large percentage of high-grade gasoline and asphalt. A *paraffin-based crude oil* contains little or no asphalt. It is usually greenish. When refined, it yields a large percentage of paraffin wax, high-quality lubricating oil, and kerosene. A *mixed-base crude oil* is a combination of both types.

Table 1–2

Average and range of hydrocarbon series molecules in crude oil

	weight percent	percent range
paraffins	30	15 to 60
naphthenes	49	30 to 60
aromatics	15	3 to 30
asphaltics	6	remainder

(modified from Levorsen, 1967 and Bruce and Schmidt, 1994)

°API

Crude oils are compared and described by density. The most commonly used density scale is *°API*. API stands for the American Petroleum Institute, based in Washington, D.C. It standardizes petroleum industry equipment and procedures. The formula for computing °API is:

$$°API = [(141.5 ÷ \text{specific gravity at } 60°F) - 131.5].$$

Fresh water, for example, has an °API of 10. The °API of crude oils varies from 5 to 55. Average weight crude oils are 25 to 35. Light oils are 35 to 45. Light oils are very fluid, often transparent, rich in gasoline, and are the most valuable. Heavy oils are below 25. They are very viscous, dark-colored, contain considerable asphalt, and are less valuable.

Sulfur

Sulfur is an undesirable impurity in fossil fuels such as crude oil, natural gas, and coal. When sulfur is burned, it forms sulfur dioxide, a gas that pollutes the air and forms acid rain. During the refining process, the refiner must remove the sulfur as the crude oil is being processed. If not, the sulfur will harm some of the chemical equipment in the refinery. Crude oils are classified as sweet and sour on the basis of their sulfur content. *Sweet crudes* have less than 1% sulfur by weight, whereas *sour crudes* have more than 1% sulfur. The refiner usually pays a US $1 to $3 per barrel premium for sweet crude. In general, heavy oils tend to be sour, whereas light oils tend to be sweet. At a refinery, *low sulfur crude* has 0 to 0.6% sulfur. *Intermediate sulfur crude* has 0.6 to 1.7% sulfur, and *high sulfur crude* has above 1.7% sulfur. Most of the sulfur in crude oil occurs bonded to the carbon atoms. A very small amount can occur as elemental sulfur in solution and as H_2S gas.

Benchmark Crude Oils

A *benchmark crude oil* is a standard against which other crude oils are compared, and prices are set. In the United States, West Texas Intermediate (WTI) is 38 to 40 °API and 0.3% S, whereas West Texas Sour, a secondary benchmark, is 33 °API and 1.6% S. Brent, the benchmark crude oil for the North Sea is very similar to WTI and is 38 °API and 0.3% S. Dubai is the benchmark crude oil for the Middle East. It is 31 °API and 2% S.

Pour Point

All crude oils contain some paraffin molecules. If the paraffin molecules are 18 carbon atoms or longer in length, they are waxes. Waxes are solid at surface temperature. A crude oil that containing a significant amount of wax is called a *waxy crude oil.* In the subsurface reservoir where it is very hot, waxy crude oil occurs as a liquid. As it is being brought up the well, it cools, and the waxes can solidify. This can clog the tubing in the well and flow-lines on the surface. The well then has to be shut in for a workover to clean out the wax.

The amount of wax in crude oil is indicated by the pour point of the oil. A sample of the oil is heated in the laboratory. It is then poured from a container as it is being cooled. The lowest temperature at which the oil will still pour before it solidifies is called the *pour point.* Crude oil pour points vary between +125° to -75°F (+52° to -60°C). Higher pour points reflect higher oil wax content. *Cloud point* is related to pour point. It is the temperature at which the oil first appears cloudy as wax forms when the temperature is lowered. It is 2° to 5°F (1° to 3°C) above the pour point. Very waxy crude oils are yellow in color. Slightly waxy crude oils can have a greenish color. Low or no wax oils are black.

In the North Sea, Ekofisk oil has a pour point of +10°F (-12°C). Brent oil has a pour point of +27°F (-3°C) whereas oil from the Statfjord field is +40°F (+4.5°C) and has a higher wax content. Crude oils from the Altamont area in the Uinta basin of Utah have very high pour points between +65° and +125°F (+18° to +52°C) and range from heavy (19 °API) to light (54 °API) oil.

Properties

The color of crude oil ranges from colorless through greenish-yellow, reddish, and brown to black. In general, the darker the crude oil, the lower the °API. The smell varies from gasoline (sweet crude) to foul (sour crude) to fruity (aromatic crude). Crude oil has a calorific heat value of 18,300 to 19,500 Btu/lb.

Crude Streams

A *crude stream* is oil that can be purchased from an oil-exporting country. It can be from a single field or a blend of oils from several fields. Table 1–3 describes some crude streams.

Table 1–3

Properties of selected crude streams

crude stream	country	°API	S%	pour point
Arabian light	Saudi Arabia	33.4	1.80	-30°F
Bachequero	Venezuela	16.8	2.40	-10°F
Bonny light	Nigeria	37.6	0.13	+36°F
Brass River	Nigeria	43.0	0.08	-5°F
Dubai	Dubai	32.5	1.68	-5°F
Ekofisk	Norway	35.8	0.18	+15°F
Iranian light	Iran	33.5	1.40	-20°F
Kuwait	Kuwait	31.2	2.50	0°F
North Slope	USA	26.8	1.04	-5°F

Measurement

The English unit of crude oil measurement is a *barrel* (*bbl*) that holds 42 U.S. gallons or 34.97 Imperial gallons. Oil well production is measured in barrels of oil per day (*bopd* or *b/d*). The metric units of oil measurement are metric tons and cubic meters. A metric ton of average crude oil (30 °API) equals 7.19 barrels. A metric ton of heavy oil (20 °API) equals 6.75 barrels whereas a metric ton of light oil (40 °API) equals 7.64 barrels. A cubic meter (m^3) of oil equals 6.29 barrels of oil.

Refining

During the refining process, various components of crude oil are separated by their boiling points. In general, the longer the hydrocarbon molecule, the higher its boiling temperature. At the refinery, crude oil is first heated in a furnace until most is vaporized. The hot vapor is then sprayed into a distilling column. Gasses rise in the distilling column and any remaining liquid falls. In the distilling column are bubble trays filled with liquid (Fig. 1–4). The rising vapors bubble up through the trays and are cooled. The cooling vapors condense into liquid on the trays where they are then removed by sidedraws. Each liquid removed by cooling is called a cut (Fig. 1–5). Heavy cuts come out at high temperatures, whereas light cuts come out at low temperatures. In order of cooling temperatures, the cuts are heavy gas oil, light gas oil, kerosene, naphtha, and straight run gasoline.

Gasoline is the refining product in most demand. A process called cracking is used to make gasoline from the other cuts. Gasoline is composed of short molecules with 5 to 10 carbon atoms. The longer, less valuable, molecules of other cuts are used as cracking stock. Cracking stock is put into cracking towers at the refinery where high temperatures and pressures and caustic chemicals split the longer molecules to form gasoline.

Refineries also produce pure chemicals, called feedstocks, from crude oil. Some common feedstocks are methane, ethylene, propylene, butylene, and naphthene. These feedstocks are sold to the petrochemical industries, where the molecules are reformed, and a large variety of products are made. Plastics, synthetic fibers, fertilizers, Teflon, polystyrene, drugs, dyes, explosives, antifreeze, and synthetic rubber are examples.

The average percent yield of crude oil in a refinery is shown in Table 1–4.

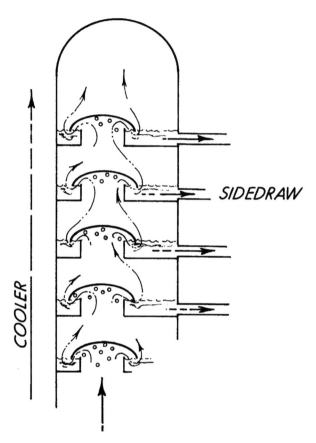

Fig. 1–4 Bubble trays in a distilling column

Natural Gas

Composition

Natural gas is composed of hydrocarbon molecules that range from one to four carbon atoms in length. The gas with one carbon atom in the molecule is *methane* (CH_4), two is *ethane* (C_2H_6), three is *propane* (C_3H_8), and four is *butane* (C_4H_{10}). All are paraffin-type hydrocarbon molecules. A typical natural gas composition is shown in Table 1–5.

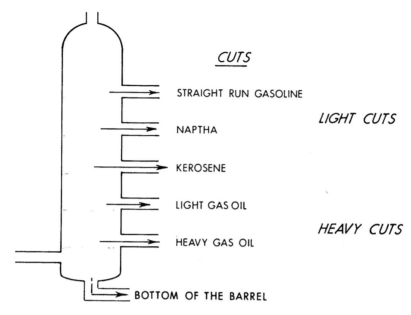

Fig. 1–5 Distilling column cuts

Table 1–4	
Percent yield of crude oil	
gasoline	46%
fuel oil	27%
jet fuel	10%
coke	5%
liquefied gasses	4%
petrochemical feedstocks	3%
asphalt	3%
lubricants	1%
kerosene	1%
(American Petroleum Institute)	

Table 1–5

Typical natural gas hydrocarbon composition

methane	70 to 98%
ethane	1 to 10%
propane	trace to 5%
butane	trace to 2%

These percentages vary from field to field, but methane gas is by far the most common hydrocarbon. Many natural gas fields contain almost pure methane. The gas from pipelines that is burned in homes and industry is methane gas. Propane and butane burn giving off more heat than methane. They are often distilled from natural gas and sold separately. *Liquefied petroleum gas* (*LPG*) is made from propane gas.

The nonhydrocarbon, gaseous impurities that don't burn in natural gas are called *inerts*. A common inert is water vapor (steam). Another inert is carbon dioxide (CO_2), a colorless, odorless, gas. Because it doesn't burn, the more carbon dioxide natural gas contains, the less valuable the gas is. In some gas reservoirs, carbon dioxide is greater than 99% of the gas. Large fields of almost pure carbon dioxide probably formed by the chemical reaction of volcanic heat on limestone rock. Carbon dioxide can be used for inert gas injection, an enhanced oil recovery process, in depleted oil fields. Nitrogen (N), another inert, is also a colorless, odorless gas that can be used for inert gas injection. Helium is a light gas used in electronic manufacturing and filling dirigibles. Gas from the Hugoton gas field in western Texas, Oklahoma, and Kansas contains 0.5 to 2% helium. It is thought to have formed by the radioactive decay of K^{40} in granite. Amarillo, Texas, near the giant gas field, is called the "helium capital of the world."

Hydrogen sulfide (*H_2S*) is a gas that can occur mixed with natural gas or by itself. It is not an inert and is a very poisonous gas that is lethal in very low concentrations. The gas has the foul odor of rotten eggs and can be smelled in extremely small amounts. It is associated with the salt domes of the Gulf of Mexico and ancient limestone reefs of Mexico, West Texas, and Louisiana. Hydrogen sulfide is common in Alberta, the overthrust belt of Wyoming, offshore Southern California, Utah, and the Middle East. Hydrogen sulfide gas is very corrosive. When it occurs mixed with natural gas, it causes corrosion of the metal tubing, fittings, and valves in the well. Hydrogen sulfide must be removed before the natural gas can be delivered to a pipeline. *Sweet natural gas* has no detectable hydrogen sulfide, whereas *sour natural gas* has detectable amounts of hydrogen sulfide.

Occurrence

Because of high pressure in the subsurface reservoir, a considerable volume of natural gas occurs dissolved in crude oil. The *formation, dissolved* or *solution gas/oil ratio* is the cubic feet of natural gas dissolved in one barrel of oil in that reservoir under subsurface conditions. The volume measurements are reported under surface conditions. In general, as the pressure of the reservoir increases with depth, the amount of natural gas that can be dissolved in crude oil increases. When crude oil is lifted up a well to the surface (Fig. 1–6), the pressure is relieved, and the natural gas, called *solution gas*, bubbles out of the oil. The *producing gas-oil ratio (GOR)* of a well is the number of cubic feet of gas the well produces per barrel of oil.

Nonassociated natural gas is gas that is not in contact with oil in the subsurface. A nonassociated gas well produces almost pure methane. *Associated natural gas* occurs in contact with crude oil in the subsurface. It occurs both as gas in the free gas cap above the oil and gas dissolved in the crude oil. Associated gas contains other hydrocarbons besides methane.

Condensate

In some subsurface gas reservoirs, at high temperatures, shorter-chain liquid hydrocarbons, primarily those with five to seven carbon atoms in length, occur as a gas. When this gas is produced, the temperature decreases, and the liquid hydrocarbons condense out of the gas. This liquid, called *condensate*, is almost pure gasoline, is clear to yellowish to bluish in color and has 45 °API to 62 °API. Condensate is commonly called *casinghead gasoline, drip gasoline, white gas,* or *natural gasoline.* Condensate can be added to

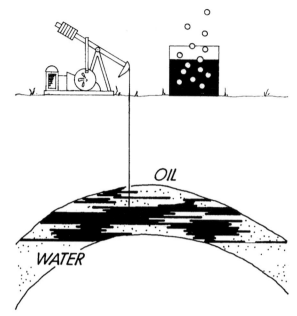

Fig. 1–6 Solution gas

crude oil in the field in a process called *spiking* to decrease the °API and increase the volume of the oil. Condensate removed from natural gas in the field is classified as crude oil by regulatory agencies.

Refiners pay almost as much for condensate as crude oil. It doesn't have a high octane and must be mixed with high-octane gasoline made by cracking in the refinery. Because of the low octane, the posted price for condensate is usually slightly less than that for crude oil. Natural gas that contains condensate is called *wet gas*, whereas natural gas lacking the condensate is called *dry gas*. The condensate along with butane, propane, and ethane that can be removed from natural gas is called *natural gas liquids* (*NGL*).

Measurement

The English unit of volume measurement for natural gas is a *cubic foot* (*cf*). Because gas expands and contracts with pressure and temperature changes, the measurement is made under or is converted to standard conditions defined by law. It is usually 60°F and 14.65 psi (15°C and 101.325 kPa) and is called *standard cubic feet (scf)*. The abbreviation for 1000 cubic feet is *Mcf*, a million cubic feet is *MMcf*, a billion cubic feet is *Bcf* and a trillion cubic feet is *Tcf*. Condensate content is measured in barrels per million cubic feet of gas (*BCPMM*).

The unit used to measure heat content of fuel in the English system is the *British thermal unit (Btu)*. One Btu is about the amount of heat given off by burning one wooden match. Pipeline natural gas ranges from 900 to 1200 Btus per cubic foot and is commonly 1000 Btu. The heat content varies with the hydrocarbon composition and the amount of inerts in the natural gas. Natural gas is sold to a pipeline by volume in thousands of cubic feet, by the amount of heat when burned in Btus or by a combination of both. If the pipeline contract has a *Btu adjustment clause*, the gas is bought at a certain price per Mcf, and the price is then adjusted for the Btu content of the gas.

In the metric system, the volume of gas is measured in cubic meters (m³). A cubic meter is equal to 35.315 cf. Heat is measured in kilojoules. A kilojoule is equal to about 1 Btu.

The Btus in one average barrel of crude oil are equivalent to the Btus in 6040 cubic feet of average natural gas and is called *barrel of oil equivalent (BOE)*. Different companies often have a slightly different BOE numbers depending on the oil and gas composition of their production.

Reservoir Hydrocarbons

Chemists classify reservoir hydrocarbons into 1) black oil, 2) volatile oil, 3) retrograde gas, 4) wet gas, and 5) dry gas. Laboratory analysis of a sample is used to determine the type.

Both black and volatile oils are liquid in the subsurface reservoir. *Black oil* or *low-shrinkage oil* has a relatively high percentage of long, heavy, non-volatile molecules. It is usually black but can have a greenish or brownish color. Black oil has an initial producing gas-oil ratio of 2000 scf/bbl or less. The °API is below 45.

Volatile oil or *high-shrinkage oil* has relatively more intermediate size molecules and less longer size molecules than black oil. The color is brown, orange, or green. Volatile oil has an initial producing gas-oil ratio between 2000 and 3300 scf/bbl. The °API is 40 or above.

Retrograde gas is a gas in the reservoir under original pressure but liquid condensate forms in the subsurface reservoir as the pressure decreases with production. The initial gas-oil ratio is 3300 scf/bbl or higher.

Wet gas occurs entirely as a gas in the reservoir, even during production, but produces a liquid condensate on the surface. It often has an initial producing gas-oil ratio of 50,000 scf/bbl or higher.

Dry gas is pure methane. It does not produce condensate either in the reservoir or on the surface

two

THE EARTH'S CRUST—WHERE WE FIND IT

Rocks and Minerals

The earth is composed of *rocks*, which are aggregates of small grains or crystals called minerals (Fig. 2–1). *Minerals* are naturally occurring, relatively pure chemical compounds. Examples of minerals are quartz (SiO_2) and calcite ($CaCO_3$). Rocks can be composed of numerous grains of several different minerals. The rock granite, for example, is composed of the minerals quartz, feldspar, hornblende, and biotite. Rocks can also be composed of numerous grains of the same mineral. The rock limestone consists only of calcite mineral grains.

Rocks have been formed throughout the billions of years of earth's history. The same chemical and physical process that form rocks today formed rocks throughout geological time. The molten lava flowing out a volcano in Hawaii or Italy today is forming lava rock similar to lava rock formed millions and billions of years ago. Ancient sandstone rock composed of sand grains was formed the same way sand is deposited today: along beaches, in river channels, and on desert dunes. There is nothing unusual about ancient rocks. They formed the same way rocks are forming today.

Types of Rocks

Three types of rocks make up the earth's crust: igneous, sedimentary, and metamorphic. *Igneous rocks* have crystallized from a hot, molten liquid. *Sedimentary rocks* are composed of sediments that were deposited on the surface of the ground or bottom of the ocean or salts that precipitated out of water. *Metamorphic rocks* have been recrystallized from other rocks under high temperatures and pressures.

Igneous Rocks

Igneous rocks are formed when a molten melt is cooled. Two types of igneous rocks are plutonic and volcanic, depending on where they formed.

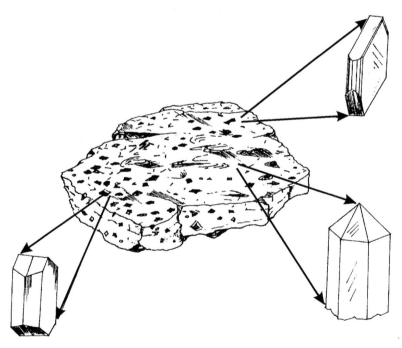

Fig. 2–1 *Mineral grains in a rock*

Plutonic igneous rocks crystallized and solidified while still below the surface of the earth. Because the rocks that surround the cooling plutonic rocks are good insulators, plutonic rocks often take thousands of years to solidify. When a cooling melt is given a long time to crystallize, large mineral crystals are formed. Plutonic igneous rocks are easy to identify because the mineral crystals are all large enough to be seen by the naked eye. Plutonic rocks formed as hot liquids that were injected into and displaced preexisting rocks in the subsurface. Because of this, plutonic rock bodies are called *intrusions*.

Volcanic igneous rocks crystallize on the surface of the earth as lava. As the lava flows out of a volcano, it immediately comes in contact with air or water and rapidly solidifies. The rapid crystallization forms very small crystals that are difficult to distinguish with the naked eye.

Sedimentary Rocks

Sedimentary rocks are composed of sediments of which there are three types. *Clastic sediments* are whole particles that were formed by the breakdown of rocks and were transported and deposited as whole particles. Boulders, sand grains, and mud particles are examples. *Organic sediments* are formed biologically such as seashells. *Crystalline sediments* are formed by the precipitation of salt out of water. As sediments are buried in the subsurface, they become solid, sedimentary rocks. Sedimentary rocks are the rocks that are drilled to find gas and oil. They are the source and reservoir rocks for gas and oil.

Loose sediments (*unconsolidated sediments*) become relatively hard sedimentary rocks (*consolidated sediments*) in the subsurface by the processes of cementation and compaction. No matter how some sediments such as sand grains are packed together, there will be *pore spaces* between the grains (Fig. 2–2). Once the grains have been buried in the subsurface, the pore spaces are filled with water that can be very salty. Under the higher temperatures and pressures of the subsurface, salts often precipitate out of the subsurface waters to coat the grains. These coatings grow together to bridge the loose grains. This process, called *cementation*, bonds the loose grains into a solid sedimentary rock. Two common cements are the minerals calcite ($CaCO_3$) and quartz (SiO_2). Also, as the sediments are buried deeper, the increasing weight of overlying rocks exerts more pressure on the grains. This compacts the sediments that also solidify the rock.

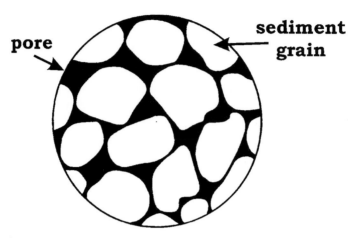

pore

sediment
grain

Fig. 2–2 Pores between sediment grains

Clastic sedimentary rocks often consist of three parts when examined under a microscope (Fig. 2–3). First, there are sediment grains. These are composed of minerals such as quartz or feldspar or seashells. Second, there are natural cements coating and bonding the grains together. Third, there are pore spaces. The pores are filled with fluids (water, gas, or oil) in the subsurface.

There is an enormous amount of water below the surface of the ground, called *ground water*, in the pores of the sedimentary rocks (Fig. 2–4). Ground water is described by salt content in parts per thousand (*ppt*). Fresh water contains so little salt (0–1 ppt) that it can be used for drinking water. *Brines* are subsurface waters that contain more salt than seawater (35–300 ppt). Brackish waters are mixtures of fresh waters and brines (1–35 ppt). Below the surface is a boundary called the *water table* between the dry pores above and pores that are filled with water below. The water table can be on the surface or very deep depending on how much rain falls in that area. Just below the water table, the ground water is usually fresh because of rain water that percolates down from the surface. Deep waters, however, are usually brines. When a well is drilled, completed, and producing, near-surface fresh waters must be protected. *Meteoric water* is fresh, subsurface water. *Connate*

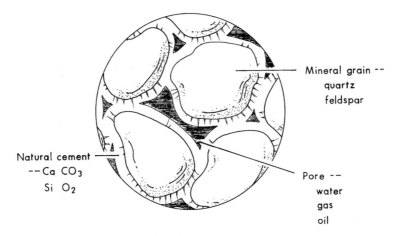

Mineral grain --
quartz
feldspar

Natural cement
--Ca CO₃
Si O₂

Pore --
water
gas
oil

Fig. 2–3 *Clastic sedimentary rock under a microscope*

water is saline, subsurface water that has been out of contact with the atmosphere for a long time. Connate water is often water that was originally trapped in the sediments when they were deposited.

The sizes of the clastic grains that make up an ancient sedimentary rock are important. The rock is often classified according to the grain size. Sandstones are composed of sand-sized grains whereas shales are composed of fine-grained (clay-sized) particles. Also, the size of the grains controls the size of the pore spaces and the quality of the oil or gas reservoir. Larger grains have larger pores between them. It is easier for fluids, such as gas and oil, to flow through larger pores. Clastic grains in sedimentary rocks are classified by their diameters in millimeters (Fig.2–5). They are called *boulder, cobble, pebble, granule, sand, silt,* and *clay-sized* particles. The finest grains (*i.e.*, sand, silt, and clay-sized) are the most common.

Sorting is the range of particle sizes in the rock (Fig. 2–6). A *well-sorted* rock is composed of particles of approximately the same size (Fig. 2–6a). A *poorly-sorted* rock is composed of particles with a wide range of sizes (Fig. 2–6b). Sorting is the most important factor in determining the amount of original pore space in a clastic sedimentary rock. Finer-sized particles in a poorly sorted rock

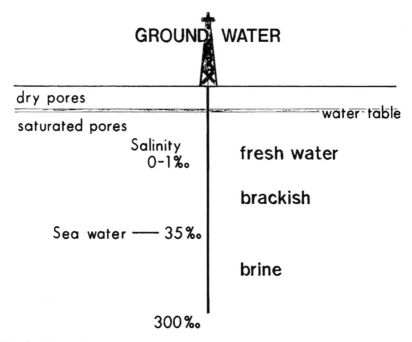

Fig. 2–4 *Ground water*

occupy the spaces between the larger-sized particles and reduce the volume of the pores. Poorly sorted rocks can hold less fluids and are lower-quality reservoir rocks than well-sorted rocks. Well-sorted sandstones are called *clean sands.* Because sand grains are light in color, clean sandstones are usually light in color. Poorly sorted sandstones with significant amounts of silt- and clay-sized grains are called *dirty sands.* Because silt- and clay-sized particles are usually dark in color, dirty sandstones are dark colored.

Sedimentary rocks are identified by their layering, called *stratification* or *bedding* (Plate 2–1). As the sediments are deposited, there are frequent variations in the amount and composition of sediment supply and the level of the ocean that cause the layering. Sediment layers are originally deposited horizontal in water.

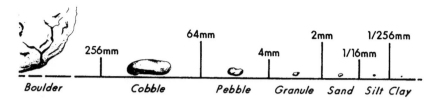

Fig. 2–5 *Grain sizes in millimeters (1mm = 1/25 in.)*

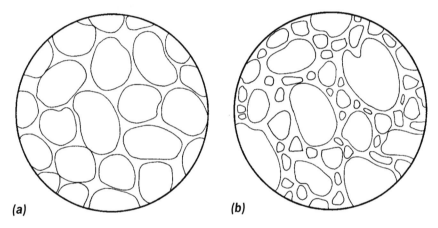

Fig. 2–6 *Particle sorting (a) well sorted (b) poorly sorted*

Geologists can interpret how sedimentary rocks were deposited. *Lithology* (rock composition) is an important clue as to how a sedimentary rock was formed. Sand grains, mud particles, and shell beds each form different sedimentary rocks. Each is originally deposited in a very different environment. *Sedimentary structures* such as ripple marks, mud cracks, and flow marks help to visualize the environment in which the rock was deposited. Another aid to interpretation is *fossils*, preserved remains of plants and animals.

Plate 2–1 *Layering in sedimentary rocks, Dead Horse Point, Utah*

Metamorphic Rocks

Metamorphic rocks are any rocks that have been altered by high heat and pressure. Marble ($CaCO_3$), a metamorphic rock, is metamorphosed limestone ($CaCO_3$), and quartzite (SiO_2) is metamorphosed quartz sandstone (SiO_2). Since temperatures and pressures become greater with depth, a rock often becomes metamorphosed when buried deep in the earth.

Structure of the Earth's Crust

The earth is estimated to be about 4.5 billion years old. Even the sedimentary rocks that generated and hold the gas and oil are millions to hundreds of millions of years old. During that vast expanse of geological time, sea level has not been constant. Sea level has been rising and falling. During the rise and fall of sea level, sediments were deposited in layers. Sands were deposited along the ancient beaches. Mud was deposited in the shallow seas offshore. Seashells were deposited in shell beds. These ancient sediments form the sedimentary rocks that are drilled to find gas and oil. The rise and fall of sea level has occurred in numerous cycles (Fig. 2–7). The largest cycles occurred every few hundreds of millions of years. There are shorter cycles on the large cycles and even shorter cycles on them. At lease five orders of sea level cycles have occurred, with the shortest occurring every few tens of thousands of years. The shorter cycles are thought to be caused by the freezing and melting of glaciers.

In a typical section of the earth's crust such as Tulsa, Oklahoma, about 5000 ft (1500 m) of well-layered sedimentary rocks are underlain by very old metamorphic or igneous rocks (Fig. 2–8). There are about one hundred layers of sedimentary rocks. Sands form the rock sandstone. Mud forms the rock shale. Sea shells form the rock limestone. The unproduc-

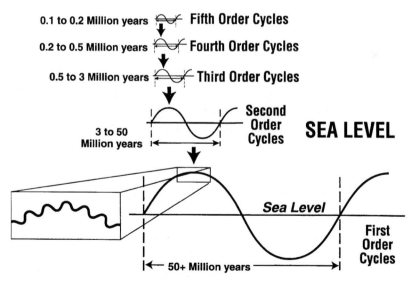

Fig. 2–7 *Sea level cycles (Hyne, 1995)*

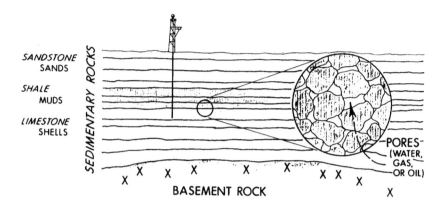

Fig. 2–8 *Cross section of earth's crust*

tive rocks, usually igneous and metamorphic rocks underlying the sedimentary rocks, are called *basement rocks*. When drilling encounters basement rock, the drilling is usually stopped.

In some areas of the earth, there are no sedimentary rocks, and the basement is on the surface. These areas are called *shields*, and there is no gas or oil. Every continent of the world has at least one shield area (Fig. 2–9). A shield, such as the Canadian shield in eastern Canada, tends to be a large, low-lying area. Ore minerals such as iron, copper, lead, zinc, gold, and silver are mined from the basement rock in shield areas. The southwest portion of Saudi Arabia is a shield. All the Saudi Arabian oil fields are located in sedimentary rocks to the northeast of the Arabian Shield.

In other areas called *basins*, the sedimentary rocks are very thick. The Caspian basin (Caspian Sea) has about 85,000 ft (26,000 m) of sedimentary rock cover. However, 20,000 to 40,000 ft (6,000 to 12,000 m) of sedimentary rocks is typical of many basins. Basins such as the Gulf of Mexico, the Anadarko basin of southwestern Oklahoma, and the Denver-Julesburg basin of Colorado are large areas that are often more than 100 miles (160 km) across.

Fig. 2–9 *Map of world showing location of shields and areas where very old (Precambrian age) rocks occur on the surface in black*

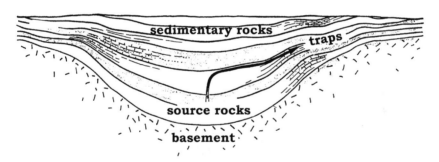

Fig. 2–10 *Cross section of sedimentary rock basin*

It is in the sedimentary rock basins that the most gas and oil is found and produced. Because of the thick sedimentary rock, most basins have source rocks that have been buried deep enough in the geological past to generate gas and oil (Fig. 2–10). The deep part of the basin where the gas and oil forms is called the *kitchen* or *oven*. After the gas and oil is generated, it flows upward in the overlying rocks. If it intersects a layer of reservoir rock, the gas and oil then migrates through the interconnected pores of the reservoir rock layer up the flanks of the basin where it can be trapped and concentrated. The trap, such as an anticline, is a relatively small feature compared to the basin. Numerous traps can occur along the flanks of the basin.

There are about 600 sedimentary rock basins in the world. Of the basins that have been explored and drilled, about 40% are very productive. About 90% of the world's oil occurs in only 30 of those basins. The other 60% of the explored basins are barren. The unproductive basins either have no source rocks, the source rocks have never been buried deep enough to generate gas and oil, or the basin has was overheated, and the oil was destroyed.

three

IDENTIFICATION OF COMMON ROCKS AND MINERALS

Just a few rocks and minerals make up the bulk of the earth's crust. All of these are readily identifiable by simple tests that can be made in the field such as at the drillsite without elaborate equipment.

Identification of Minerals

Minerals occur as crystals and grains in rocks. Color is the first property that is noticed in a mineral. Many colors, such as the brassy yellow of pyrite and the steel gray of galena, are diagnostic. Some transparent minerals are tinted by slight impurities such as gas bubbles, iron, or titanium. Rose quartz, milky quartz, and smoky quartz are examples. *Luster* is the appearance of light reflected from the surface of a mineral. Two common lusters are metallic and nonmetallic. Nonmetallic lusters have descriptive names as greasy, glassy, and earthy. A few minerals are transparent in thin sheets, and others are translucent (they transmit light but not an image), but most are opaque.

The form that a mineral crystal takes such as cubes or pyramids can also be diagnostic. Other minerals have no crystal form and are called *amorphous*. The tendency for some minerals to break along smooth surfaces is called *cleavage*. Cleavage is described by three properties (Fig.

3–1). The first is the number of cleavage surfaces of different directions. The second is the quality of the surfaces, such as poor or excellent. The third is the angle between the surfaces. Fracture is the breakage of the mineral along an irregular surface.

The hardness of a mineral is quantified by *Moh's scale* that ranges from 1 to 10. The mineral talc is the softest (1), and diamond is the hardest (10). A mineral that is higher on the scale can scratch a mineral that is lower on the scale. Some common objects that are used for hardness comparisons are a fingernail (2.5), a copper penny (3.5), a knife or steel key (6), and glass (7).

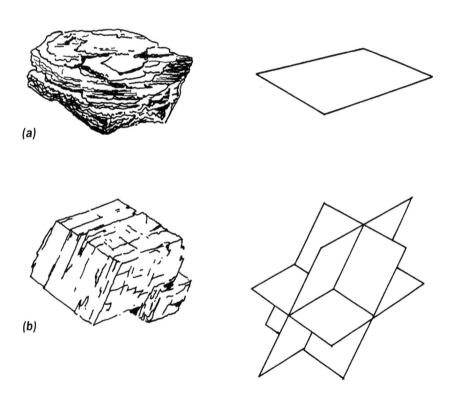

(a)

(b)

Fig. 3–1 Cleavage (a) one, perfect cleavage (b) three, perfect cleavages not at right angles

Specific gravity is the relative weight of a mineral compared to the weight of an equal volume of water. A specific gravity of 2.5 means the mineral weighs 2.5 times an equal volume of water. The specific gravity of an average rock or mineral is about 2.5, whereas metallic ore minerals often have specific gravities above 3.5. Ore minerals mined for metals, such as iron, copper, or nickel, are heavy.

The mineral halite (common table salt) can be identified by its taste. A very important test is the application of cold, dilute hydrochloric acid to a sample. Only the mineral calcite or the sedimentary rock limestone, formed by calcite grains, will bubble.

Minerals

Mica is a common mineral that breaks along one perfect cleavage plane, forming very thin, elastic flakes. Two types of mica are white mica and black mica. *White mica, (muscovite)* is composed of $KAl_3Si_3O_{10}(OH)_2$. It is colorless and is transparent in thin flakes (Plate 3–1a). *Black mica (biotite)* is composed of $K(Mg,Fe)_3AlSi_3O_{10}(OH)_2$. It is brown to black in color (Plate 3–1b).

Quartz (SiO_2) is a very common mineral (Plate 3–1c) that is colorless when pure. It is often, however, tinted by impurities. Common varieties include rose, cloudy, milky, and smoky quartz. Quartz is the hardest of the common minerals (7 on the scale). It will scratch all other common minerals and a knife. It forms six-sided, prismatic crystals but can occur as amorphous grains. Most sand grains on a beach or in sandstone rock are composed of quartz.

Calcite $(CaCO_3)$ is a common mineral that is either colorless or white. Calcite breaks along three perfect cleavage planes that are not at right angles to form rhombs (Plate 3–1d). Calcite is relatively soft (3) and can be scratched by a knife. Calcite crystals are dog-tooth or flat shaped. Calcite will bubble in cold, dilute acid. Most sea shells are composed of calcite.

Halite (NaCl) is common table salt. It is colorless to white (Plate 3–1e). Halite forms a granular mass or crystallizes in cubes. It breaks along three perfect cleavage planes at right angles, forming rectangles and cubes. Halite tastes salty. The mineral halite forms from the evaporation of sea water. It is very common in ancient salt deposits.

Gypsum ($CaSO_4 \cdot 2H_2O$) is colorless to white (Plate 3–1f). It forms tabular crystals and has one perfect cleavage plane. Gypsum is very soft (2) and can be scratched by a fingernail. It has a specific gravity of 2.3. Gypsum is also called *selenite* or *alabaster*. Gypsum and a similar mineral, *anhydrite* ($CaSO_4$), form by the evaporation of sea water.

Pyrite (FeS_2) is known as fool's gold. It has a brassy yellow color and a metallic luster (Plate 3–1g). Pyrite forms either cubes or an earthy mass and is relatively heavy, with a specific gravity of 5. Pyrite is an iron ore and can sometimes be found as grains in river sands.

Galena (PbS) is a lead ore. It has a silvery-gray color with a metallic luster (Plate 3–1h). Galena forms cubic crystals. It has a hardness of 2.5. Galena is very heavy, with a specific gravity of 7.5.

Table 3–1 lists the properties of these minerals.

Each of these minerals has one or two characteristic tests that readily distinguish it from other minerals. For example, quartz is the hardest of the common minerals and cannot be scratched by a knife. Calcite is relatively soft, can be scratched with a knife, and will bubble in cold, dilute acid.

Table 3–1
Mineral properties

name	hardness	specific gravity	luster
white mica	2 to 3	3	pearly to vitreous
black mica	2.5 to 3	3	pearly to vitreous
quartz	7	2.65	vitreous to greasy
calcite	3	2.72	vitreous to dull
halite	2 to 2.5	2.1	vitreous
gypsum	2	2.3	vitreous, pearly, oily
pyrite	6 to 6.5	5	metallic
galena	2.5	7.5	metallic

Identification of Rocks

Rocks are classified and identified by their textures and mineral compositions. Igneous rock textures are based on the size of the mineral crystals. The grain sizes range from large enough to see with the naked eye to glassy with no crystals. Metamorphic rock textures are based on the size and orientation of the mineral crystals. A *foliated* metamorphic rock has parallel, platy crystals (Fig. 3–2a). *Nonfoliated* metamorphic rock has either uniform-sized crystals or a nonparallel orientation of platy crystals (Fig. 3–2b). Sedimentary rock textures are based on the nature, size, and shape of the grains and how they are bound together.

Rocks

Igneous Rocks

Granite is the most common intrusive, igneous rock. It has the coarse-grained texture characteristic of all plutonic rocks. Granite is composed of the minerals quartz, feldspar, biotite, and hornblende. Quartz grains are most common, giving granite a light color (Plate 3–2a). Dark-colored min-

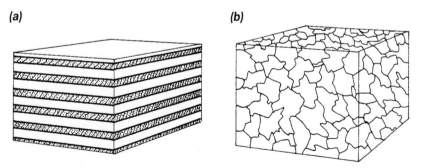

(a) **(b)**

Fig. 3–2 *Metamorphic rock textures (a) foliated (b) nonfoliated*

eral grains give it a speckled texture. Some granites are reddish or pinkish from iron impurities. Granite is commonly used for building stone.

Basalt is the most abundant volcanic rock. It has the fine-grained texture characteristic of lava. Basalt is black to gray in color (Plate 3–2b). In some instances, the basalt coming out of the volcano cooled so rapidly that gas bubbles were frozen in the basalt. Fragments of basalt with numerous gas bubbles are called *scoria*.

Metamorphic Rock

Gneiss is a product of intense metamorphism. It is easily identified by its foliated texture of alternating, wide bands of light- and dark-colored coarse mineral grains (Plate 3–2c). Feldspar and quartz are the most common light-colored minerals. Amphibole, garnet, and mica are the most common dark-colored minerals.

Sedimentary Rocks

Conglomerate is a clastic rock with a wide range of pebble- to clay-size grains (Plate 3–2d). The coarse grains distinguish it from other clastic sedimentary rocks. The particles are all well-rounded. A conglomerate is commonly deposited in a river channel or on an alluvial fan formed where a mountain stream empties into the desert. If the particles are angular, the rock is called *breccia*.

Sandstone is composed primarily of sand grains (Plate 3–2e) that have been naturally cemented together. The sand grains can be broken off if the rock is loosely cemented. Sandstone is rough to the touch. The rock can be white to buff to dark in color. Sandstones are commonly deposited on beaches, river channels, or dunes. It is a common reservoir rock for gas and oil and is the most important reservoir rock in North America.

Shale is composed of clay-sized particles (Plate 3–2f) and is the most common sedimentary rock. It is usually well-layered and relatively soft. Shale breaks down into mud when exposed to water. The color of shale ranges from green and gray to black, depending on the organic content. The darker the shale, the higher the organic content. Shale is commonly deposited on river floodplains and on the bottom of oceans, lakes, or lagoons. Black shales are common source rocks for gas and oil. A gray shale can be a caprock on a reservoir rock in a petroleum trap. *Mudstone* is similar to shale but is composed of both silt- and clay-sized grains.

Limestone is composed of calcite mineral grains that range in size from very fine to large, sparkling crystals (Plate 2–2g). The rock is commonly white or gray in color. The calcite mineral grains are soft enough to be scratched by a knife and will bubble in cold, dilute acid. Limestones often have fossil fragments that are also usually composed of calcite. Limestone is a common reservoir rock and is the most important reservoir rock in the oil and gas fields of the Middle East. An organic-rich, dark-colored limestone can also be a source rock for gas and oil.

Coal is brown to black in color and very brittle (Plate 3–2h). It has few, if any layers. Coal is composed of plant remains that were buried in the sub-surface and transformed by heat and time. *Lignite, bituminous,* and *anthracite* are varieties of coal formed by increasing heat that causes the coal to become harder and change in texture and composition.

Chert or *flint* is amorphous quartz (Plate 3–2i). It is very hard and cannot be scratched by a knife. Being amorphous (without crystals), chert breaks along smooth, curved surfaces, forming sharp edges and points. American Indians used chert to make arrowheads. Colored varieties of chert include jasper, chalcedony, and agate. Chert can be formed by precipitation directly out of ground water or by recrystallization of fossil shells composed of SiO_2 by heat and pressure. Chert is the hardest of all sedimentary rock to drill.

Ninety-nine percent of the sedimentary rocks that make up the earth's crust are shales, sandstones, and limestones. Many sedimentary rocks are a combination of these three types. Sedimentary rock mixtures are described as sandy, shaly, and limey or calcareous (Fig. 3–3).

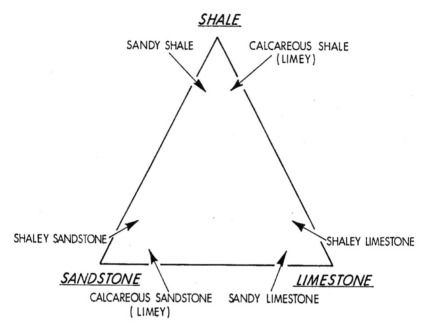

Fig. 3–3 Common sedimentary rocks

PLATE 3-1: MINERALS

3–1a white mica

3–1b black mica

3–1c quartz

3–1d calcite

3–1e halite

3–1f gypsum

3–1g pyrite

3–1h galena

PLATE 3-2: ROCKS

3–2a granite

3–2b basalt

3–2c gneiss

3–2d conglomerate

3–2e sandstone

3–2f black shale and gray shale

3–2g limestone

3–2h coal

3–2i chert

four

GEOLOGICAL TIME

Two methods for dating the formation of the rocks and events in the earth's crust are absolute and relative age dating. *Absolute age dating* puts an exact time (e.g., 253 million years ago) on the formation of a rock or an event. *Relative age dating* arranges the rocks and events into a sequence of older to younger.

Radioactive Age Dating

Exact dates for the formation of rocks are made by radioactive analysis. *Radioactivity* is the spontaneous decay of radioactive atoms that occur naturally in rocks (Fig. 4–1). The atoms decay by giving off atomic particles and energy. For example, uranium (U^{238}) decays by giving off atomic particles to form lead (Pb^{206}). The original atom, uranium (U^{238}), is called the *parent*. The product of radioactive decay, lead (Pb^{206}), is the *daughter*. There are four relatively abundant radioactive atoms that occur in rocks. These are two isotopes of uranium (U^{238} and U^{235}), potassium (K^{40}), and rubidium (Rb^{87}). Each decays at a different rate. The rate of radioactive decay is measured in *half-lives*. A half-life is the time in years that it takes one-half of the parent atoms to decay into daughter atoms (Fig. 4–2). The half-lives are shown in Table 4–1.

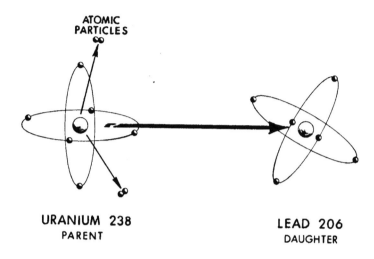

URANIUM 238
PARENT

LEAD 206
DAUGHTER

Fig. 4–1 *Radioactive decay*

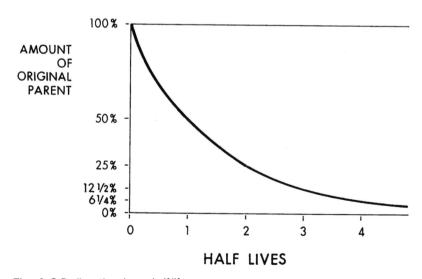

Fig. 4–2 *Radioactive decay half-life*

Table 4–1	
Half-lives of common radioactive atoms	
atom	*years*
U^{238}	4.5×10^9
U^{235}	0.7×10^9
Rb^{87}	4.7×10^{10}
K^{40}	1.3×10^9

One daughter atom is formed by the decay of each parent atom. As time goes on, the amount of radioactive parent atoms decreases, and the amount of daughter atoms increases. By measuring the amount of parent atoms left and daughter atoms created, the age of the mineral grains in a rock can be determined. This parent-daughter technique is used in the potassium-argon method. Potassium is a common element. A potassium isotope (K^{40}) decays into argon (Ar^{40}) with a half-life of 1.3 billion years. The assumption is made that when a mineral crystal forms, only potassium is accepted into the crystal structure, never argon because it is an inert gas. Any argon that is found in the crystal today could have come only from the radioactive decay of potassium. By measuring the amount of K^{40} and Ar^{40} in the mineral, the ratio can be applied to the radioactive decay curve, and the age of the mineral determined. For example, if the ratio of K^{40} to Ar^{40} is 3 to 1, the age of the mineral grain is 2 half-lives, or 2.6 billion years (Fig. 4–3).

Carbon (C^{14}) decays very quickly and is not useful for most rocks. The half-life is only 5710 years. After about 10 half-lives or 60 thousand years, there is not enough parent left to date the material. It is used for archeology where ages are much younger.

Radioactive age dating is used primarily on igneous and metamorphic rocks and cannot be used directly on sedimentary rocks. Sedimentary rocks are derived from the erosion of pre-existing rocks. Absolute age dating of sedimentary rock grains will tell the age of the formation of the mineral grains but not

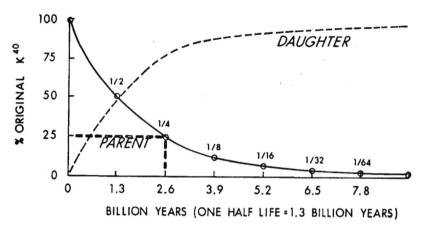

Fig. 4–3 Using half-lives to determine age

the time the sediments were deposited. However, igneous intrusions that cut sedimentary rocks or lava flows that cover sedimentary rocks can be dated. In that way, the age of the affected sedimentary rocks can be estimated. A lava flow will be older than any sedimentary rock that covers it. An intrusion will be younger than any sedimentary rock that it intrudes.

Relative Age Dating

In sedimentary rock sequences, relative age dating is used. Sedimentary rocks and events are put in order from oldest to youngest. In a sequence of undisturbed sedimentary rock layers, the youngest rock is on top, and the oldest is on the bottom. Events such as faulting, folding, intrusions, and erosion, can also be relative age dated. If any one of those events affects a sedimentary rock layer, the event must be younger than the affected rock. The sequence in Figure 4–4, from oldest to youngest is a) deposition of sedimentary rocks *1, 2,* and *3,* b) faulting, c) erosion (unconformity), and d) deposition of sedimentary rocks *4* and *5.*

Fossils

An important tool in relative age dating of sedimentary rocks is fossils. *Fossils* are the preserved remains of plants and animals in the rocks (Fig. 4–5). There are several ways in which fossils are preserved. Many animals that live in the sea such as corals and clams secrete shells of lime ($CaCO_3$). The shells can be preserved unaltered in sedimentary rocks. Other animals have bones that can be preserved. Plants can be preserved as films of carbon in mud, which becomes shale. Sometimes the pore spaces of bone or shell are filled with minerals deposited by groundwater in a process similar to the cementation of clastic grains into sedimentary rock. Other fossils are preserved when the original matter is completely replaced by another mineral in the subsurface. Petrified wood is formed by the replacement of wood by silicon dioxide that preserves the grain structure of the wood. *Trace fossils*, such as burrows, tracks, or trails, are indirect evidence of ancient life.

Certain species of plants and animals lived during certain geologic times. They eventually became *extinct* (disappeared from the earth) and replaced by newer plants and animals. This continuous succession of

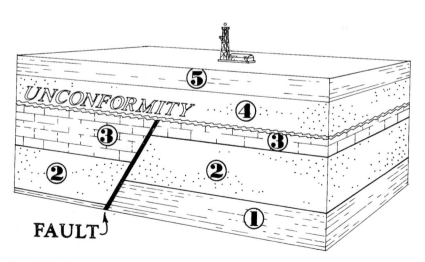

Fig. 4–4 *Relative age dating*

organisms throughout geologic time is known as *evolution*. Vertical sequences of sedimentary rock layers that have been relative age dated can be used to determine the relative ages of the fossils in those rock layers (Fig. 4–6). Geologists have collected and established the relative ages of most fossils. The evolutionary sequence of the fossils can be used to relative age date any sedimentary rocks that contain those fossils. In Figure 4–7, the rocks labeled A are older than those labeled B.

A *guide* or *index fossil* is a distinctive plant or animal that lived during a relatively short span of geologic time. This fossil species identifies the age of any sedimentary rock in which it occurs. A *fossil assemblage* is a group of fossils found in the same sedimentary rocks. It identifies that zone of rocks and the geologic time during which those rocks and fossils were deposited.

Fossils can also be used to determine the depositional environment of the sediments. Different plants and animals live in different environments such as beach, marsh, and deep ocean.

Fossils, like sedimentary rocks, can be indirectly dated by radioactivity (Fig. 4–8). If volcanic ash layers that occur above and below the fossil are dated, the fossil is younger than the underlying ash layer and older than the overlying ash layer.

Microfossils

A well is drilled down to a *drilling target*, a potential reservoir rock. But as the well is drilled, it penetrates hundreds of sedimentary rock layers consisting of shales, sandstones and limestones that look very similar. How can each sedimentary rock layer and the drilling target be identified while drilling through them?

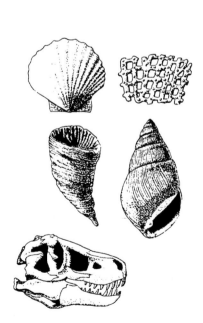

Fig. 4–5 *Fossils*

SEDIMENTARY ROCKS

FOSSILS

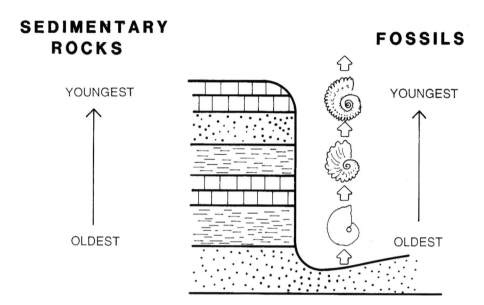

YOUNGEST

OLDEST

YOUNGEST

OLDEST

Fig. 4-6 *Relative age dating of fossils*

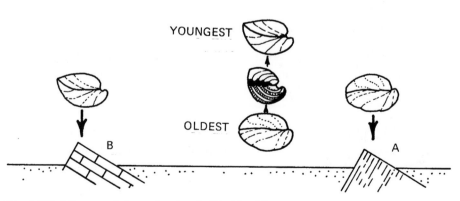

YOUNGEST

OLDEST

B

A

Fig. 4-7 *Relative age dating using fossils (A) older (B) younger*

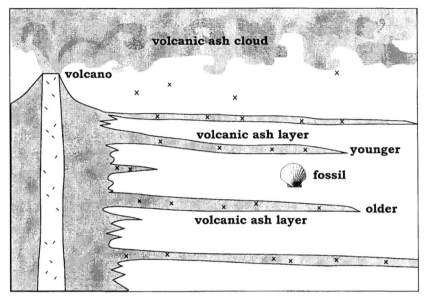

Fig. 4-8 *Indirect age dating of a fossils by radioactive age dating volcanic ash layers*

Each sedimentary rock layer was deposited during a different time and has different fossils that are the key to identifying the subsurface rock layers. Large fossils such as clams and corals, however, are broken into small chips (well cuttings) by the drill bit. It is almost impossible to identify the pieces when they are finally flushed to the surface by the drilling mud. The sedimentary rocks, however, also contain abundant microfossils.

Microfossils are fossils that are so small that they can be identified only with a microscope (Plate 4–1). They are often undamaged by the drill bit and are flushed unbroken in the well cuttings up to the surface. Microfossils are used in *biostratigraphy* to identify the rocks and their ages. Rock layers that contain a characteristic microfossil are often named after that microfossil and called a *zone* or *biozone*. The Siphonia davisi zone is composed of sedimentary rock layers that contain the microfossil Siphonia davisi (Fig. 4–9). A horizon in a well that is identified by the first appearance, most abundant occurrence, or last appearance of a microfossil is called a *paleo pick*. Paleo picks can also be used to determine if the sedimentary rock layers in a well are higher or lower in elevation than those in a well that has already been drilled (Fig. 4–10).

Many microfossils are shells of single-celled plants and animals that live in the ocean. *Foraminifera* (*forams*) are single-celled animals with shells (Plate 4–1a) composed primarily of $CaCO_3$. They live by either floating in the ocean or growing attached to the bottom. More than 30,000 species of forams have existed throughout geologic time.

Radiolaria are single-celled, floating animals that live in the ocean and have shells (Plate 4–1c) of SiO_2. Some cherts were formed by alteration of radiolarian shells deposits.

Coccoliths are plates from the spherical $CaCO_3$ shell (coccolithophore) of algae that floats in the ocean. They are so small that they can be identified only with a scanning electron microscope. Ancient, relatively pure $CaCO_3$ deposits of coccoliths or foraminifera microfossils are called *chalks*.

Diatoms are single-celled plants that float in water and have shells of SiO_2 (Plate 4–1b). Ancient deposits of relatively pure diatom microfossils are called *diatomaceous earth*. Some cherts formed from ancient diatom shell deposits. Spores and pollens given off by plants to reproduce also are good microfossils.

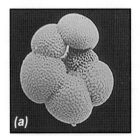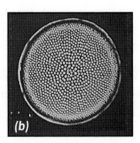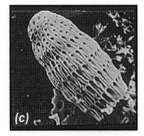

Plate 4-1 *Magnified microfossils (a) foraminifera, (b) diatom, and (c) radiolaria*

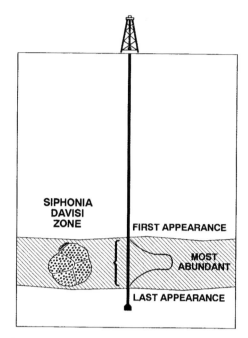

Fig. 4-9 *Microfossil zone*

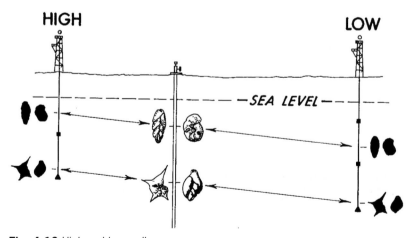

Fig. 4-10 *High and low wells*

Scientists who study fossils are called *paleontologists*, and those who specialize in microfossils are *micropaleontologists*. Because they pick microfossils (bugs) from well cuttings to examine them under a microscope, they are often called *bug pickers*. Micropaleontologists specializing in spores and pollens are called *palynologists* or *weed and seed people*.

The Geologic Time Scale

The geologic time scale was developed during the early 1800s by relative age dating sedimentary rocks and fossils in Europe. Large divisions of geologic time are called *eras*. Eras are subdivided into *periods*, and periods into *epochs*. The geologic time scale is presented in Table 4–2. In Europe, the Mississippian and Pennsylvanian periods are combined into the Carboniferous Period.

Earth History

It is known by radioactive age dating that the earth is about 4.5 billion years old. For the first part of the Precambrian Era, there is no fossil evidence that life existed on earth. The first life, probably bacteria followed later by algae floating in the ocean, appeared approximately 3.5 billion years ago. The fossil record throughout the later part of the Precambrian is sparse.

A great abundance of diverse plants and animals were living in the ocean at the start of the Paleozoic Era (543 million years ago) as shown in the fossil record. All the major animal phyla that we know today in the oceans, except the vertebrates, were present. Nothing, however, existed on the land. During the Ordovician Period, fish, the first vertebrates, came into existence. Plants and animals finally adapted to life on the land in the next period, the Silurian. During the Pennsylvanian Period, swamps covered large areas of the land. Primitive plants, such a ferns and horsetail rushes, grew to great heights. These Pennsylvanian swamp deposits formed many of the world's coal deposits. During the last period of the Paleozoic, the Permian, the climate was very dry and warm, and the lands were covered with deserts.

Table 4–2

Geological time scale

Era	Period	Epoch	Absolute age (years)
- -			0
		Holocene	
	Quaternary	————	10,000 thousand
		Pleistocene	
Cenozoic	————————		1.8 million
		Pliocene	
		————	5.3 million
		Miocene	
	Tertiary	————	24 million
		Oligocene	
		————	34 million
		Eocene	
		————	55 million
		Paleocene	
- -			65 million
	Cretaceous		
	————————		144 million
Mesozoic	Jurassic		
	————————		206 million
	Triassic		
- -			248 million
	Permian		
	————————		290 million
	Pennsylvanian		
	————————		323 million
	Mississippian		
Paleozoic	————————		354 million
	Devonian		
	————————		417 million
	Silurian		
	————————		443 million
	Ordovician		
	————————		490 million
	Cambrian		
- -			543 million
Precambrian			
- -			4.5 billion

During the Permian Period, egg-laying reptiles appeared. At the end of the Permian, the greatest extinction (90%) of plants and animals in the history of the earth occurred. There have been four great extinctions during geological time but this was the most extensive

The Mesozoic Era, starting about 248 million years ago, is known as the age of reptiles. These animals, which include the dinosaurs, dominated the earth. They filled a great diversity of ecological niches for more than 150 million years. Most dinosaurs were plant eaters, but some were carnivores. During this time, great reptiles lived in the oceans, while others flew through the air with wingspans more than 50 ft (16 m). During the Jurassic Period, the middle period of the Mesozoic, mammals and birds appeared. They were small and were dominated by the reptiles throughout the remainder of the Mesozoic.

At the very end of the Mesozoic Era (65 million years ago), one of the most complete and sudden extinction of life in the history of the earth occurred. All the dinosaurs died out, along with the flying reptiles, most swimming reptiles, and 75% of all species of plants and animals. The extinction was remarkable because reptiles apparently dominated their environment until the very end. They then disappeared in an instant of geologic time called the *great killing*.

This extinction was caused when a comet or meteorite hit the earth. It was at least 10 miles (16 kms) in diameter and hit the earth head on with a speed of about 100,000 miles per hour (160,000 km/hr). In every location throughout the world, the thin clay layer separating the Mesozoic and Cenozoic sedimentary rocks has an abnormally large concentration of iridium 121 (Ir^{121}). This rare, radioactive element is also found in abundance in meteorites. The Ir^{121} layer is thickest in the Caribbean, indicating that the impact occurred in that area. A crater, now filled with sediments and buried 3000 ft (1000 m) deep, occurs under the fishing village of Chicxulub on the Yucatan peninsula of Mexico (Fig. 4–11). It is 130 miles (208 km) in diameter and is the same age as the great killing.

Soot is common in the iridium 121 layer, indicating that the surface of the earth was consumed in a fire storm caused by the heat of the impact after the collision. The soot and dust caused by the collision and ejected into the atmosphere must have thrown the earth into total darkness for several months. Evidence in Texas and other areas, indicates that the waters of the Gulf of Mexico were thrown up into a great wave which swept across North America. Large amounts of sulfur dioxide were vaporized during the impact. It mixed with water in the atmosphere to form sulfuric acid that rained down on the earth.

Some mammals survived the great killing at the end of the Mesozoic Era. They flourished after their reptilian competition was eliminated, and the Cenozoic Era is known as the age of mammals. Grasses evolved during the Cenozoic and became an important food source for the mammals. Late in the Cenozoic, during the Pleistocene Epoch, or Ice Age, the climate was colder than it is today. Extensive ice sheets, thousands of feet thick, covered approximately one-third of the land. During four separate times, the glaciers advanced across the land and then retreated. The last ice sheet did not retreat until the end of the Pleistocene, just 10,000 years ago.

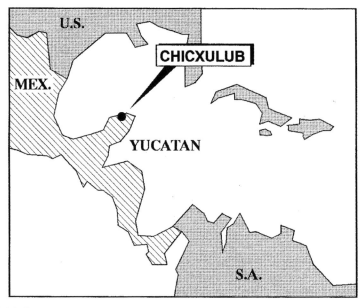

Fig. 4-11 Cretaceous-Tertiary impact site

five

DEFORMATION OF SEDIMENTARY ROCKS

Sedimentary rocks are originally deposited in horizontal layers. A type of oil and gas trap, called a *structural trap*, is formed by the deformation of these rock layers.

Weathering, Erosion, and Unconformities

Weathering is the breakdown of solid rock. Once a rock is exposed on the surface of the earth, either to the atmosphere or ocean bottom, it will eventually be mechanically broken into particles or chemically dissolved by the forces of weathering. Some sedimentary rocks are more resistant to weathering (*e.g.*, sandstones or even limestones in arid climates). Other rocks (*e.g.*, shales) readily break down. *Erosional processes* are those that transport and deposit sediments. These processes include gravity (landslides), wind, glaciers, waves, and rivers.

Sea level has been rising and falling throughout geological time. Whenever sea level was lower, the land was exposed to erosion, and some of the sedimentary rocks were stripped off the surface of the land. Buried, ancient erosional surfaces are called *unconformities*. Two types are disconformities and angular unconformities.

A *disconformity* is an erosional channel in which the sedimentary rock layers above and below the erosional surface parallel (Fig. 5–1). It is an ancient river channel that is usually filled with sand that has become sandstone.

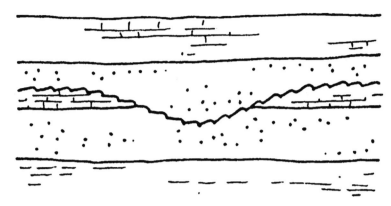

Fig. 5-1 Disconformity

An *angular unconformity* is an ancient erosional surface in which the sedimentary rock layers below the unconformity are tilted at an angle to the layers above the unconformity (Plate 5–1). The formation of an angular unconformity starts with the deposition of horizontal sediment layers as ancient seas cover the earth. After the seas have retreated, exposing the earth, the sedimentary rocks are tilted to form hills and mountains. The hills and mountains are then eroded down, leaving an erosional surface. The seas again cover the land, depositing sedimentary rock layers on the erosional surface and bury it in the subsurface. An angular unconformity represents a time of mountain building followed by erosion. It often covers a large subsurface area.

Angular unconformities can form gas and oil traps (Fig. 5–2). One of the sedimentary rock layers below it must be a reservoir rock that can store gas and oil; usually a sandstone or limestone. The sedimentary rock layer above it must be a caprock that acts as a seal; usually a shale or salt layer. The gas and oil forms below the unconformity in a source rock such as black shale. It migrates up into and then through the pore spaces of the reservoir rock until it reaches the angular unconformity surface where it is trapped below the caprock.

Because angular unconformities can cover large subsurface areas, they can form giant gas and oil fields. The two largest oil fields in North America, the East Texas field and the Prudhoe Bay field in Alaska, are both

Plate 5-1 *Angular unconformity showing flat sedimentary rocks above and sedimentary rocks tilted at an angle below (sea cliff England)*

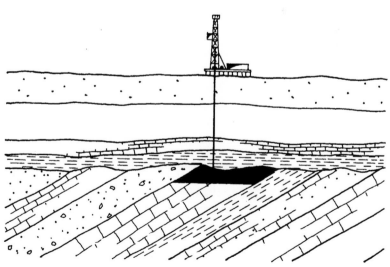

Fig. 5–2 *Angular unconformity trap*

in angular unconformity traps. In both fields, the horizontal rock layers on the surface of the ground don't give any indication of the subsurface angular unconformities and their giant oil accumulations.

The East Texas field originally contained more than 7 billion bbls (1.1 billion m³) of oil. The oil is located in the Woodbine Sandstone below the angular unconformity (Fig. 5–3) at a depth below about 3500 ft (1070 m). The Austin Chalk, the caprock, directly overlies the angular unconformity. The Woodbine Sandstone was originally deposited as a horizontal layer of sands when shallow seas covered East Texas about 100 million years ago (Fig. 5–4a). The sandstone was then buried in the subsurface as it was covered with other sediments (Fig. 5–4b). Later, the Sabine uplift, along the Texas-Louisiana border, arched up and exposed the Woodbine Sandstone (Fig. 5–4c). Erosion removed the Woodbine Sandstone from the top of the arch (Fig. 5–4d). After that, the seas invaded the area, depositing the Austin Chalk and other sediments, covering the angular unconformity (Fig. 5–4e). The oil formed in the Eagle Ford Shale source rock below and migrated up into the Woodbine Sandstone. It then flowed along the porous sandstone toward the east until it was trapped under the angular unconformity, unable to flow into the Austin Chalk.

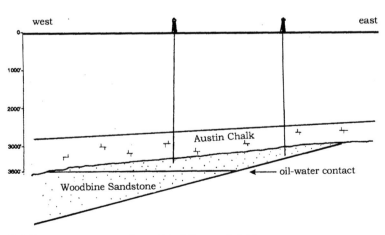

Fig. 5–3 *East-west cross section of East Texas oil field*

The discovery of the East Texas oil field in 1930 is a classical story of petroleum history. Oil companies explored this area in the early 1900s. There were no oil seeps in the area, and they became discouraged after drilling some dry holes. They abandoned this area by the mid-1920s to drill in the newly discovered giant West Texas oil fields. Because of this, Charlie Joiner, a driller and promoter, was able to obtain leases for drilling on a large area of eastern Texas at a very low cost.

He started to drill in the area in the late 1920s with what is best described as "random drilling." His only geological help was from a veterinarian by the name of Dr. A. D. Lloyd. It was poor farm land, and the local

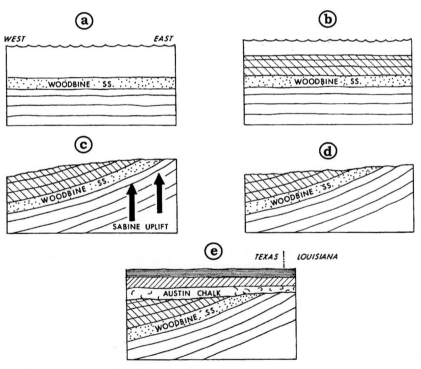

Fig. 5–4 Formation of East Texas oil field trap

cotton farmers would often volunteer to help drill the well. Because Charlie Joiner had little money, he traded shares in the well for room and board, repairing equipment, buying supplies and hiring help.

The No. 3 Daisy Bradford well finally reached below the angular unconformity at 3725 ft (1135 m) after 16 months of drilling and blew in the East Texas oil field. The well initially tested 6800 bbl/day (1080 m³/day) and was completed to produce 300 bbl/day (48 m³/day). Unfortunately, Charlie

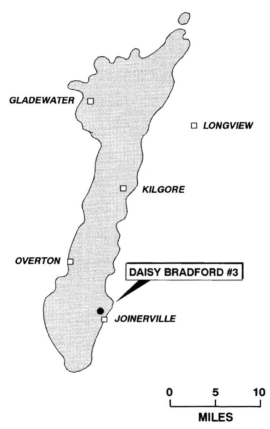

Fig. 5–5 Map of East Texas oil field (modified from Halbouty, 1991)

Joiner in his financial need during drilling, had sold 300% of the Daisy Bradford well and was in legal trouble. H.L. Hunt negotiated a deal with Charlie to settle his legal problems and pay him US $1,335,000, mostly in future oil production for 5000 acres of prime leases in the field.

The East Texas oil field is 45 miles (72 km) long and 5 miles (8 km) wide (Fig. 5-5). Almost instantly, many poor farming families that had land in the field became Texas millionaires. More than 30,000 wells were drilled in the field that has now produced more than 5 billion bbls (800 million m^3) of light, sweet crude oil.

Anticlines and Synclines

An *anticline* is a large, upward arch of sedimentary rocks (Fig. 5–6), whereas a *syncline* is a large, downward arch of rocks. Anticlines (but not synclines) form gas and oil traps. Folds such as anticlines expose the rocks to erosion. If the anticlines are relatively young, they haven't been eroded and appear as topographic ridges on the surface. A series of young rising anticlines that are also prolific petroleum producers occur as a line of hills that cross the Los Angeles basin (Fig. 5–7). These trend from Beverly Hills in the north, through the Inglewood (Baldwin Hills) and Dominguez fields, southward to Long Beach, and offshore into the Huntington Beach field.

Most anticlines and synclines are not level and are tilted with respect to the surface of the earth. These are called *plunging anticlines* (Fig. 5–8) and *synclines*. During and after folding, erosion rapidly levels the folds. The pattern of an eroded plunging anticline or syncline is lobate-shaped, called a *nose* (Fig. 5–9).

The formation of anticlines and synclines results in shortening of the earth's crust (Fig. 5–10). Forces that shorten the earth's crust are compressional. If an area of the earth's crust is compressed, the rocks will be folded into anticlines and synclines. If folds are present in the rocks of the earth's crust, that area probably has been compressed some time in the past.

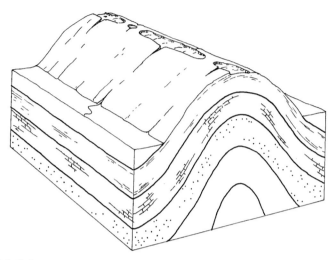

Fig. 5–6 Anticline

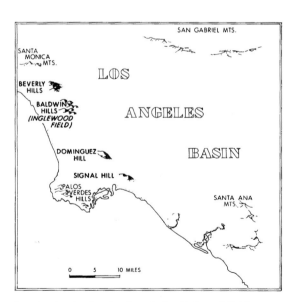

Fig. 5–7 Los Angeles basin showing trend of anticline oil fields

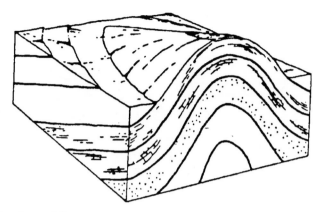

Fig. 5–8 *Plunging anticline*

Domes

A *dome* is a circular or elliptical uplift. When eroded flat, a dome forms a bull's eye pattern of concentric rock layers with the oldest rock in the center (Fig. 5–11). Domes also form gas and oil traps. Oil was first discovered in the Middle East in Bahrain, an island in the Persian Gulf, in 1932. The traps in Bahrain were domes with a low hill on the surface above each of them. A similar low hill above a dome in Saudi Arabia was drilled to find the first oil field there in 1937.

Anticlines and domes were the first type of petroleum trap recognized. They form many of the giant oil and gas fields of the world. Most of the Middle East oil fields are in anticline and dome traps. The Cushing oil field of Oklahoma, discovered in 1912, is located to the southwest of Tulsa. The trap is an anticline with three domes superimposed on it (Fig. 5–12). The reservoir rock is the Bartlesville Sandstone along with several other sandstones and a limestone. The best producing wells are on the domes. The Cushing oil field will produce 450 million bbls (72 million m^3) of oil. It was the largest oil field in the world during World War I.

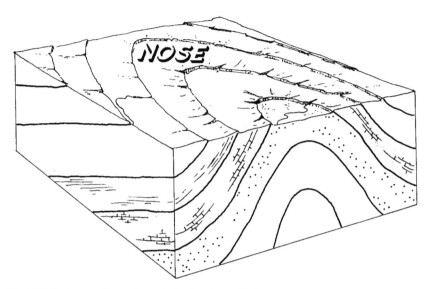

Fig. 5–9 *Surface pattern of eroded, plunging anticline*

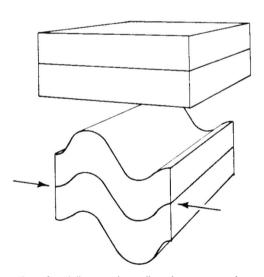

Fig. 5–10 *Formation of anticlines and synclines by compression*

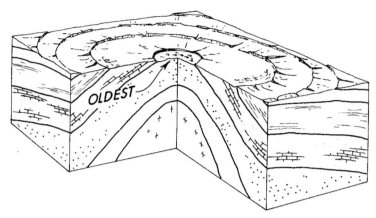

Fig. 5–11 Surface pattern of eroded dome

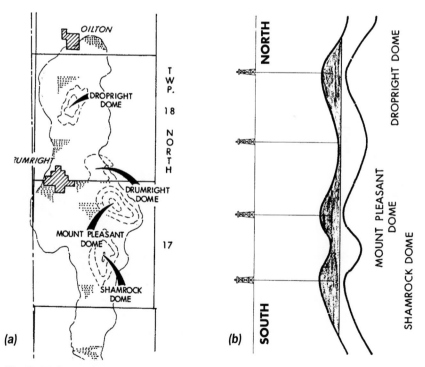

Fig. 5–12 Cushing oil field, Oklahoma (a) map (b) cross section

Homoclines

Sedimentary rocks dipping uniformly in one direction are known as a *homocline* (Fig. 5–13). Alternating layers of hard and soft sedimentary rocks in a homocline will be eroded to form a series of parallel ridges on the surface of the ground.

Fractures

Joints

A *joint* is a fracture in the rocks without any movement of one side relative to the other (Plate 5–2). Joints in sedimentary rocks improve the reservoir qualities of the rock. They increase the fluid storage capacity of the rock (porosity) and the ability of the fluid to flow through the rock (permeability). Any naturally fractured rock is a potential reservoir rock.

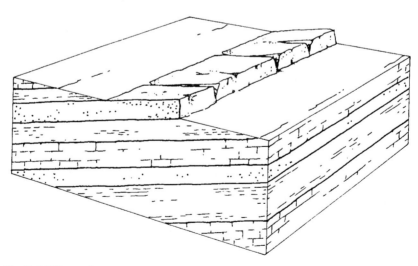

Fig. 5–13 Homocline

Plate 5–2 *Joints in a sandstone, Winding Stair Mountains, Oklahoma*

Faults

Faults are breaks in the rocks along which one side has moved relative to the other (Plate 5–3). The relative movement of each side is used to classify faults (Fig. 5–14). *Dip-slip* faults move primarily up and down, whereas *strike-slip* faults move primarily horizontal. *Oblique-slip* faults have roughly equal dip-slip and strike-slip displacements.

The side of a fault that extends under the fault plane is called the *footwall* (Fig. 5–15), and the side that protrudes above the fault plane is the *hanging wall*. *Throw* (Fig. 5–16) is the vertical displacement on a dip-slip fault. The side of the dip-slip fault that goes down is called the *downthrown* side, and the side that goes up is called the *upthrown* side.

Plate 5–3 *Fault showing displacement of sedimentary rock layers, Austin Chalk, Texas*

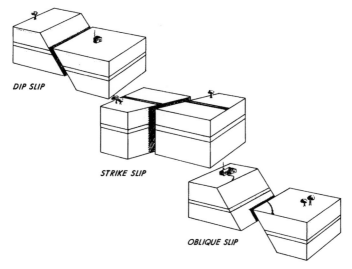

Fig. 5–14 *Types of faults*

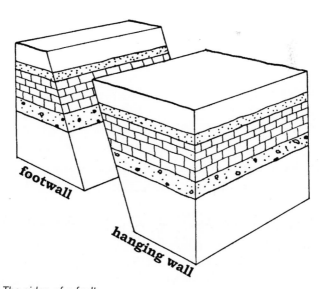

Fig. 5–15 *The sides of a fault*

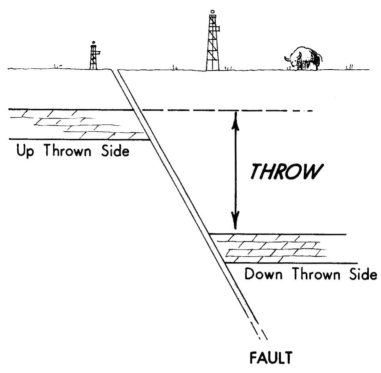

THROW

Up Thrown Side

Down Thrown Side

FAULT

Fig. 5–16 Dip-slip fault terminology

Two types of dip-slip faults are normal and reverse. If the hanging wall has moved down relative to the footwall, it is a *normal dip-slip fault* (Fig. 5–17). In a normal dip-slip fault, the beds are separated and pulled apart. A normal dip-slip fault is identified in the subsurface by a *lost section*, a missing layer, or layers of rocks (Fig. 5–18).

A series of parallel, normal dip-slip faults forms a structure called horst and graben (Fig. 5–19). A *graben* is the down-dropped block between two normal faults. A *horst* is the ridge left standing between two grabens. These can range in size from inches to tens of miles across.

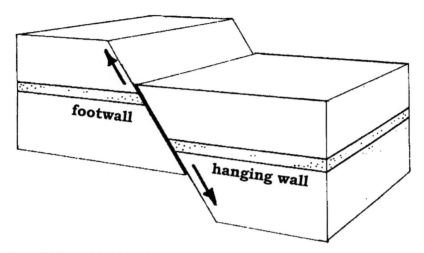

Fig. 5–17 Normal dip-slip fault

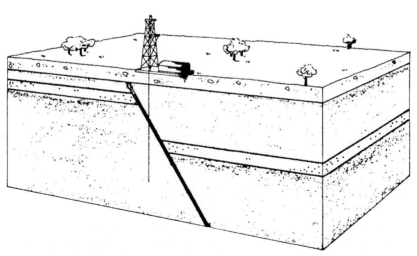

Fig. 5–18 Lost section on a normal dip-slip fault

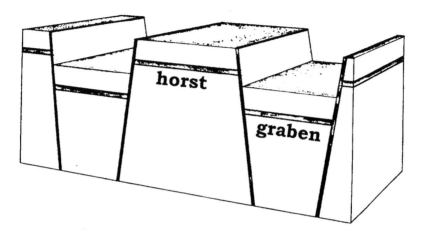

Fig. 5–19 Horst and graben

If the hanging wall has moved up relative to the footwall, it is a *reverse dip-slip fault.* In a reverse dip-slip fault, some subsurface beds overlap. It is possible to drill through this fault and encounter the same rock layers twice in a *double section* (Fig. 5–20).

A *thrust fault* is a reverse fault with a fault plane less than 45° from horizontal (Fig. 5–21). On a thrust fault, the upper part (the hanging wall) has been thrust up and over the lower part (the footwall). There are some thrust faults in the earth's crust where the hanging wall has been thrust horizontally tens of miles over the footwall. A series of large thrust faults, called the Rocky Mountain Overthrust Belt, occurs in a band along the eastern side of the Rocky Mountains. A series of large gas and oil traps are located in the overthrust belt.

A normal dip-slip fault is formed when the rocks are pulled apart by tensional forces (Fig. 5–22). A reverse dip-slip fault is formed by shortening the rocks with compressional forces (Fig. 5–22). When the earth's crust is pulled apart, normal dip-slip faults with horsts and grabens are formed. When the earth's crust is squeezed, reverse dip-slip and thrust faults and folds, such as anticlines and synclines, are formed.

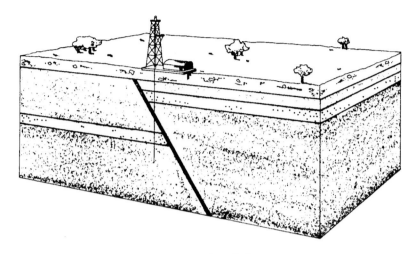

Fig. 5–20 *Double section on a reverse dip-slip fault*

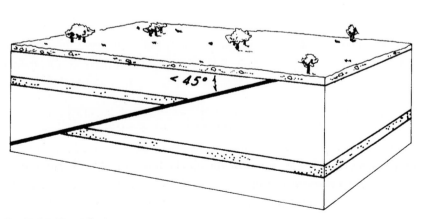

Fig. 5–21 *Thrust fault*

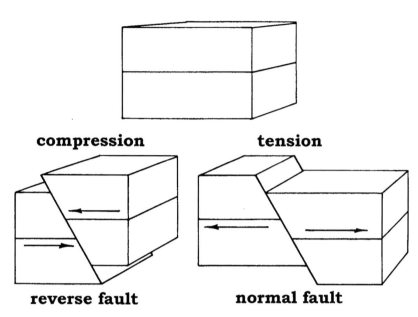

Fig. 5–22 *Forces that form a reverse dip-slip fault and normal dip-slip fault*

Faults can be both active and inactive. When a fault moves, it can produce shock waves called an *earthquake*. Many faults, however, moved a long time ago and are inactive today. Two very large faults occur in Oklahoma, the Seneca and Nemaha faults. Both were active hundreds of millions of years ago but are inactive today.

Dip-slip faults form traps by displacing the reservoir rock (Fig. 5–23). The fault must be a *sealing fault*, which means it prevents fluid flow across or along the fault. Any gas and oil migrating up a reservoir rock will be trapped under the sealing fault. Fault traps are common but are usually relatively small because a single fault is linear and doesn't have sides to contain large volumes of petroleum. The Bridge pool oil field of Santa Paula,

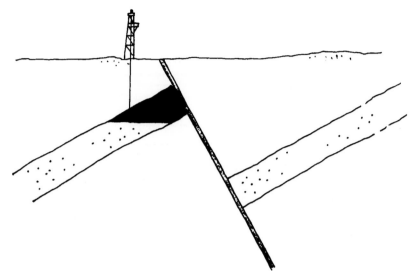

Fig. 5–23 Fault trap

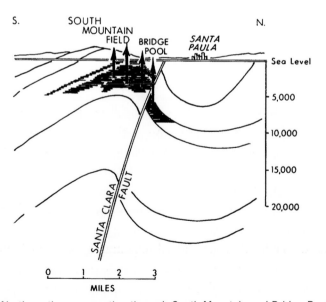

Fig. 5–24 North-south cross section through South Mountain and Bridge Pool oil fields, California

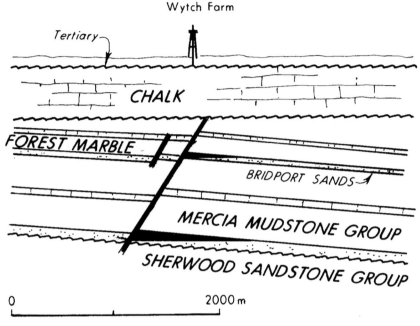

Fig. 5–25 *Wytch Farm oil field, England (modified from Colter and Harvard, 1981)*

California is a fault trap (Fig. 5–24). The reservoir rock is the Oligiocene age sandstones in the Sespe Formation that is terminated against the Santa Clara Fault. The largest oil field on land in England is the Wytch Farm field located southwest of London on the South Dorset coast. There are natural oil seeps along the coast, and the field was discovered in 1973. The trap was formed by a fault cutting the Sherwood Sandstone reservoir rocks (Fig. 5–25). It contains 286 million bbls (45 million m³) of oil.

A strike-slip fault is described by the horizontal movement of one side relative to the other (Fig. 5–26). If the opposite side of the fault moves to the right, it is a *right-lateral strike-slip fault*. If it moves to the left, it is a *left-lateral strike-slip fault*. The San Andreas Fault of California is an active right-

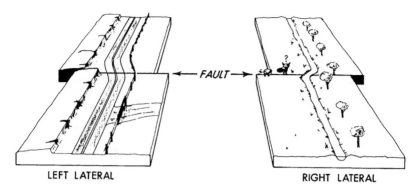

LEFT LATERAL   RIGHT LATERAL

Fig. 5–26 *Strike slip faults*

lateral strike-slip fault. It is hundreds of miles long and has moved many tens of miles over a long time.

The Potrero oil field in California (Fig. 5–27) is formed by an anticline on sandstone reservoir rocks. The crest of the anticline is displaced 1200 ft (365 m) by the Potrero fault, a right-lateral strike slip fault.

Buried, tilted fault blocks can form large petroleum traps. During the geological past, horizontal sedimentary rocks (Fig. 5–28a) were broken by normal faults into large, tilted fault blocks (Fig. 5–28b). Some of the blocks can contain reservoir rocks. Later, the seas covered the tilted fault blocks and deposited a caprock of shale or salt on them (Fig. 5–28c). The oil and gas then formed and migrated up along the reservoir rock to below the sealing fault or caprock.

The Statford oil field, the largest oil field in the North Sea, is located in both the United Kingdom and Norway sectors. It was discovered by seismic exploration and drilled in 1974. The trap is a westward-tilted fault block (Fig. 5–29). The two, very porous and permeable sandstone reservoirs rocks (Brent and Statfjord sandstones) are several hundred feet thick. The seal is the overlying shale. The ultimate production will be 3 billion bbls (500 million m^3) of light, sweet crude oil.

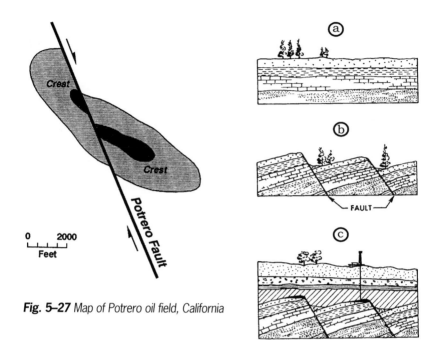

Fig. 5–27 Map of Potrero oil field, California

Fig. 5–28 Formation of tilted fault block traps

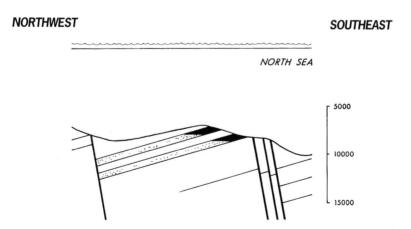

NORTHWEST

SOUTHEAST

NORTH SEA

5000

10000

15000

Fig. 5–29 Cross section of Statfjord field, North Sea (modified from Kirk, 1980)

six

SANDSTONE RESERVOIR ROCKS

The two most common petroleum reservoir rocks are sandstones and carbonates (limestones and dolomites). Most sandstones and limestones, however, are not reservoir rocks. Sandstones and limestones show a wide variety of textures and were deposited in a variety of environments. Some reservoir rocks, such as limestone reefs and river channel sandstones, were deposited completely encased in shale. Because the shale below is a source rock for gas and oil and the shale above is a caprock, the reef or channel contains gas and oil. This is called a *primary stratigraphic trap*.

Dune Sandstones

Sand dunes are formed by wind in both desert and coastal environments. The wind picks up and suspends only clay- and silt-sized particles in the air. The only other size particle moved by the wind is fine sand that is bounced along the ground and deposited locally. Coarser grained sediments cannot be moved by the wind. Because of this, sand dunes are composed of very well-sorted fine sand and can be excellent reservoir rock.

Sand dunes have a distinctive shape and internal structure. During periods of low wind velocity, wind and bouncing fine sand flow smoothly over any surface irregularity such as a hill (Fig. 6–1a). As the wind velocity increases, an eddy current forms downwind from the crest of the hill (Fig. 6–1b). The eddy current flows opposite the wind direction, causing the fine sand to pile up on the crest of the hill. Eventually, the sand becomes unstable and avalanches down the back side of the hill depositing a layer of sand on the avalanche or slip face of the dune (Fig. 6–1c). This process continues as long as the wind is blowing fast enough to maintain the eddy current.

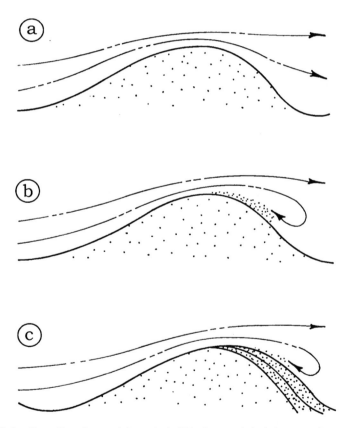

Fig. 6–1 *Formation of a sand dune a) wind blowing at relatively low speed across a sand dune, b) wind blowing at a faster speed, forming an eddy current on the backside of the dune, and c) sand avalanching down the backside of the dune*

The sand layers formed by the avalanching sands have a characteristic shape and are called *crossbeds* (Fig. 6–2). They are steep at the top and are almost horizontal at the bottom. As the wind blows, sand is eroded off the windward side of the dune and is deposited on the opposite side of the dune. This causes the dune to migrate in the direction of the slip face and the wind. Dune sand is characterized by internal crossbeds.

Ancient dune sandstones form extensive subsurface sandstone reservoirs. One example is the Jurassic age Nugget Sandstone, the primary reservoir in the Rocky Mountain overthrust belt oil and gas fields of the United States. It was deposited as sand dunes in a desert that covered much of the western part of the United States during Jurassic time.

During the Permian time, sea level fell to expose the bottom of the North Sea, and a desert climate occurred. A large salt lake surrounded by sand dunes formed in the center of the North Sea. These sand dunes are now buried by other sediments on the bottom of the North Sea and form the Rotliegend Sandstone. It underlies the southern portion of the North Sea and parts of England and Europe, where it is the reservoir for several North Sea gas and oil fields and the giant Groningen gas field in the Netherlands.

Shoreline Sandstones

Beaches are long, narrow deposits of well-sorted sand. Waves wash the finer-grained silt and clay particles out of the beach sands. The muddy water is carried offshore where the mud settles out of the water and is deposited in deep water. Waves shape the beach into a long strip of sand.

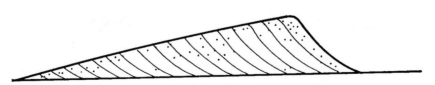

Fig. 6–2 Crossbeds

In the South Texas coastal plain (Fig. 6–3). there is a series of buried shoreline sandstones that are oil and gas reservoirs. The Yegua-Jackson sands of Eocene age are located inland, and the Frio-Vicksburg sands of Oligocene age are located on the Gulf of Mexico side.

The Clinton oil and gas fields run north to south through the state of Ohio (Fig. 6–4). The reservoir rock is the Silurian Age Clinton Sandstone. It was deposited as a shoreline sandstone and is completely encased in shale, which is the source rock below the sandstone and the caprock above the sandstone. Drilling into the Clinton sandstone started in the 1890s with cable tool drilling rigs.

Beach sands, called *buttress sands*, can be deposited on an angular unconformity during rising seas (Fig. 6–5) and form giant oil and gas field reservoirs. The Bolivar Coastal fields of Lake Maracaibo, Venezuela (Fig. 6–6a), containing more than 30 billion bbls (5 billion m^3) of recoverable oil, are examples. A cross section (Fig. 6–6b) from west to east shows

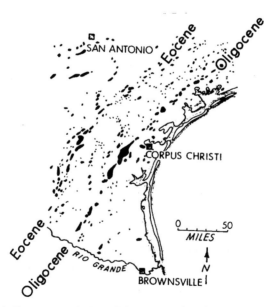

Fig. 6–3 *South Texas coastal plain oil fields (modified from Landes, 1970)*

Fig. 6–4   *Clinton Sandstone gas fields, Ohio (modified from State of Ohio, 1966)*

the Oligiocene age angular unconformity dipping down to the west under Lake Maracaibo. Miocene age reservoir sandstones directly overlay the unconformity under Lake Maracaibo. Some of the oil leaks out along the unconformity to form a line of oil seeps along the eastern shore of Lake Maracaibo. The Bolivar Coastal fields were discovered in 1917 and produce from wells on platforms in the shallow lake.

Pembina, the largest oil field in Canada (Fig. 6–7) also produces from a beach sandstone and conglomerate reservoir, the Cretaceous age Cardium Formation, overlying an unconformity. The field was discovered in 1953 by drilling to a deeper seismic anomaly that was thought to be a Devonian age reef. The reef was not there, but the wellsite geologist tested the 20 feet (6 m) of oil sandstone encountered higher in the well and discovered the field. The field covers 900 sq miles (2330 sq km), making it one of the largest fields in the world by surface area. There are 7.5 billion bbls (1.2 billion m³) of oil in place. Pembina will eventually yield 1.56 billion bbls (250 million m³) of sweet, 37 °API gravity oil. There is no gas cap, and the average net pay is only about 20 ft (6 m).

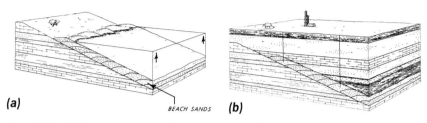

Fig. 6–5 (a) Rising seas depositing beach sands (b) buttress sands

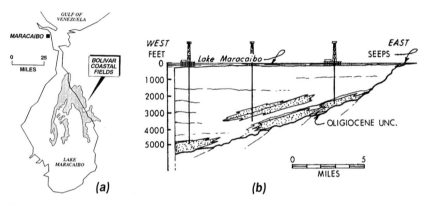

Fig. 6–6 Bolivar Coastal oil fields, Lake Maracaibo, Venezuela (a) map (b) cross section (modified from Martinez, 1970)

River Sandstones

Most rivers flow through loops and bends (Fig. 6–8) called *meanders*. These are caused by friction of water flowing along the bottom of the channel. In a meander, water must flow further along the outside of the meander bend to keep up with the water on the inside of the meander. Water on the

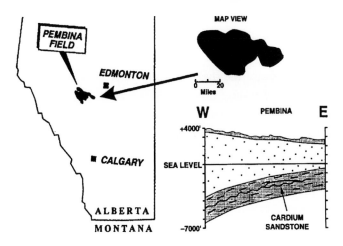

Fig. 6–7 Map and cross section of Pembina oil field, Alberta

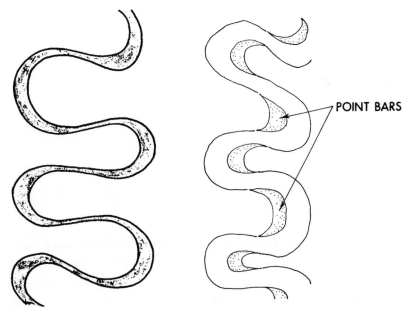

Fig. 6–8 Meandering river

Fig. 6–9 Point bar sands

outside of the meander bend flows faster and erodes the outer bank. Water
on the inside of the meander slows down, depositing the coarsest sediment
(sand) that was suspended in the water. The pattern of erosion and deposition on a meander causes the meander form to grow. The crescent-shaped,
well-sorted sand bars deposited on the inside of the river meanders (Fig. 6–9)
are called *point bars*. After the river abandons a meander, a clay plug that will
become shale is deposited in the channel.

Buried point bar sandstones are often good oil and gas reservoirs. The
Miller Creek field in the Powder River basin of Wyoming (Fig. 6–10) is an
example. It produces out of the Cretaceous age Fall River Sandstone.
Expected production is 5 million bbls (0.8 million m^3) of 33 °API oil. The
sandstone is tilted 2° to the southwest. The oil is trapped in the sandstone
against the shale channel fill to the east. There is 35 ft (11 m) of pay with
good porosity and permeability.

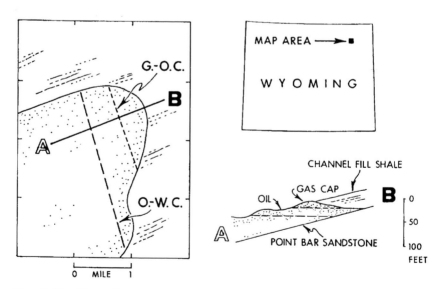

Fig. 6–10 *Map and cross section of Miller Creek field, Wyoming (modified from
Berg, 1968)*

The Bush City oil field of eastern Kansas (Fig. 6–11) was formed in a river channel segment. The Pennsylvanian age sandstone reservoir rock is 13 miles (21 km) long, about ¹/₄ of a mile (0.4 km) wide and up to 50 ft (15 m) thick. The entire channel segment, which is only about 600 ft (180 m) deep, is productive.

Most river channel sandstones are deposited and preserved as *incised valley fills* during a fall and rise of sea level. During sea level fall, the river incises (erodes) a valley; and during sea level rise, the valley is filled with sands. If the sands are overlain by a caprock, it can form a gas or oil trap.

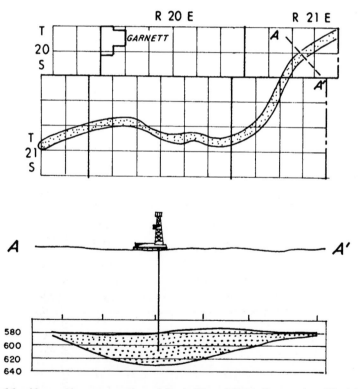

Fig. 6–11 *Map and cross section of Bush City oil field, Kansas (modified from Charles, 1941)*

The Stockholm Southwest field in Kansas, discovered in 1979, is an example of an incised-valley fill (Fig. 6–12). The Pennsylvanian age Stockholm Sandstone is the reservoir rock at a depth of about 5000 ft (1500 m). The caprock is the overlying shale, and the source rock is a deeper shale. The field contains 27 million bbls (4 million m³) of oil and will eventually produce 11 million bbls (2 million m³) of light, sweet oil.

A *braided river* forms with interconnecting channels separated by sand and gravel bars (Fig. 6–13). Braided rivers are caused by an overload of sediments that the river cannot transport. The Triassic age Sadlerochit Sandstone, the primary reservoir rock in the Prudhoe Bay Field, Alaska, was deposited partially by a braided river.

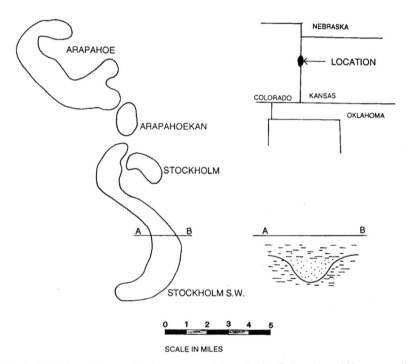

Fig. 6–12 *Map of Pennsylvanian age sandstone fields, Colorado and Kansas and cross section of Stockholm SW field (modified from Shumard, 1991)*

Fig. 6–13 *Braided river*

Delta Sandstones

A *delta* is a mass of sediments deposited by a river flowing into a body of water, such as a lake or ocean. There are several environments on a delta (Fig. 6–14). The river often bifurcates or divides into numerous channels, called *distributaries*, on the delta. The distributaries are located on low-lying swamps and marshes that are covered with river water during floods.

Two important processes occur on a delta. The river deposits sediments, which is a constructive force. Waves erode the sediments, which is a destructive force. The geometry of the delta that forms is a result of the relative importance of river deposition and wave erosion. A *constructive delta* (Fig. 6–15a) is shaped by river deposition. Wave erosion is relatively minor. A constructive delta has lobes of sediments that protrude into the ocean. The Mississippi River Delta is an example. A *destructive delta* (Fig. 6–15b) is shaped by wave erosion. It hardly protrudes from the shoreline. Wave erosion forms well-developed beaches in front of a destructive delta. The Niger River Delta and the Nile River Delta are examples.

Deltas can be good environments for the formation and accumulation of gas and oil (Fig. 6–16). Nutrient-rich, river water flowing into the ocean causes large, offshore algal blooms. The organic matter eventually falls to the sea bottom, forming an organic mud that is preserved as black shale in front

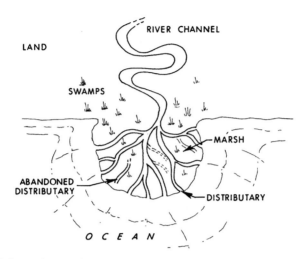

Fig. 6–14 Delta environments

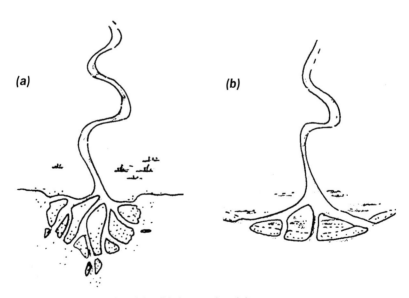

Fig. 6–15 (a) constructive delta (b) destructive delta

of the delta. Sediments cover the black shale source rock as the delta is deposited out into the ocean. The overlying delta sediments contain beach, and river channel sandstone reservoir rocks. As the loose shales compact, the delta surface subsides and is covered with marsh, swamp and river deposits. The gas and oil forms in the underlying black shale source rock and migrates up into the sandstone reservoir rocks.

The Pennsylvanian age Booch Sandstone in Oklahoma (Fig. 6–17) is a good reservoir rock. It was deposited as a southward-flowing river channel and distributary channel sandstone on a constructive delta in the Arkoma basin. The Booch Sandstone is productive where the sandstone crosses anticlines.

The Bell Creek oil field reservoir in Montana is a destructive delta. During the Cretaceous, the area bordered by Montana, Wyoming, and South Dakota was occupied by the water-filled Powder River basin. The Black Hills uplift was located to the east. A meandering river flowed out of the Black Hills and emptied into the Powder River basin, forming a destructive delta (Fig. 6–18a). The beach and channel sands, called the Muddy Sandstone, are encased in shale and form the reservoir rock for the Bell Creek oil field (Fig. 6–18b). The producing sandstone at a depth of 4500 ft (1400 m) is 20 to 40 ft (6 to 12 m) thick. The field will ultimately produce more than 200 million bbls (32 million m^3) of oil.

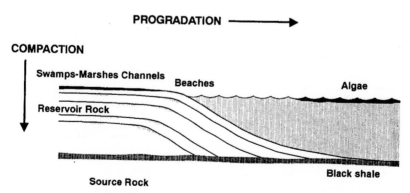

Fig. 6–16 *Cross section of delta showing source and reservoir rocks*

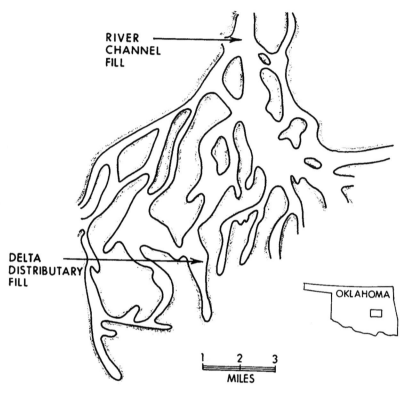

Fig. 6–17 *Map of the Booch Sandstone of Oklahoma (modified from Busch, 1974)*

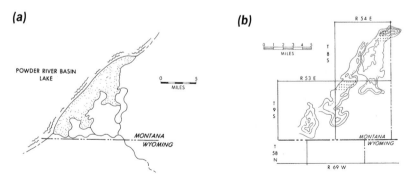

Fig. 6–18 *(a) shoreline of Powder River basin, Montana during Cretaceous time (b) map of the Bell Creek oil field, Montana (modified from McGreger and Biggs, 1970)*

A process that occurs on constructive deltas is *delta switching* (Fig. 6–19a). After a delta progrades out into a basin, the river becomes more inefficient as the river must flow over the flat delta to empty into the ocean. During a flood, the river can break through to a shorter, more efficient route to the sea. It will then abandon its older, less efficient route. A new delta will form to the side of the old delta, and the old delta will be eroded back by waves and covered with shallow water. The old delta will also subside due to sediment compaction. The process of delta switching by a river can deposit a thick mass of deltaic sediments along the margin of a basin. On a destructive delta, switching is confined to the major river channel that jumps about the shoreline (Fig. 6–19b).

The Mississippi River Delta province of Louisiana will produce more than 22 billion bbls (3.5 billion m^3) of oil. Most of the offshore gas and oil production in the Gulf of Mexico is located in shallow water to the southwest of

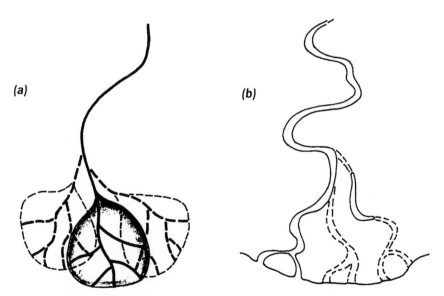

(a)

(b)

Fig. 6–19 (a) delta switching on a constructive delta (b) distributary switching on a destructive delta

the modern Mississippi River Delta. Several million years ago, sea level was lower by about 300 ft (90 m). At that time, the Mississippi River Delta switched back and forth off southwestern Louisiana (Fig. 6–20). Rising sea level covered these Miocene, Pliocene, and Pleistocene Age deltas. Much of the offshore production is from sandstone reservoirs in the ancient deltas.

By using carbon 14 (C^{14}) age dating, it has been determined there have been six previous deltas (Fig. 6–21) of the Mississippi River during the past 5000 years. Each prograded out into the Gulf of Mexico and was abandoned by delta switching. A new route, the Atchafalaya River route by Morgan City, is forming today. This route is only 140 miles (225 km) to the Gulf. The present route by Baton Rouge and New Orleans is inefficient and 300 miles (480 km) long. Some time in the future, the Mississippi River will abandon its present route and switch down the Atchafalaya River route.

A map of the Niger River Delta, a destructive delta in central western Africa shows the present and ancient shorelines (Fig. 6–22). The deposition of the delta by the Niger River into the Atlantic Ocean is shown by the progression of shorelines with time. Ancient shorelines (5 and 2 million years ago) located offshore from the present shoreline occurred when sea level was lower. The Niger River Delta is one of the world's greatest oil producing areas with more than 41 billion bbls (6.5 billion m^3) of recoverable oil in ancient beach and river channel sandstones.

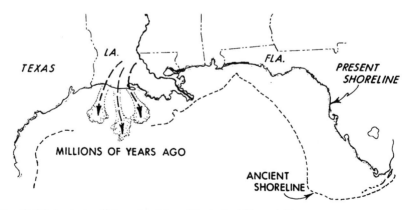

Fig. 6–20 *Ancient, offshore Louisiana Mississippi River deltas*

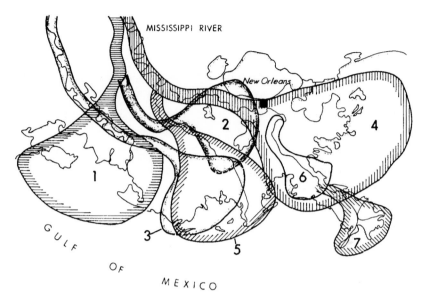

Fig. 6–21 *Last 5000 years of Mississippi River Delta history (modified from Kolb and Van Lopik, 1996)*

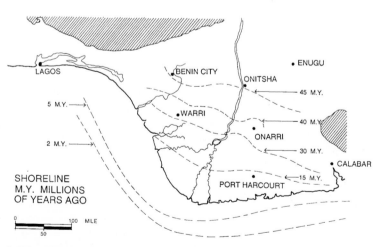

Fig. 6–22 *Present day and ancient shorelines of the Niger River Delta, Nigeria (modified from Burke, 1972)*

seven

CARBONATE RESERVOIR ROCKS

The sedimentary rock limestone is common in the ancient rock record. This is because lime ($CaCO_3$) is secreted as shells by many animals and plants that live in the ocean. Most limestones were deposited on the bottom of shallow, tropical seas that covered the land many times in the geological past. *Carbonates* are limestones and dolomites.

Reefs

Reefs are mounds of shells. All reefs have a wave-resistant, calcium carbonate framework of overlapping organic branches formed by a plant or animal. Other plants and animals live in the protection of the framework. Modern reefs often have corals as the framework. During the Paleozoic and Mesozoic eras, corals were not as important as they are today, and sponges, calcareous algae, clams called rudistids, and other organisms formed the reefs. Modern reefs grow only in clear, shallow, warm waters.

There are several types of reefs based on shape (Fig. 7–1). *Barrier reefs* grow parallel to a shoreline but are separated from the land by a lagoon. Both modern and ancient barrier reefs tend to be large. The Great Barrier Reef off eastern Australia is more than 1000 miles (1600 km) long. An *atoll* is a circular or elliptical reef surrounding a central lagoon. Atolls are common in the Pacific Ocean. *Pinnacle reefs* are smaller, steep-sided reefs.

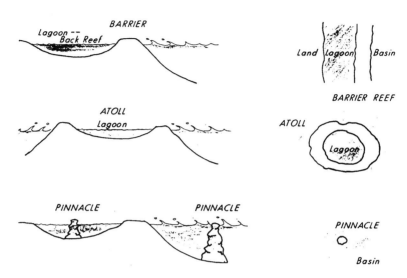

Fig. 7–1 *Reef types in cross section and map view*

Ancient barrier reefs and atolls are prolific petroleum reservoirs, especially in North America. The reef rock has the most original pore spaces. These spaces are often enhanced in the subsurface when fresh waters percolate through the pores and dissolve the limestone. In the lagoon, limestone mud called *micrite* was deposited. It is not reservoir rock. If the ancient reef is covered with a shale or salt caprock, it forms a gas and oil trap.

During the Permian Period, the climate of the world was hot and dry, a desert climate. The eastern and central portions of North America were exposed above sea level. Large areas were covered with sand dunes and salt flats. The last area covered with water was three, deep, tropical-water basins in west Texas, New Mexico and Mexico (Fig. 7–2). They are the Midland, Marfa and Delaware basins, known as the *Permian basins*. During the late Pennsylvanian and throughout most of the Permian time, thick limestones were deposited in and between these basins.

A barrier reef grew along the margin of the Delaware basin. The Capitan reef, more than 600 ft (180 m) thick, is exposed today in the Guadalupe Mountains of west Texas (Plate 7–1). To the north of the Capitan reef, is the

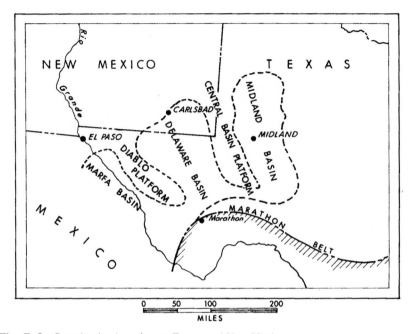

Fig. 7–2   *Permian basins of west Texas and New Mexico*

Abo reef (Fig. 7–3). Along the top of the buried Abo reef are numerous oil fields. One of the largest of these fields is the Empire Abo field (Fig. 7–4) that will produce more than 250 million bbls (40 million m³) of oil. The caprock is shale. It is bounded on the north by impermeable micrite limestones of the former lagoon.

During the Silurian Period, two large basins, the Illinois and Michigan basins, occurred in north central United States (Fig. 7–5). A barrier reef grew along the northern edge of the Michigan basin. This barrier reef, now exposed on the surface of the ground, has been unproductive. However, the hundreds of smaller, pinnacle reefs that grew basinward of the barrier reef are still buried in the subsurface and are productive (Fig. 7–6). The pinnacle reefs are hundreds of feet to several miles across and hundreds of feet thick. Each is overlain by a salt layer caprock. More than 1200 of the pinnacle reefs have been drilled in Michigan. Of these, more than 900 are productive for gas or oil.

Plate 7–1 *Capitan reef on Guadalupe Mountains, Midland basin, Texas*

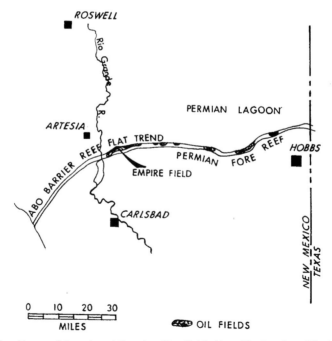

Fig. 7–3 *Abo reef trend and Empire Abo field, New Mexico (modified from Le May, 1972)*

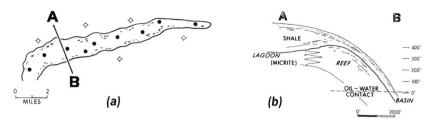

Fig. 7–4 *Empire Abo field, New Mexico (a) map (b) cross section*

Fig. 7–5 *Map of Illinois and Michigan basin*

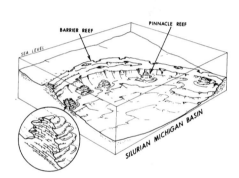

Fig. 7–6 *Pinnacle reefs growing basin-ward from the barrier reef, Michigan basin*

The Belle River Mills gas field near Detroit (Fig. 7–7a) is an example of a Silurian pinnacle reef field. It is 3 miles (5 km) long, 1 mile (1.6 km) wide and 400 ft (120 m) high (Fig. 7–7b). This field will eventually produce 50 bcf (1.4 billion m^3) of natural gas.

A large portion of the oil production in Canada comes from buried Devonian age atolls and barrier reefs, located primarily in Alberta (Fig. 7–8). Leduc was the first of these discovered in 1947. It was located by seismic exploration but was originally thought to be an anticline until it was drilled.

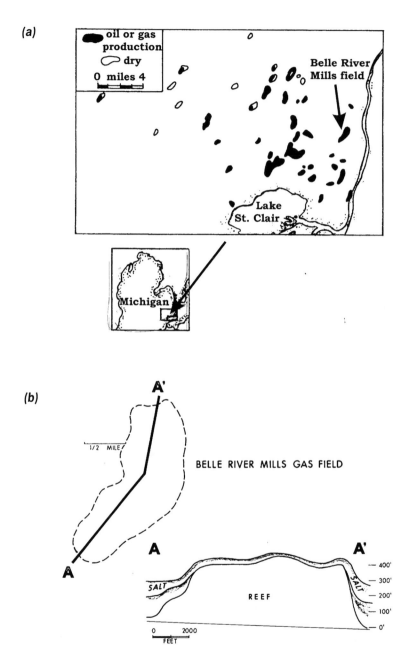

Fig. 7-7 (a) Location of pinnacle reefs near Detroit, Michigan (b) map and cross section of Belle River Mills gas field (modified from Gill, 1985)

The Redwater oil field of Alberta, the second largest oil field in Canada, was discovered in 1948. A map (Fig. 7–9a) shows the elliptical shape of the large, subsurface atoll that covers 200 sq miles (520 sq km) and is 15 miles (24 km) across. The reef rock, the reservoir rock, is located on the outside, whereas the lagoonal micrite limestone is located on the inside of the reef (Fig. 7–9b).The reef is tilted down to the west. The oil occurs in the highest part of the reservoir rock located to the east. The field will eventually yield 850 million bbls (135 million m^3) of oil.

Limestone Platforms

A *limestone platform* is a large area covered with shallow tropical seas where limestones are being deposited. The Bahaman banks, a modern example of a limestone platform, is located to the southeast of Florida. Sand-, silt- and clay-sized limestone particles are being deposited and reworked by waves and currents.

Fig. 7-8 *Devonian reefs of Alberta (modified from Procter, Taylor, and Wade, 1983)*

On some limestone platforms, strong tidal currents flow back and forth. The tropical water is saturated with calcium carbonate. As it flows over the limestone platform, calcium carbonate precipitates out of the water forming sand- and silt-sized spheres called *oolites*. Limestones composed of oolites are called *oolitic limestones*. They have excellent original porosity and can be good reservoir rock.

The Magnolia oil field in Arkansas (Fig. 7–10) was discovered in 1938 by surface mapping and seismic exploration. The trap is an anticline about 6 miles (10 km) long and $1\,^1/_2$ ($2\,^1/_2$ km) miles wide. It produces from the Reynolds Oolite Member of the Jurassic age Smackover Formation that is 300 ft (91 m) thick. Production has totaled 140 million bbls (22 million m^3) of oil.

(a)

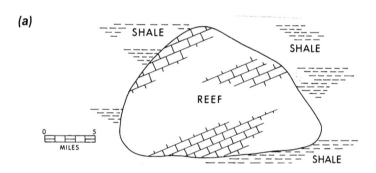

(b)

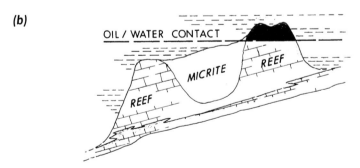

Fig. 7-9 *Redwater oil field, Alberta (a) map (b) east-west cross section (modified from Jardine, Andrews, Wishart, and Young, 1977)*

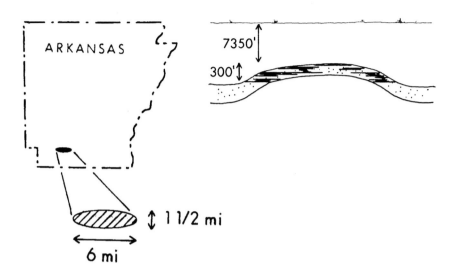

Fig. 7-10 *Map and East-west cross section of Magnolia oil field, Arkansas*

A *rimmed platform* is formed by a beach of broken sea shells and/or oolites deposited by waves along the platform margin. The rim can be good reservoir rock. During the Permian time, the Midland, Marfa, and Delaware basins in west Texas and New Mexico were separated by the Diablo and Central Basin platforms. A rim was deposited along the eastern margin of the Central Basin platform (Fig. 7–11). This rim forms the reservoir rock for many of the major West Texas oil fields such as Yates, McElroy, and Means (Fig. 7–12). The McElroy field will produce 601 million bbls (96 million m^3) of oil.

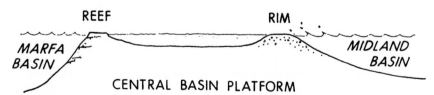

Fig. 7-11 *East-west cross section of Central basin platform, west Texas during Permian time*

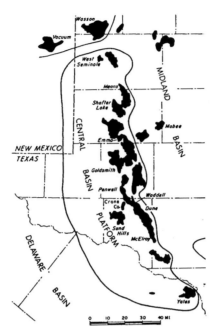

Fig. 7-12 *Map of oil fields along eastern rim of Central basin platform, west
Texas (modified from Harris and Walker, 1990)*

Karst Limestone

Limestone is very soluble in fresh water, especially in warm, humid climates. This is because rain and soil water absorb carbon dioxide, a common gas, to form carbonic acid that dissolves limestones. A highly dissolved limestone is called *karst limestone* and can have excellent porosity and permeability. Some karst limestones have caves but more often they have solution pores up to several inches in diameter called *vugs*. The great Middle East oil fields are primarily in limestone reservoir rocks. Their excellent porosity and permeability come from vugs in the karst limestone.

The prolific Golden Lane oil fields, both onshore and offshore, near Tampico, Mexico (Fig. 7–13) produce from karst limestone reservoir rock. The reservoir rock is a large, Cretaceous age atoll. The limestone atoll

(El Abra reef) was buried in the subsurface by sediments. Later, the sedimentary rocks covering the reef were eroded, exposing the reef. The top of the reef was dissolved, forming cavernous pores. After that, the reef was again buried in the subsurface by sediments. Finally, the oil migrated up into the karst limestone. The fields were discovered in the early 1900s by drilling on oil seeps. One field on the trend, Cerro Azul, has 1 $^1/_4$ billion bbls (200 million m^3) of recoverable oil. The El Abra reef limestone is extremely porous and permeable. One well (Potero del Llano # 4) on the reef produced a total of more than 115,000,000 bbls (18 million m^3) of oil in 40 years.

The Dollarhide pool oil field, located north of Odessa, Texas, is in a limestone cave that ranges from 3 to 16 ft (1 to 5 m) high. The reservoir rock is the Silurian Age Fusselman Limestone. Fifteen wells have been drilled into this cave.

Limestones caves cannot be detected from the surface using the seismic method and are usually found by chance. Karst limestone, however, often occurs directly below a subsurface angular unconformity. An angular unconformity is an ancient erosional surface that was exposed on

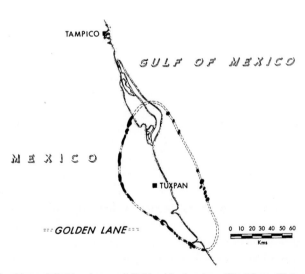

Fig. 7–13 *Map of Golden Lane oil fields, Mexico (modified from Viniegra and Castillo-Tejero, 1970)*

the surface. Any limestone directly below it is often highly dissolved. In the Sooner trend of oil fields to the southwest of Oklahoma City, Oklahoma, one of the reservoir rocks is the Hunton Limestone, a karst limestone directly under an angular unconformity.

Chalk

Chalks are extremely fine-grained limestones composed of microfossil shells. Single-celled animals called foraminifera and plants called coccolithophores grow floating in tropical seas. Both have shells of calcium carbonate. When they die, their shells fall to the ocean bottom. If they are not dissolved or diluted by other sediments, a chalk is formed. During the Cretaceous time, several chalks were deposited that include the Austin Chalk of Texas and Louisiana, the Niobrara Chalk of South Dakota, Nebraska and Kansas, the Selma Chalk of Alabama, the chalk cliffs of Dover, England, and the Ekofisk Chalk underlying the North Sea.

Dolomite

Dolomite [$CaMg(CO_3)_2$] is a sedimentary rock that forms only by the alteration of a pre-existing limestone. Magnesium-rich waters percolating through subsurface limestone ($CaCO_3$) replace calcium atoms with magnesium (Mg) to form dolomite. Often entire beds of permeable limestone have been transformed into dolomite. In impermeable limestones such as micrite or crystalline limestone, dolomite formation is restricted to areas along fractures through which the magnesium waters were able to flow.

Dolomite is difficult to distinguish from limestone in the field. Dolomite and limestone have a similar crystal shape, color, and hardness. Limestone will bubble in cold, dilute acid. Dolomite, however, will bubble only in hot, concentrated acid. Rhomb-shaped crystals that will not bubble in cold, dilute acid are the best field indication of dolomite. In the laboratory, they are easy to identify.

Dolomite is often a good reservoir rock. Limestones tend to lose porosity when they are buried deeper due to compaction. Dolomites, however, are harder and less soluble than limestones and retain porosity. The dolomite crystals are often larger then the limestones particles that they replace, increasing the permeability of the rock. The dolomitized oil field reefs of western Canada have an average of 1% more porosity and 10 times greater permeabilities than the limestone reefs.

The Jay field in northwest Florida is located on an anticline at 15,000 ft (4500 m) deep. Only about one-half of the structure is productive. The original Jurassic age Smackover Limestone of dense micrite is not reservoir rock on the unproductive side of the anticline (Fig. 7–14). On the productive side, dolomite has replaced micrite. There are 730 million bbls (116 million m^3) of 51 °API oil in place. Ultimately, more than 346 million bbls (55 million m^3) of oil will be produced from the Jay field. The average net pay is 95 ft (29 m).

The Scipio-Albian field, discovered in 1957, is the largest oil field in Michigan. It is located in dolomitized Trenton Limestone at a depth of about 3550 ft (1080 m). A large strike-slip fault in the basement (Fig. 7–15) folded

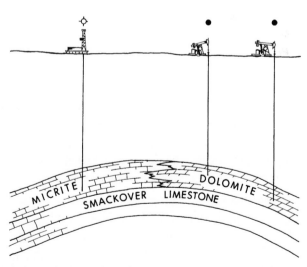

Fig. 7-14 *Cross section of Jay field, Florida (modified from Ottman, Hayes, and Ziegler, 1976)*

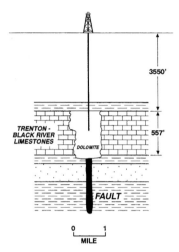

Fig. 7-15 *East-west cross section of Scipio field, Michigan*

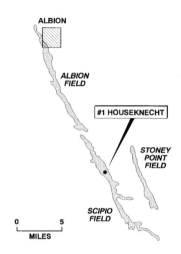

Fig. 7-16 *Map of Scipio-Albion trend, Michigan (modified from Hurley and Biedros, 1990)*

and fractured the Trenton Limestone above it. The fractures allowed magnesium-rich waters to flow into the limestone and change it into dolomite. Because of this, the field is no wider than a mile (1.6 km) as it follows the subsurface fault (Fig. 7–16). The caprock is the Utica Shale that overlies the dolomite. The Scipio-Albian field will ultimately produce more than 150 million bbls (24 million m³) of oil and 200 billion cf (6 billion m³) of gas.

The discovery well (#1 Houseknecht) was drilled by the landowner, Ferne Bradford (maiden name of Houseknecht), on the advise of a psychic friend, "Ma" Zulah Larkin. The nearest oil production was more than 50 miles (80 kms) away when the well was spudded in 1955. Twenty months later, in 1957, the well encountered an oil reservoir and came in at the rate of 140 bbls of oil/day with considerable gas. Continental Oil Co. had drilled and tested the Scipio oil reservoir in 1943 but decided to plug the well and complete in a shallower gas zone. The adjacent Stoney Point field, located five miles to the east, was not discovered until 1982.

The Yates oil field in West Texas just to the west of the Pecos River, is a giant oil field that produces from both karst and dolomite. The primary reservoir rock is the Permian age Grayburg Dolomite that contains caverns that are up to 21 ft (6 m) high. These were formed when the area was a tropical island, and fresh waters percolated through the limestone. The dolomite formed later. The caprock is salt in the overlying Seven Sisters Formation.

There were oil seeps on the Yates Ranch, and a large anticline was mapped on the surface (Fig. 7–17). The field was discovered in 1926 by drilling to 997 ft (304 m) where the well blew out. The well was later deepened to 1032 ft (315 m) where it tested at 72,000 bbls (11,500 m³) of oil per day. One well on the Yates oil field had an initial production test that yielded more than 8500 bbls (1350 m³) of oil per hour. The field has already produced more than 1 billion bbls (160 million m³) of oil and is expected to produce another billion barrels with a recovery of 50% of the oil in the reservoir.

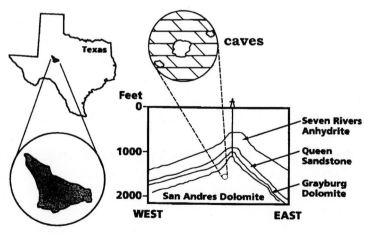

Fig. 7–17 *Yates field, west Texas (a) map (b) cross section (modified from Craig, 1988)*

eight

SEDIMENTARY ROCK DISTRIBUTION

Petroleum is not evenly distributed throughout the world. Sweden has virtually none, whereas the Middle East has more than one-third of the world's known oil supply. Sedimentary rocks are both source and reservoir rocks for petroleum. Where the sedimentary rocks are thick, a large amount of petroleum is likely to occur. Where there are no sedimentary rocks, there is no petroleum.

Basin Formation

A *basin* is a large area with thick sedimentary rocks. It is where most gas and oil is found. There are several ways to form a basin.

A basin can be formed by subsidence of the basement rock. The depression is originally filled with ocean water and, eventually, with sediments. The Michigan basin is an example of this type of basin (Fig. 7–5 and Fig. 8–1). Subsidence occurred during the Paleozoic time and very thick carbonates and evaporates accumulated during the Silurian time. Petroleum occurs in reefs that surrounded the basin during that time.

The southern California oil basins were formed by grabens. Some basins are located on land (Los Angeles and Ventura basins), whereas others are off-shore (Santa Barbara and San Pedro basins). An east to west cross section

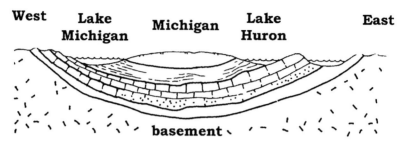

Fig. 8–1 *East-west cross section of Michigan basin*

(Fig. 8–2) of the basins shows that the area was subjected to tensional forces that formed a series of parallel normal dip-slip faults with north-south horsts and grabens in the basement rock. The grabens were originally filled with ocean water, and many of the horsts stood above sea level to form islands.

Sediments filling the grabens came primarily from the east, where land was being eroded. Because of the easterly source of sediments, the eastern grabens (Los Angeles and Ventura basins) were the first to be filled and are now dry land. The city of Los Angeles is located on the Los Angeles basin. To the northwest is the Ventura basin. Presently, the sediments are filling the next basins to the west (San Pedro and Santa Barbara basins). The basins furthest to the west will be the last to be filled. The sedimentary rocks, rich in source rocks and reservoir rocks, are tens of thousands of feet thick in the eastern basins. Numerous traps were formed by anticlines, domes, and faults.

The Los Angeles basin (Fig. 8–3) is the most prolific petroleum basin on earth. There is more known gas and oil per cubic mile of sedimentary rocks here than anywhere else in the world. The 61 fields in the Los Angeles basin have produced a total of 7.8 billion bbls (1.2 billion m^3) of oil and 7.1 Tcf (200 billion m^3) of gas.

Half-graben basins are formed by subsidence along one side of a normal fault (Fig. 8–4). These basins are common and are productive in the North Sea, offshore western Africa, and offshore Brazil.

Intermontane basins form when mountain ranges are created. The basin is located between the mountain peaks and is often occupied by a lake.

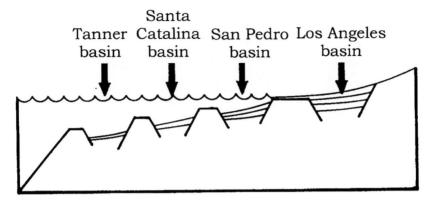

Fig. 8–2 *East-west cross section of Southern California graben basins*

Fig. 8–3 *Oil fields of the Los Angeles basin*

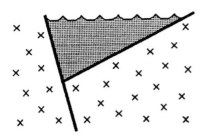

Fig. 8-4 *Half-graben basin*

Algae growing in the lake contribute organic matter to the bottom sediments for source rocks. Streams, eroding the surrounding mountains, deposit numerous channel and beach sandstone reservoir rocks in the basin. When the Rocky Mountains were uplifted during the Cretaceous time, several intermontane basins were formed (Fig. 8–5). Many of these basins such as the Big Horn, Powder River, Green River, and Uinta basins are good petroleum producers today.

Fig. 8-5 *Rocky Mountain intermontane basins*

Basins also form along the edges of mountains. As the mountains are eroded by streams, sediments fill in the areas adjacent to the mountains. The Alberta, Denver-Julesburg, and Raton basins formed in this manner (Fig. 8–5).

Coastal plains are formed by thick sediments deposited adjacent to an ocean (Fig. 8–6). They originate when mountains are uplifted adjacent to a coast. As erosion lowers the mountains, streams deposit sands along the beaches, and waves carry the silts and clays offshore. The sandy beaches are deposited (prograde) out into the ocean, forming the coastal plain. The Gulf of Mexico coastal plain was created by this process. Because the sediments on the surface are young and have never been buried, they are loose and uncemented (unconsolidated).

The oldest sediments on the surface of a coastal plain are furthest inland from the modern beach. A well drilled on a modern coastal plain beach encounters older and older sedimentary rocks with depth. Near the surface it will penetrate beach and near-shore sands. At deeper depths, the well penetrates sedimentary rocks that were deposited in deeper waters.

The Atlantic and Gulf coastal plains of the United States were deposited by sediments eroded from the Appalachian and Ouachita mountains, which rose during the Pennsylvanian Period. Underlying the Gulf of Mexico coastal plain is 40,000 to 60,000 ft (12,000 to 18,000 m) of sedimentary rocks.

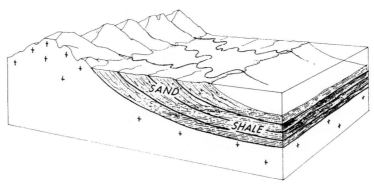

Fig. 8–6 *Coastal plain*

This area is one of the most prolific petroleum producing areas of the world. The Atlantic coastal plain is barren for gas and oil, possibly due to the lack of source rocks.

Most mountain ranges on land were formed by the compression of sedimentary rocks. They display large compressional features such as anticlines, synclines, reverse faults, and thrust faults. Mountain ranges, however, are relatively unproductive because most petroleum reservoirs have been breached by erosion, and the oil and gas has leaked out. In many areas, the sedimentary rocks have been eroded away, exposing the basement rock. Many of the remaining sedimentary rocks in the mountains have been metamorphosed by the high heat and pressure that occurred during the compression. Only intermontane basins are good areas to explore for oil and gas.

Sedimentary Rock Facies

Sedimentary rock layers are not deposited uniformly. A single layer can be composed of several different rock types (Fig. 8–7). Each is a *facies*, a distinctive portion of the rock layer. The change between rock types is called a *facies change*.

Sediments being deposited along the margin of an ocean have both a sandstone and shale facies. As waves wash ashore, silt and clay are suspended in the water. The mud is carried offshore and deposited in deep water to become shale. Sand, too heavy to be suspended in the water by the waves, is deposited on the beach to become sandstone. This forms a layer of sediments with a shale (deep-water) and sandstone (beach) facies.

The Jurassic age Smackover Limestone occurs in the subsurface of the northern Gulf of Mexico coastal plain. It is 15,000 to 20,000 ft (4600 to 6000 m) deep and in some areas is a good reservoir rock. A facies map the Upper Smackover (Fig. 8–8) illustrates the different depositional environments and textures of the limestone. The best reservoir rock is the oolite facies that was deposited on a tropical, shallow-water shelf and has 20 to 25% porosity in Arkansas. In Texas, parts of the oolite facies have been dolomitized, increasing the porosity to 30%. The salt facies, in contrast, is impermeable and acts as a caprock.

Fig. 8-7 *Sedimentary rock facies*

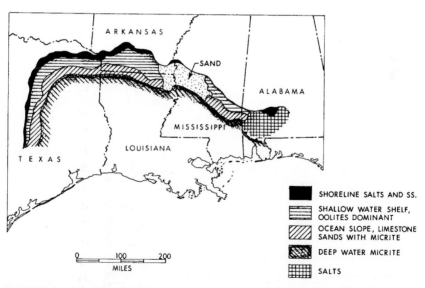

Fig. 8-8 *Facies of the Smackover Limestone in the Gulf Coast (modified from Bishop, 1968)*

The largest gas field on the North American continent, the Hugoton field of Texas, Oklahoma, and Kansas (Fig. 8–9a) was formed by a facies change. The field covers an enormous area that is 275 miles (443 km) long and from 8 to 57 miles (13 to 92 km) wide. It will eventually produce 70 Tcf (2 trillion m^3) of gas. In Texas there is also oil production along the eastern margin of the field. This is called the Panhandle field and will eventually produce 1.4 billion bbls (200 million m^3) of oil.

The reservoir rocks are primarily limestones and dolomites known as the Chase Group of Permian age. The Wichita Formation, containing salt layers, lies directly over the reservoir rocks and forms the seal. The Chase Group was deposited as limestone reservoir rocks to the east and impermeable, red-colored shales and sands to the west in a facies change (Fig. 8–9 b–1). A later uplift to the west, the Stratford arch completed the trap (Fig. 8–9 b–2). Enormous volumes of gas formed in the deep Anadarko basin to the southeast. The gas was trapped as it migrated updip to the west by the change from permeable into impermeable Chase Group limestones.

Subsurface Rock Layers

Few sedimentary rock layers are uniform in thickness and rock type. A rock layer often thins in one direction and thickens in another (Fig 8–10). It can grade from one facies into another facies. Sometimes the boundary between two facies is sharp, and the rocks *interfinger* or *wedge out* into each other. Often a single rock layer will *pinch* or *wedge out* (Fig. 8–11) in another rock layer. Sandstone wedges, deposited as beaches or river channels, are common in shale layers.

An updip pinch out of a reservoir rock in a shale or salt layer can form a petroleum trap. Pinch outs of dolomite reservoir rocks in salt occurs along the northern edge of the Midland basin in west Texas (Fig. 7–2 and Fig. 8–12) to trap oil.

The Glenn Pool oil field, located just south of Tulsa, Oklahoma, is formed by an updip pinch out of a sandstone wedge in a shale layer (Fig. 8–13). The Pennsylvanian age reservoir rock, the Bartlesville Sandstone, is an incised valley fill that is tilted up to the east at 1°. The wells are only about 1500 ft (500 m) deep. The field has already produced 327 million bbls (53 million m^3) of sweet, 36 to 41 °API oil. The Glenn Pool was discovered in 1905 and started the Oklahoma oil boom.

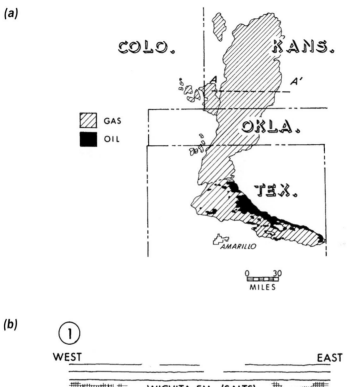

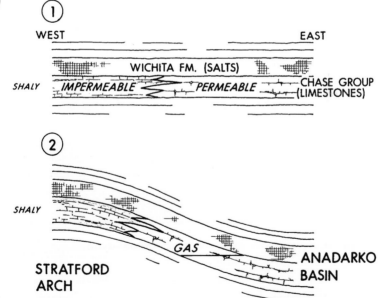

Fig. 8-9 Hugoton gas field and Panhandle oil field, Kansas, Oklahoma, and Texas (a) map (b) formation (modified from Pippen, 1970)

115

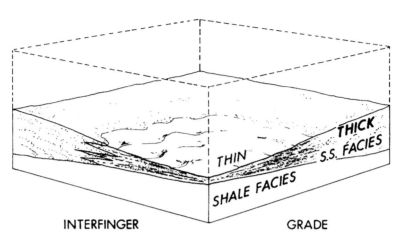

Fig. 8-10 *Variations in a sedimentary rock layer*

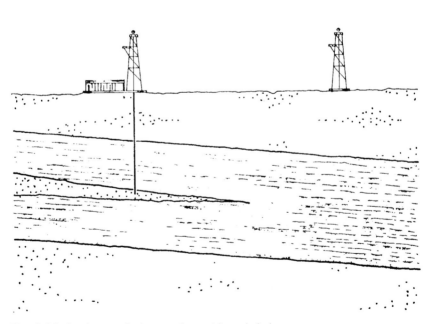

Fig. 8-11 *Sandstone pinch or wedge out in a shale layer*

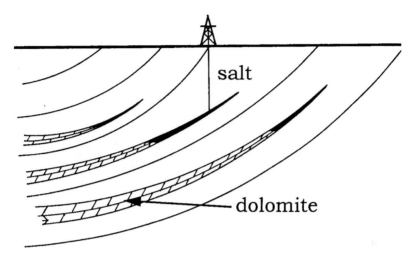

Fig. 8-12 *Pinch out of dolomite in salt along the edge of a basin*

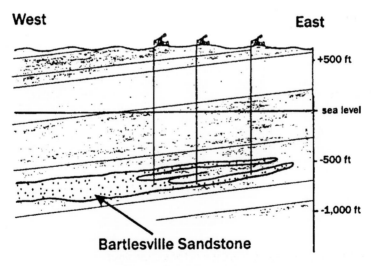

Fig. 8-13 *East-west cross section of Glenn Pool oil field, Oklahoma*

nine

MAPPING

An important tool that geologists use to find gas and oil is maps of both the surface and the subsurface. All maps are oriented with north to the top, south to the bottom, east to the right and west to the left.

Topographic Maps

A *topographic map* shows the elevation of the earth's surface (Fig. 9–1). To illustrate the third dimension (elevation) on a flat, two-dimensional map, contour lines are used. A *contour line* is a line of equal value on a map, and a contour line on a topographic map is a line of equal elevation. A contour line is always labeled with an elevation that is above or below sea level. All along that contour line, the elevation is exactly the same. For example, any-where along the +400 ft contour line on a topographic map, the elevation is exactly 400 ft above sea level. The *contour interval* of a topographic map is the difference in elevation between two adjacent contour lines. The contour interval of the topographic map in Figure 9–1 is 100 ft.

If the contour line elevations increase in a direction, the slope is rising (Fig. 9–2). If the contours are spaced relatively close together, the elevation is changing rapidly, and the slope is steep. If the contours are relatively far apart, the slope is gentle. There are some important characteristics of contours on a topographic map. Contour lines never cross. Contour lines are single lines; they never branch. Contour lines are continuous; they always close or run off the map and never end on the map.

Elevations can be accurately estimated from a topographic map. If a point is on the +300 ft contour, it must be, by definition, exactly 300 ft above sea level. If the point is about half way between the +300 and +400 ft contour, an elevation of +350 ft is a good estimate. The shape of the contours is characteristic for many topographic features such as hills, ridges, and canyons.

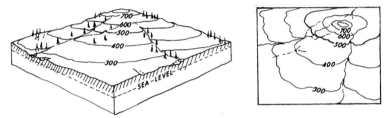

Fig. 9–1 *Land and a topographic map of the land*

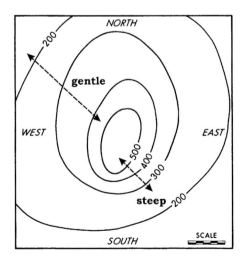

Fig. 9–2 *Contoured topographic map showing steep and gentle slope*

A topographic map (or any contoured map) cannot be drawn without some accurately surveyed points. After the elevations or values are located on a map, contours can be drawn between the points. Contouring of any map can be done either by hand or computer. The position of a contour line between two data points can be accurately located by using proportions. For example, the 400 contour must run between data points of 402 and 399 (Fig. 9–3). A straight line is drawn between the two data points. Because there is a difference of 3 between the data points (402 and 399), the line is divided into three equal segments. The 400 contour is located one segment from the 399 point and two segments from the 402 point. Anything that can be expressed by mathematics can be programmed into a computer, and computer-generated contour maps can be made.

Geologic Maps

A *geologic map* (Fig. 9–4) shows where each rock layer crops out on the surface of the earth. Each rock layer is given a different pattern, color, and symbol on the map. The basic sedimentary rock layer used for geologic mapping is a *formation*. A formation is a mapable rock layer with a definite top and bottom. Geologists have divided all sedimentary rocks into formations.

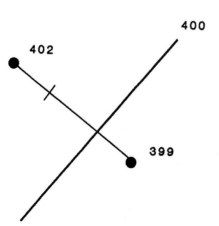

Each formation has a two-part name. The first part is a town, where the layer crops out on the surface. The second part is the dominant rock type, such as sandstone or limestone. San Andres Limestone, Bartlesville Sandstone, and Mancos Shale are formation names. If the sedimentary rock layer is a mixture of rock types, the word formation is used. The Coffeyville Formation, composed of both sandstone and shale, is an example.

Fig. 9–3 *Locating a contour using proportions*

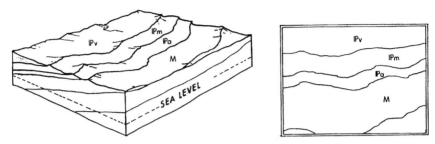

Fig. 89–4 Geologic map

Formations can be subdivided into smaller units called *members*. A member is a distinctive but local bed in a formation. It is also given a formal, two-part name. For example, the Layton Sandstone Member is part of the Coffeyville Formation. If the member is found only in the subsurface, it is given a letter and number such as the H5 Sands Member. Adjacent formations of similar rocks can be joined to form a *group* and given a geographic name (*i.e.*, the Chase Group).

A geologic map is a flat, two-dimensional representation of the earth's surface. The orientation of rock layers, the third dimension, is shown with a strike-and-dip symbol. *Strike* (Fig. 9–5a) is the horizontal orientation of a plane, such as a sedimentary rock layer or a fault. It is measured with a compass orientation, such as North 30° East. Strike is shown as a short line on the geological map (Fig. 9–5b) that is oriented in the measured compass direction. *Dip* is the direction and vertical angle of the plane. It is measured perpendicular to the strike (Fig. 9–5a). The dip symbol on the map is a small bar attached to the middle of the strike line (Fig. 9–5b). It points in the direction that the plane goes down into the earth. The angle in degrees is often on the dip symbol.

The dip of a rock layer is the angle and direction it goes into the subsurface. Drilling *updip* means that the drillsite will be up the angle (dip) of

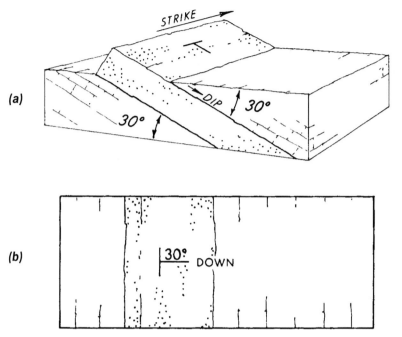

Fig. 9-5 *(a) strike and dip of a sedimentary rock layer (b) strike and dip symbol on a geologic map*

the rock layer from the last drillsite. Updip in a gas or oil reservoir is usually a favorable position as gas and oil rises in a reservoir filled with water.

A stratigraphic column (Fig. 9–6) is a convenient method for presenting the vertical sequence of rocks on a geologic map. Any deformation of the rocks, such as faulting, has been removed. The youngest formation in the geologic map area is at the top of the column, and the oldest is located at the bottom. The column is drawn as a cliff of weathered rocks with the weaker rock types (*e.g.*, shales) indented. Stronger rock types, such as sandstones, protrude outward as they would weather in nature.

Common geological symbols (Fig. 9–7) are used for rocks, structures, and wells on a geological map.

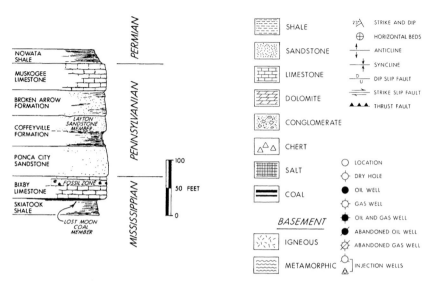

Fig. 9-6 *Stratigraphic column*   **Fig. 9-7** *Common geological symbols*

Base Maps

A *base map* is a map that shows the position of all the wells that have been drilled in an area. *Spotting a well* involves the location of a wellsite and placing the well symbol (Fig. 9–7) on a base map. Base maps can also include seismic lines and other data.

Subsurface Maps

Three important types of subsurface maps are structural, isopach, and percentage. All three maps use contours to describe the subsurface rock.

Structural Map

A *structural map* uses contours to show the elevation of the top of a subsurface sedimentary rock layer (Fig. 9–8). The contours are usually in feet below sea level, as most rocks are located below sea level. An important structural map would be one contoured on the top of a potential reservoir rock or drilling target.

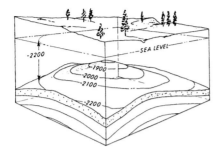
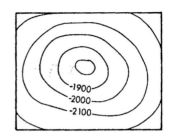

Fig. 9-8 *Structural map*

Domes, anticlines, and faults can be identified on structural maps. Both a hill on a topographic map and a dome on structural map have a bull's eye pattern (Fig. 9–9) with the highest elevation in the center. Both a ridge on a topographic map and an anticline on a structural map have a concentric but oblong pattern (Fig. 9–10) with the highest elevation in the center.

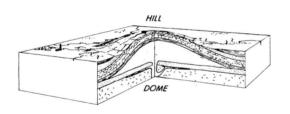

Fig. 9-9 *Topographic map of a hill and structural map of a dome*

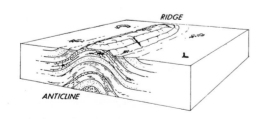
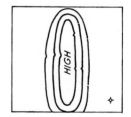

Fig. 9-10 *Topographic map of a ridge and structural map of an anticline*

Faults are characterized by a rapid change in elevation along a relatively straight line (Fig. 9–11). A normal dip-slip fault that causes a lost section in the rock layer being mapped (Fig. 5-18) is seen on the map as two lines separating the contours (Fig. 9–12).

Fig. 9-11 *Fault on a structural map*

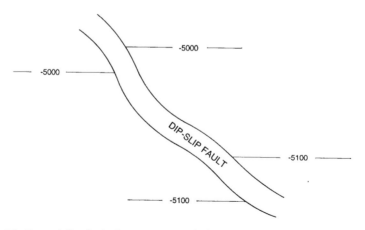

Fig. 9-12 *Normal dip-slip fault on a structural map*

Isopach Map

An *isopach map* (Fig. 9–13) uses contours to show the thickness of a sub-surface layer. If an oil or gas field has been drilled, an isopach map can be made of the pay zone. The *pay zone* is the vertical distance in a well that produces gas and/or oil. *Gross pay* contours the entire reservoir thickness including nonproductive water-bearing and tight zones. *Net pay* contours only the productive thickness of the reservoir. A net pay isopach map of a reservoir is used to calculate the reservoir volume and the oil and gas reserves.

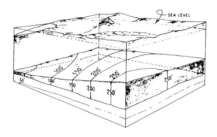 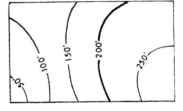

Fig. 9-13 *Isopach map*

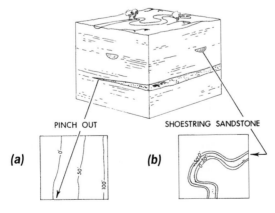

Fig. 9-14 *Isopach map of (a) a sandstone pinch out and (b) a shoestring sandstone*

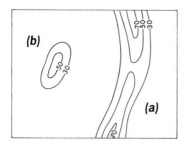

Fig. 9-15 *Isopach map of a barrier
reef (a) and a pinnacle reef (b)*

An isopach map can be used in exploration to delineate a sandstone pinch or wedge out (Fig. 9–14a) where the isopach contour becomes zero. The aerial patterns of beach and channel sandstones (shoestring sandstones) are seen on an isopach map (Fig. 9–14b).

An isopach map of a limestone layer can be used to locate a reef. A reef is a mound and is shown by thick contours (Fig. 9–15). Barrier reefs, which are long, can be distinguished from pinnacle reefs, which are more circular.

Percentage Map

A *percentage map* (Fig. 9–16) plots the percentage of a specific rock type such as sandstone in a formation. Higher percentages of reservoir quality rocks, such as sandstones and carbonates, imply a larger reservoir net pay.

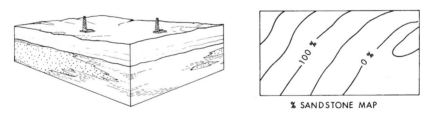

% SANDSTONE MAP

Fig. 9-16 *Sandstone percentage map of a formation composed of sandstone and shale*

ten

OCEAN ENVIRONMENT AND PLATE TECTONICS

Ocean Topography

Shallow Water

The continents are surrounded by a shallow, almost flat platform, the *continental shelf* (Fig. 10–1). It extends from the beach out with a slope of less than 1° to the shelf break. At the *shelf break*, the ocean bottom sharply increases in the slope. The shelf break is located in an average water depth of 450 ft (140 m). The width of the continental shelf varies from $^1/_2$ mile (0.8 km) to more than 500 miles (800 km) with an average width of 50 miles (80 km).

The continental shelf is geologically part of the continents. Sedimentary rocks that are encountered in drilling along the beach extend out under the continental shelf. Many large structures such as faults and folds continue from the land onto the continental shelf. The giant Wilmington oil field is formed by an anticline that lies partially under land (Long Beach, California) and partially offshore. The San Andres fault extends offshore onto the continental shelf in northern California. Throughout geologic time, sea level has been rising and falling, and the seas are now covering this part of land, the continental shelves. The continental shelf is a very active petroleum exploration and production area. The same source rocks, reservoir rocks, and traps that produce on land are found on the continental shelves.

Seaward of the continental shelf and slope break is the *continental slope.* The continental slope leading down to the bottom of the ocean has a slope of about 4° and is the geological edge of the continents.

Eroded into the continental shelves and slopes in many areas are *submarine canyons* (Fig. 10–2). They often extend from shallow depths off the beach down to the bottom of the continental slope, thousands of feet deep. Submarine canyons are relatively common throughout the world and often occur offshore from rivers. The Mississippi, Amazon, Ganges, Niger, Nile, and many other rivers have submarine canyons offshore.

Submarine canyons are eroded, and sediments are transported down submarine canyons by turbidity currents. *Turbidity currents* are masses of water with suspended sediments such as sand, silt, and clay. The turbidity current is denser than the surrounding sea water and is pulled by gravity down the submarine canyon similar to river water being pulled by gravity down a river channel on land. Turbidity currents can originate from rivers with a large sediment load flowing into the ocean. They are thought to erode the submarine canyons.

A turbidity current will continue to flow down a submarine canyon as long as a slope exists. When the turbidity current flows onto the relatively flat ocean bottom, it stops, and the sediments settle out of the water. The coarsest

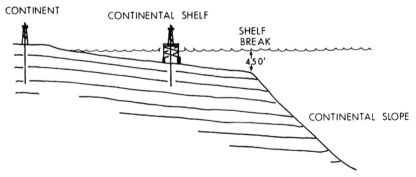

Fig. 10-1 *Cross section of a continental shelf*

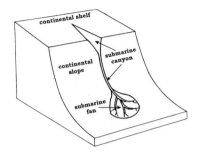

Fig. 10-2 *Submarine canyon and fan*

sediments (usually sand) settle out first. The finest sediments (silt and then clay) settle out last. This deposits a *graded bed* (Fig. 10–3) with the coarsest sediments on the bottom and the finest on the top. Accumulations of turbidity current sediments at the base of the submarine canyon form a large sedimentary deposit called a *submarine fan* (Fig. 10–2). A channel usually leads out of the submarine canyon and divides into smaller distributary channels on the submarine fan.

Turbidity current deposits are called *turbidites* and can be sandstone reservoirs (Fig. 10–4). Relatively thin sandstones separated by shales are characteristic of submarine fan reservoirs. Relatively thick sandstones were deposited in submarine canyons and channels where currents eroded away the finer-grained sediments. Deep-water production on the continental slopes of the Gulf of Mexico, western Africa, and Brazil are from turbidite sands deposited in submarine canyons and submarine fan distributary channels now buried in the subsurface. The sands are relatively young and haven't been buried too deep. Because of this, they often have porosities of 20 to 30%, thousands of millidarcy permeabilities, and are excellent reservoir rocks. They yield oil and gas at a high rate to wells to justify the cost of drilling in deep water.

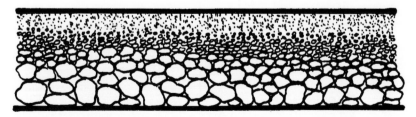

Fig. 10-3 *Graded bed*

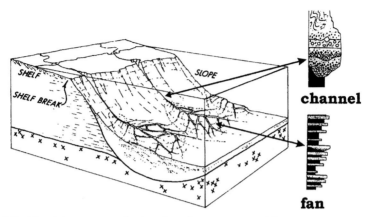

channel

fan

Fig. 10–4 Sandstone reservoirs in submarine canyons and fans

The Santa Fe Springs oil field in the Los Angeles basin, California is a symmetrical dome (Fig. 10–5). It produces gas and oil from 25 sandstones, each located at the base of Miocene and Pliocene age graded beds. The field has 622 million bbls (99 million m³) of recoverable oil and 0.87 Tcf (28 million m³) of gas.

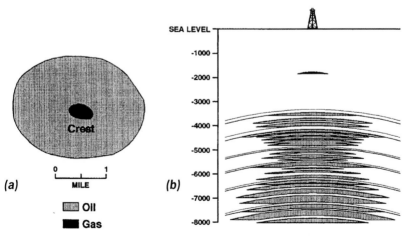

Fig. 10–5 Santa Fe Springs field, Los Angeles basin, California (a) map and (b) cross section

During the Miocene time, the San Joaquin Valley of California was occupied by a deep ocean. Streams carried sediments off the high Sierra Nevada Mountains to the east. They emptied into the deep ocean, eroding submarine canyons and depositing a series of submarine fans. Today these sand-rich submarine fans are buried in the subsurface. They are called the Stevens sands of Miocene age, a prolific petroleum producer.

The largest, producing offshore gas field in the world is the Frigg field located in both the United Kingdom and Norwegian sectors of the North Sea (Fig. 10–6a). A map of the reservoir rock (Fig. 10–6b) shows it produces from an ancient submarine canyon and fan sandstone (the Frigg Sandstone of Paleocene age) at a depth of about 6000 ft (1800 m) below the bottom of the North Sea. The sandstone is completely encased in shale, which is both the source rock for the gas and the caprock for the reservoir (Fig. 10–7). The field will eventually produce 7 Tcf (200 million m^3) of natural gas.

When sediments prograde out into a basin, shallow-water sediments are deposited on deep-water submarine canyon and fan sediments. In the northeastern Texas Gulf Coast, the Hackberry Sandstone Member of the Oligocene age Frio Formation is an excellent reservoir rock (Fig. 10–8) that occurs at a

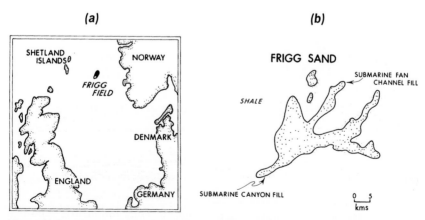

Fig. 10–6 *Frigg gas field, North Sea (a) location map and (b) map of reservoir rock (modified from Heritier, Lassel, and Wathne, 1980)*

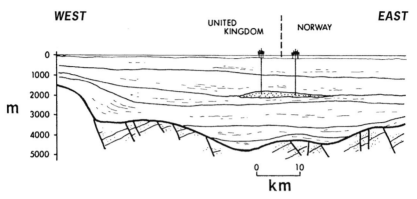

Fig. 10–7 *Frigg gas field, North sea cross section (modified from Heritier, Lossel, and Wathne, 1980)*

depth of about 10,000 ft (3000 m). It was deposited as sands filling submarine canyons and distributary channels on submarine fans millions of years ago. Salt structures, anticlines, faults, and pinch outs form many traps in the sandstone. The Port Acres field in this trend will eventually produce 326 Bcf

Fig. 10–8 *Oil and gas field map of Hackberry Sandstone Member of Frio Formation, Texas Gulf Coast (modified from Jackson, 1991)*

(9 billion m³) of natural gas and 10.5 million bbls (1.7 million m³) of condensate. There the Hackberry Sandstone has an average net pay of 26 ft (8 m).

River channels and deltas are closely related to submarine canyons and fans. During relatively high sea levels, such as today, the sediments are deposited in deltas, and the submarine canyons and fans are inactive. This is evident by recent Mississippi River Delta switching (Fig. 6–21). During relatively low sea level, the sediments are deposited on submarine fans, and little is deposited in deltas.

Deep Water

The deepest parts of the seafloor are *ocean trenches*. Ocean trenches are long, narrow depressions usually located along the margins of the oceans. Adjacent to many deep ocean trenches are active volcanic islands.

Located almost in the very center of the Atlantic Ocean, is a segment of the *mid-ocean ridge*. The mid-ocean ridge is the world's longest mountain chain that can be traced for 50,000 miles (80,000 km). It extends down the Atlantic Ocean, around south Africa, into the Indian Ocean and continues between Australia and Antarctica and up into the eastern Pacific Ocean (Fig. 10–9).

The ridge bifurcates in several locations. In the Indian Ocean, one segment extends into the Gulf of Aden and the Red Sea. The ridge is very wide, averaging 1000 miles (1600 km) and rises about 1 to 2 miles (1.6 to 3 km) above the adjacent ocean floor. The center of the ridge typically has a rift valley (graben). The ridge protrudes through the surface of the ocean at Iceland and the Azores, both formed by active, basalt volcanoes. Observations from submarines have shown that active volcanoes occur all along the submerged graben on the crest of the mid-ocean ridge.

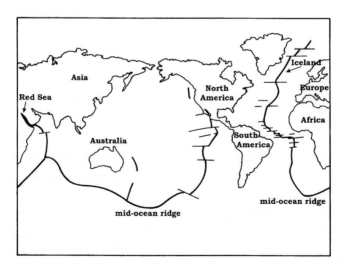

Fig. 10-9 *Map of the mid-ocean ridge*

Ocean Sediments

Seismic exploration of the ocean basins has shown that the average sediment thickness on the ocean bottom is about one-half mile (0.8 km). This is unevenly distributed. Sediments are very thick along the edge of the oceans and thin or absent on the mid-ocean ridge. The oldest sedimentary rocks in the ocean basins anywhere in the world are only Jurassic in age, about 150 million years old. Many sedimentary rocks found on land are considerably older than the oldest ocean-bottom sediments.

Earth's Interior

The basement rock that forms the crust of the earth (Fig. 10–10) is basalt lava rock. It averages 5 miles (8 km) in thickness. Continents are high in elevation because they also have granite, averaging 17 miles (27 km) thick, on the basalt basement rock. Granite is relatively light in density compared to basalt and floats higher on the rocks in the interior of the earth. Because no one has drilled or mined very deep into basement rock, there is little direct evidence of what occurs below the crust of the earth. However, temperature and pressure both increase with depth. Because of the high temperatures, rocks below the crust of the earth are partially melted and behave as thick, viscous liquids.

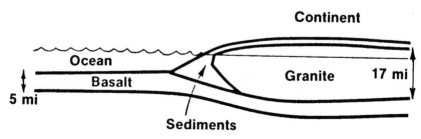

Fig. 10-10   *Cross section of the earth's crust*

Continental Drift

The theory of continental drift dates back to the early 1900s. Continental drift suggests that all the present-day continents were previously joined into one super continent, Pangaea (Fig. 10–11). During the Mesozoic Era, about 150 million years ago, Pangaea broke up. The fragments of the super continent drifted (moved) across the face of the earth into their present positions to form the modern continents.

The theory was not widely accepted at first. It was not known what process would cause Pangaea to break up and the continents to move.

Seafloor Spreading

A new theory, seafloor spreading, was presented in the early 1960s. This theory provided the processes for the breakup of Pangaea and the drifting of the continents.

Fig. 10-11 *Pangaea*

Seafloor spreading postulates that large convection currents occur in the interior of the earth (Fig. 10–12) where rocks act as viscous liquids. A convection current is a cell of flowing liquid caused by heating and cooling. Where the liquid is heated, it becomes less dense and rises. Where the liquid is cooled, it becomes more dense and sinks. Convection currents cause the interior of the earth to be constantly moving.

A rising hot current from the interior of the earth cannot penetrate the crust of the earth. It arches the crust up to form the mid-ocean ridge (Fig. 10–13). The hot, molten current then divides and flows to either side of the mid-ocean ridge. This splits the solid crust of the earth at the ridge crest and drags it to either side of the ridge. The name seafloor spreading comes from the seafloor being spread out at right angles from the crest of the mid-ocean ridge. The existence of a graben, a tensional feature that runs along the center of the mid-ocean ridge, supports this idea. The graben contains erupting volcanoes that produce basalt lava. This new basalt crust of the earth is split and spread out from the ridge crest.

The seafloor is spreading out from several mid-ocean ridges in different oceans. Areas where seafloors from two different mid-ocean ridges collide and are being destroyed are called *subduction zones.*

There are three types of subduction zones. First, if two seafloors from different mid-ocean ridges meet (Fig. 10–14), one seafloor is thrust below the other. This forms a long, narrow depression, an ocean trench. The deeper the seafloor is thrust into the interior of the earth, the hotter it becomes. When the subducted seafloor becomes too hot, it melts, and the light, molten rock rises to the surface to form a series of volcanoes adjacent to the ocean trench. The Aleutian trench and islands off Alaska is an example.

Second, where one seafloor meets another seafloor with a continent riding on it (Fig. 10–15), the seafloor without the continent is thrust under

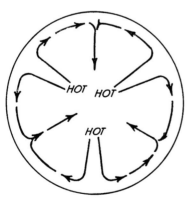

Fig. 10-12 *Convection currents in the interior of the earth*

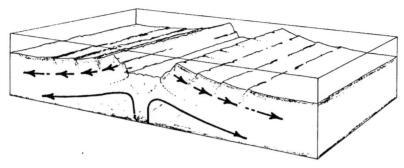

Fig. 10-13 *Formation of the mid-ocean ridge*

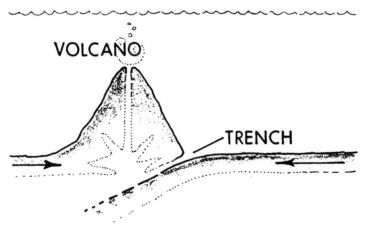

Fig. 10-14 *Subduction zone with two seafloors*

the one with the continent. This forms an ocean trench off the coast of the continent. The edge of the continent is compressed to form a coastal mountain range. Molten rock from the subducted seafloor under the edge of the continent rises to form volcanoes in the coastal mountains. The west coast of South America is an example of this. The Peru-Chile trench occurs just offshore, and the volcanic Andes Mountains occur along the coast.

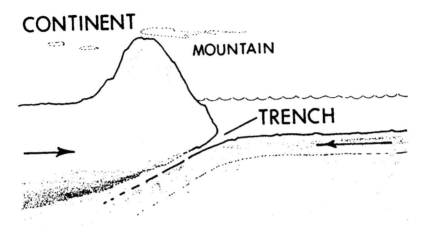

Fig. 10-15   *Subduction zone with two seafloors, one with a continent*

Third, when two seafloors meet, both carrying continents (Fig. 10–16), neither continent is subducted into the interior because both are composed of relatively light granite. The colliding continents are compressed to form a mountain range between the continents. The Himalayan Mountains between India and Asia are an example of this.

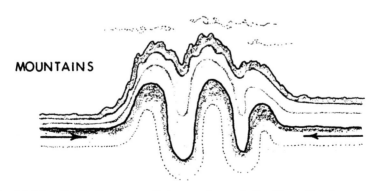

Fig. 10-16   *Subduction zone with two seafloors, both with continents*

Since the early 1960s, considerable evidence has accumulated to support the theory of seafloor spreading. By dividing the distance of the basalt crust from the mid-ocean ridge where it formed, into the basalt seafloor's age, the seafloor spreading rate can be calculated. Spreading rates vary with the location of the ridge and range from 7 in/yr (17 cm/yr) to 0.5 in/yr (1 cm/yr). These are very fast rates for geologic processes. The mid-ocean ridge in the North Atlantic Ocean is spreading at the rate of 1 in/yr (2.5 cm/yr). Because of this, the North Atlantic Ocean is getting wider by the rate of 2 in/yr (5 cm/yr). The North American continent is moving at the rate of 1 in/yr (2.5 cm/yr) to the west whereas the European continent is moving at the rate of 1 in/yr (2.5 cm/yr) to the east.

Seafloor spreading and continental drift are compatible theories. A mid-ocean ridge formed under Pangaea during the Jurassic time and caused it to break up. The continents, riding on the spreading seafloor, would have been carried to their present positions as the Atlantic Ocean became wider.

There are modern examples of a newly formed ocean and a continent that is breaking up. A segment of the mid-ocean ridge from the Indian Ocean enters the Gulf of Aden and bifurcates into two sections. One section is located on the bottom of the Red Sea (Fig. 10–9). The Red Sea is a long, narrow arm of the ocean that separates Egypt and Sudan, in Africa, from Saudi Arabia. Africa and Saudi Arabia were joined millions of years ago. A mid-ocean ridge rose beneath then, split them apart and created the Red Sea. The Red Sea is growing wider by inches each year. It is similar to the Atlantic Ocean when Pangaea first broke up.

Another section of the mid-ocean ridge underlies the Great Rift Valley of East Africa. The valley is a series of large, long grabens, with active volcanoes, earthquakes, and deep lakes. Apparently East Africa is breaking up today. A long, narrow arm of the ocean, similar to the Red Sea, will eventually occupy the rift valley in the next few thousands of years, forming two Africas.

Plate Tectonics

The modern day theory of *plate tectonics,* suggested in 1967, combines the ideas of seafloor spreading and continental drift. Plate tectonics postulates that the crust of the earth is divided into large, moving places (Fig. 10–17) that are about 60 miles (100 km) thick.

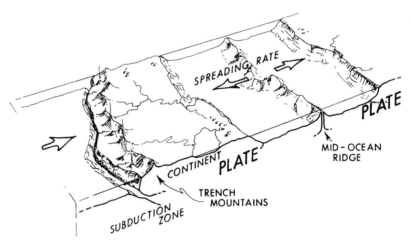

Fig. 10-17   *Cross section of plates*

Every location on the earth's surface, whether a continent or a seafloor, is on a moving plate. Each plate originates at a mid-ocean ridge where new seafloor is being formed. The plate is moving at right angles away from the crest of the ridge. Each plate is moving at the spreading rate of that ridge. At the opposite side of the plate from the mid-ocean ridge is a subduction zone, an ocean trench, and/or mountain range. Large strike-slip faults and earthquakes occur where different plates scrape against each other. Continents ride along on the moving plates.

At the present, there are 16 large plates (Fig. 10–18). The North American Plate is moving to the west at 1 in/yr (2.5 cm/yr) per year. Below California and off the west coast of the United States and Canada, the North American Plate is colliding with the Pacific Plate that is moving to the northwest. The subduction zone has formed the San Andreas Fault (Fig. 10–19), earthquakes, volcanoes such as Mount St. Helens, and a mountain range (Coastal Ranges).

The major features of the earth's surface, both modern and ancient, can be explained by moving plates. In the geologic past, the number and size of the plates have varied along with their rates and directions. Mountains are formed by the collision of plates. For hundreds of millions of years, thick sediments accumulate along continental margins on two different seafloor

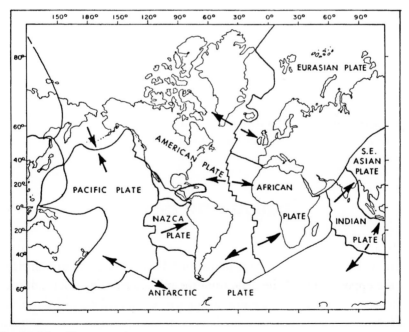

Fig. 10-18   *Plate tectonics*

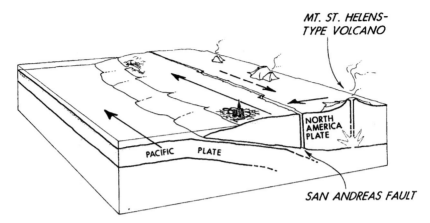

Fig. 10-19 *Cross section of California showing underlying subduction zone*

plates (Fig. 10–20). The continents eventually collide, and the sediments are compressed, forming a mountain range such as the Himalayan Mountains. Mt. Everest is composed of sedimentary rocks deposited in the seas between India and Asia before they collided. The collision between one seafloor plate with a continent and another seafloor plate (Fig. 10–21) forms a coastal mountain range. The Andes Mountains along the west coast of South America are an example.

Failed Arm Basins

The initial breakup of a continent by plate tectonics can take the form of a *triple junction*. A triple junction has three rifts (arms) that join in the center (Fig. 10–22). Usually, two of the arms unite and continue rifting to form an ocean. The other arm stops spreading and is called a *failed arm*. A failed arm is a graben that can be filled with sediments. Several are oil and gas producers.

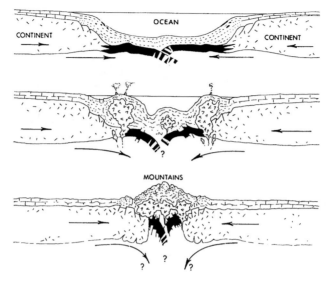

Fig. 10-20   *Formation of a mountain range by two colliding plates, each with a continent*

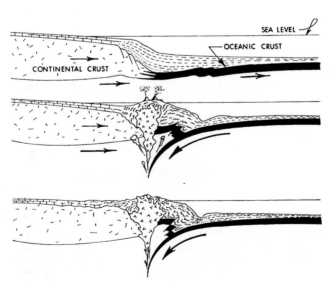

Fig. 10-21   *Formation of a mountain range by two colliding plates, one with a continent*

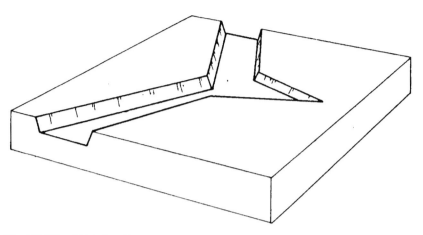

Fig. 10-22   *Triple junction*

During the Mesozoic breakup of the super continent Pangaea, several triple-junctions with failed arms formed. As North America separated from Europe, a triple junction formed near the present day North Sea. Two arms joined to become the North Atlantic Ocean. The failed arm became the Central Graben that runs down the center of the North Sea (Fig. 10–23). The Central Graben and other related grabens are now filled with sedimentary rocks that are 3000 to 6000 ft (900 to 1800 m) thicker than the sedimentary rocks in adjacent areas on the North Sea bottom. Many of the North Sea gas and oil fields are in sedimentary rocks filling this graben.

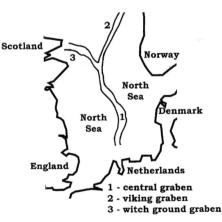

1 - central graben
2 - viking graben
3 - witch ground graben

Fig. 10-23   *North Sea grabens*

As South America pulled away from Africa, a triple junction formed under Nigeria. Two arms joined to become the South Atlantic Ocean. The other arm, the Benue Trough, failed and is located under Nigeria. The Niger River deposited its delta along the length of the Benue Trough out into the Atlantic Ocean. Most of the Nigerian oil production comes from the sedimentary rocks on the Niger River Delta filling the Benue Trough.

Middle East Oil Fields

The large petroleum traps of the Middle East (Fig. 10–24) were formed by plate tectonics. The mid-ocean ridge in the Red Sea (Fig. 10–9) is causing Saudi Arabia on the Arabian Plate to move northeastward and collide with the Eurasian Plate. The Persian Gulf area is being compressed between the two plates, forming the folded traps of the Middle East fields. The deformation becomes more intense in a northeastward direction from Saudi Arabia toward Iran and Iraq and forms the Zargos Mountains.

The Ghawar oil field in Saudi Arabia is the largest conventional oil field on earth. The trap is an anticline 145 miles (233 km) long and 13 miles (21 km) wide (Fig. 10–25a). The reservoir rock is fractured limestone, the Jurassic age Arab D Limestone with an average pay zone of 260 ft (79 m). The reservoir rock is not very deep (-1500 ft or –500 m below sea level), and the oil/water contact occurs at a depth of –1800 ft (-600 m). Salt layers in the overlying Hith Formation are the seal (Fig. 10–25b). The average production of a Ghawar well is 11,400 bbls (1800 m^3) of oil per day, and the field will eventually produce 82 billion bbls (13 billion m^3) of oil.

Fig. 10-24 *Map of Middle East oil fields (modified from Beydown, 1991)*

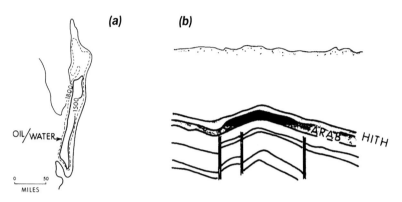

Fig. 10-25 *Ghawar oil field, Saudi Arabia (a) structural contour map on Arab D limestone and (b) cross section (modified from Arabian American Oil Company staff, 1959)*

The Gashsaran oil field in Iran is a deformed anticline (Fig. 10–26). Large thrust and reverse faults associated with the anticline are also the result of the compression between the moving plates. The reservoir rock, the Asmari Limestone of Oligocene-Miocene age, is 1000 to 1500 ft (300 to 450 m) thick. The limestone is overlain by salt that forms the seal. On the steep side of the anticline, the thick, steep-dipping limestone reservoir rock forms a 6000 ft (1800 m) net oil pay zone. Ultimate production from this field will be 8.5 billion bbls (1.35 billion m^3) of oil.

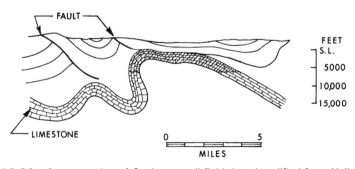

Fig. 10-26 *Cross section of Gashsaran oil field, Iran (modified from Hull and Warman, 1970)*

eleven

SOURCE ROCKS, GENERATION, MIGRATION, AND ACCUMULATION OF PETROLEUM

Source Rocks

A *source rock* is a rock that forms gas or oil. The source of gas and oil is organic matter preserved in sedimentary rocks. As sediments are deposited, both inorganic mineral grains such as sands and mud and organic matter (dead plants and animals) are mixed. Some of the organic matter is lost on the surface by decay, a process of oxidation. The decaying organic matter on land gets oxygen from the air and the decaying organic matter on the ocean bottom gets the oxygen from out of the water. Some organic matter, however, is preserved. It was either rapidly buried by other sediments before it decayed or was deposited on the bottom of a sea with stagnant, oxygen-free waters.

The black color of sedimentary rocks comes primarily from its organic content. Black-colored, organic-rich sedimentary rocks include coal, shale and some limestones.

Plant material is transformed into coal by temperature and time. Wood and coal have a specific chemistry that can generate only methane gas (CH_4). This is why coal mines are dangerous; they contain methane gas and sometimes explode. Wells are often drilled into coal beds to produce *coal bed* or *coal seam gas* that is pure methane gas. The gas occurs adsorbed to the surface of the coal along natural fractures called *clints*. A coal seam gas well first produces water for a period of time up to a year as the fractures drain. As the pressure in the coal decreases, the methane gas is released from the surface of the coal.

Shale is the most common sedimentary rock, and many are black. Black shale commonly has 1 to 3% organic matter by weight and can have up to 20%. Green or gray shale has only about 0.5% organic matter. Black shales contain a large variety of plant and animal organic matter including non-woody plant matter such as algae. They have the right chemical composition to generate both natural gas and crude oil. In some areas such as North Africa and the Middle East, organic-rich, dark limestones are also source rocks.

Some sedimentary basins are gas-prone and produce primarily natural gas. Examples of these are the Sacramento basin of northern California, the Arkoma basin of southern Oklahoma and Arkansas, and the southern portion of the North Sea. This is because the only effective source rock is coal.

Generation

The most important factor in the generation of crude oil from organic matter in sedimentary rocks is temperature. A minimum temperature of about 150°F (65°C) is necessary for oil generation under typical sedimentary basin conditions (Fig. 11–1). This temperature is obtained by burying the organic-rich source rocks. The deeper the depth: the higher the temperature.

At relatively shallow depths, the temperature is not sufficient to generate oil. There, just a couple of feet below the surface, bacterial action on the organic matter forms large volumes of *biogenic* or *microbial gas.* It is generated very fast and is almost pure methane gas. This gas is commonly known as swamp or marsh gas. Biogenic gas is rarely trapped and usually leaks into the atmosphere in enormous volumes. However, the largest gas field in the world, Urengoy in Siberia, is believed to be filled with biogenic gas. The gas is trapped below the permanently frozen ground (permafrost). The field contains 285 Tcf (8 trillion m^3) of gas. Generation of biogenic gas decreases with depth as bacterial action decreases with increasing temperature.

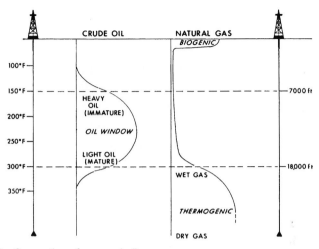

Fig. 11-1 *Generation of gas and oil*

In a typical sedimentary basin, oil generation starts at about 150°F (65°C) and ends at about 300°F (150°C). If the source rock is buried deeper, where temperatures are above 300°F (150°C), *thermogenic gas* is generated from organic matter left in the source rock. Thermogenic gas is the gas that can be trapped and is either dry or wet gas.

The zone in the earth's crust where the oil is generated is called the *oil window*. It occurs from about 7000 to 18,000 ft (2100 to 5500 m) deep. Heavy oil, called *immature oil* is generated at lower temperatures in the oil window whereas light oil, called *mature oil*, is generated at higher temperatures. At lower temperatures where thermogenic gas is generated, wet gas is formed. Under higher temperatures at deeper depths, dry gas is formed.

Time is also a generation factor. Both biogenic and thermogenic gas form very fast, but oil takes a long time to form. Because chemical reactions double in speed for each 10°F (5.5°C) temperature increase, the time at which the source rock is exposed to each temperature as it is buried deeper also influences oil formation. Oil can be generated at lower temperatures if the source rock is exposed to those temperatures for a longer time. Higher temperatures need shorter times to generate oil.

151

At temperatures higher than about 300°F (150°C), crude oil is irreversibly transformed into graphite (carbon) and natural gas. The process is similar to thermal cracking in a refinery. This temperature occurs at a depth of about 18,000 ft (5500 m) in a sedimentary rock basin. This is a floor, below which only gas can occur in the reservoir. Deep wells are drilled for natural gas. The deepest sustained oil production is from 17,192 ft (5240 m) in the Bulla-More field of offshore Baku, Caspian Sea. In several instances, a deep well has discovered a gas reservoir, and the sand grains in the sandstone reservoir rock are coated with carbon. Apparently, there was oil originally in the reservoir, but it was buried too deep and was thermally cracked.

Many sedimentary basins are unproductive. An unproductive basin might not have an organic-rich source rock that could generate petroleum. Even if an unproductive basin has a source rock, it might never have been buried into the oil window. *Maturity* is the degree to which petroleum generation has occurred in a source rock. A mature source rock has experienced the temperature and time to generate petroleum in contrast to an immature source rock. In sedimentary rock basins, between 30 to 70% of the organic matter in the source rock that has been buried deep enough generates gas and oil.

Migration

After gas and oil is generated in shale source rock, some is expelled from the impermeable shale. The generation of a liquid (crude oil) or gas (natural gas) from a solid (organic matter) causes a large increase in volume. This stresses the source rock and fractures the shale. The hydrocarbons escape upward through the fractures. After the pressure is released, the fractures close, and the shale becomes impermeable again.

Because gas and oil are light in density compared to water that also occurs in the pores of the subsurface rocks, petroleum rises (Fig. 11–2). Oil and gas can flow upward along faults and fractures. It can also flow laterally and upwards along unconformities and through carrier beds. *Carrier beds* are rock layers that are very permeable and transmit fluids.

The vertical and lateral flow of the petroleum from the source rock is called *migration*. If there is no trap on the migration route, the gas and oil will flow out onto the surface as a gas or oil seep. If there is a trap along the

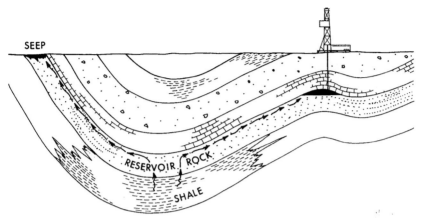

Fig. 11-2 *Migration of gas and oil in a sedimentary rock basin*

migration route, the gas and oil will accumulate in the trap. Of all the gas and oil that forms in a sedimentary rock basins, only from 0.3 to 36% is ever trapped. On the average, only 10% of the gas and oil is trapped. The rest of the gas and oil either did not get out of the source rock, was lost during migration, or seeped into the earth's surface.

Because of migration, where the petroleum formed in the deep basin and where it ends up in the trap are different both vertically and horizontally. This is why a reservoir of light oil and thermogenic gas that originally formed deep, can be found at shallow depths. In the Williston basin of Montana, the oil has migrated more than 200 miles (320 km) horizontally out from the source area in the deep basin to the traps on the flanks of the basin.

Accumulation

The trap must be in position before the gas and oil migrates through the area. If the trap forms after the migration, no gas and oil will occur in the trap (Fig. 11–3).

(a)   (b)

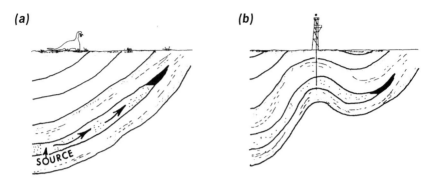

Fig. 11-3 *(a) generation, migration, and accumulation of oil in a sandstone pinch
out (b) later formation of an anticline that is barren of oil*

Once the gas and oil migrate into the trap, they separate according to
density. The gas, being lightest, goes to the top of the trap to form the *free gas
cap* where the pores of the reservoir rock are occupied by gas. The oil goes to the
middle of the trap, the *oil reservoir*. Salt water, the heaviest, goes to the bottom.

The most common trap is a *saturated pool* that always has a free gas cap on
top of the oil reservoir (Fig. 11–4a). The oil in the reservoir has dissolved all the
natural gas it can hold and is saturated. An *unsaturated pool* lacks a free gas cap
(Fig. 11–4b). The oil has some dissolved gas but it can hold more and is unsaturated. Sometimes there is only a gas reservoir on water (Fig. 11–4c).

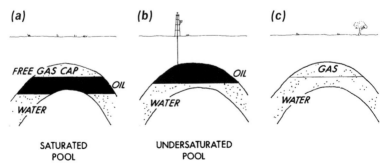

Fig. 11-4 *Types of hydrocarbon traps*

The boundary in the reservoir between the free gas cap and the oil is the *gas-oil contact* (Fig. 11–5). The boundary between the oil and water reservoir is the *oil-water contact*. The gas-oil and oil-water contacts are either relatively sharp or gradational and are usually level.

The top of the trap is called the *crest*. The first exploratory well is usually drilled *on structure*, on the crest of the structure where the probability is highest to encounter petroleum (Fig. 11–5). When a well is drilled to the side of the crest, it is drilled *off structure*. If a well is drilled too far off structure, it might not encounter commercial amounts of oil or gas and is called a *dry hole, duster,* or *wet well.*

In a trap, the reservoir rock must be overlain by a *caprock* or *seal* (Fig. 11–5), an impermeable rock layer that doesn't allow fluids to flow through it. Without a caprock or seal, the gas and oil would leak onto the surface. Two common caprocks are shales and salt layers. Well-cemented or shaly rocks, micrite, chalk, and permafrost can also be caprocks.

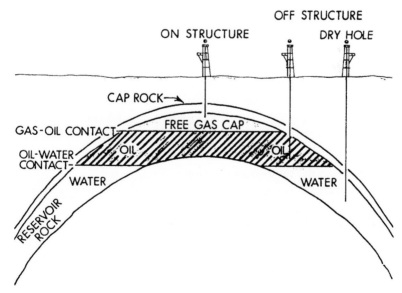

Fig. 11-5 *Details of a saturated pool*

Age

The time in which the crude oil was generated, migrated through the rocks, and accumulated in the trap can be very old to very recent. Oil in Oklahoma was generated and trapped during the Pennsylvanian time. In the North Slope of Alaska, it was during the Cretaceous time. In the Gulf of Mexico, however, some of the oils are very young, just a couple million years old, and some oil is probably still forming and accumulating today.

Reservoir Rocks

A *reservoir rock* is a rock that can both store and transmit fluids. A reservoir rock must have both porosity and permeability. *Porosity* is the percent volume of the rock that is not occupied by solids. These spaces are called *pores*. In the subsurface, the pores are filled with fluids such as water, gas and oil. Porosity measures the fluid storage capacity of a reservoir rock. There are several accurate methods to measure porosity using well cuttings, cores, and wireline well logs.

First, when a well is drilled, the rock chips (well cuttings) made by the drill bit are flushed up the well by the drilling mud. The well cuttings are sampled at regular intervals. A geologist examines them under a binocular microscope to identify the rock types and see the pores. The geologist can often accurately estimate the porosity of the rock to within 1 to 2%.

Second, a *core* is a cylinder of rock that is drilled from the well. A small plug in the shape of a cylinder, 1 to $1^1/_2$ in. (2.5 to 3.8 cm) in diameter and 1 to 3 in. (2.5 to 7.6 cm) long, is cut from the core. The plug is then dried to remove the fluids from the pores. An instrument called a *porosimeter* is used to measure the porosity of the plug.

Third, accurate porosity measurements can also be made after the well is drilled without taking samples of the rock by a service company that runs one of three wireline well logs (neutron porosity, formation density, and sonic-velocity logs).

Typical porosity values for an oil reservoir are shown in Table 11–1. Natural gas compresses and needs less porosity than an oil reservoir. Very deep gas reservoirs need very little porosity because of the very high pressure.

Table 11–1

Porosity values for an oil reservoir

0–5%	insignificant
5–10%	poor
10–15%	fair
15–20%	good
20–25%	excellent

(modified from Levorsen, 1967)

Porosity is a relatively easy and accurate measurement to make. Because of this, a *porosity cutoff*, a minimum porosity value is often used to help decide whether or not to complete an oil well. For sandstones, a typical porosity cutoff is 8 to 10%. Limestones often have less porosity than sandstones but have fractures that drain larger areas. For limestones, a typical porosity cutoff of 3 to 5% is used. These values vary depending on the depth and economics of the well.

Permeability is a measure of the ease with which a fluid can flow through a rock. Permeability is measured in units of *darcies* (D) or *millidarcies* (*md*). A millidarcy is 1/1000th of a darcy. A darcy is the permeability that will allow a flow of 1 cubic centimeter per second of a fluid with 1 centipoise viscosity (resistance to flow) through a distance of 1 centimeter through an area of 1 square centimeter under a differential pressure of 1 atmosphere. The greater the permeability of a rock, the easier it is for the fluids to flow through the rock.

The only way to make a quantitative permeability measurement is to drill a core of the reservoir rock and cut a plug. The plug is dried to remove any liquids. An instrument called a *permeameter* is used to measure the permeability of the dried plug by measuring the flow of air or nitrogen through it.

Typical permeability values of an oil reservoir rock are given in Table 11–2. Gas is more fluid than oil and needs less permeability than an oil reservoir.

Table 11–2

Permeability values for an oil reservoir

1–10md	poor
10–100md	good
100–1000md	excellent

(modified from Levorsen, 1967)

Porosity and permeability in a single sedimentary rock layer are related. In general: the higher the porosity, the greater the permeability (Fig. 11–6). Permeability, however, is also controlled by the grain size. The hardest place for the oil or gas to flow through the rock is the narrow connections (*pore throats*) between the pores (Fig. 11–7). The smaller the pore throats, the harder it is for the oil or gas to flow. Smaller grain sizes have smaller pore throats. Because of this, porous, coarse-grained rocks such as sandstones that have large pore throats are usually very permeable. A porous, fine-grained rock such as shale or chalk has small pore throats and little or no permeability.

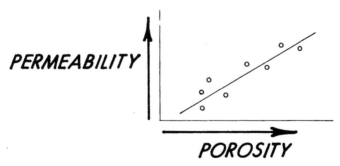

Fig. 11-6   *Porosity-permeability relationship in a single reservoir rock*

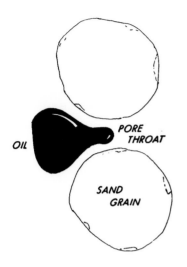

Fig. 11-7 *Magnified view of sand-stone showing two sand grains and the pore throat between them*

Most reservoir rocks are sandstone and carbonates, but most sandstones and carbonates are not reservoir rocks. A very low or no permeability sandstone or carbonate is called *tight* or *tight sands.*

The Spraberry field in the Midland basin of Texas is a series of oil fields in tight sand reservoir rocks of the Permian age Spraberry Formation at a depth of 7000 ft (2130 m). The field covers an area of 150 by 75 miles (240 by 120 km). The reservoir rock is about 1000 ft (300 m) thick with 300 ft (90 m) of pay and contains 594 million bbls (94 million m^3) of recoverable oil. Because the reservoir rock is fine-grained (*e.g.*, shale and silt-stones), it has an average permeability of only 0.5 millidarcy.

Any fractured rock can be a reservoir rock. Fractured shales, cherts, and chalks are reservoir rocks in many fields. Even fractured basement rock can be reservoir rock. When an anticline or dome is originally folded, the brittle basement rock is often fractured along the crest of the fold. Oil and gas forms in source rock along the flanks of the anticline at a lower elevation than the crest (Fig. 11–8) and migrate up into the fractured basement rock. The Wilmington Oil Field of California is formed by an anticline and produces out of seven sandstone reservoir rocks and the top of the fractured, metamorphic basement rock.

Point Arguello oil field in offshore southern California (Fig. 11–9) produces from an anticline in the Miocene age Monterey Formation. The Monterey Formation is composed of alternating layers of fractured chert and shale at a depth of 6000 to 8000 ft (1800 to 2450 m) below the seafloor. Net pay is about 700 ft (210 m). The average porosity is 15%, and the permeability varies from 0.1 to 3000 md due to fractures. Without the fractures, the Monterey formation would not be a reservoir rock. The field will eventually produce 300 million bbls (48 million m^3) of oil.

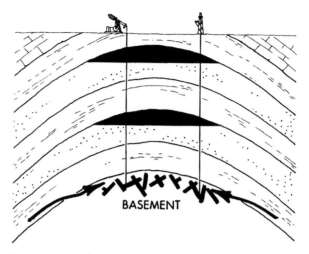

Fig. 11-8 *Accumulation of gas and oil in fractured basement rock on an anti-cline or dome*

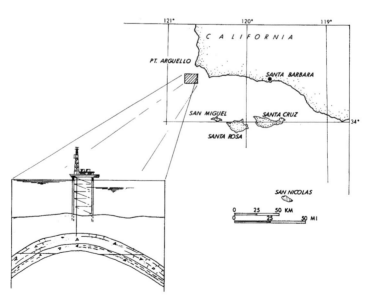

Fig. 11-9 *Map and cross section of Point Arguello oil field, offshore California
(modified from Mero, 1991)*

In the subsurface, to the east of Austin, Texas, there are several ancient volcanoes that produce oil. When the volcanoes erupted during the Cretaceous time, the area was covered with Gulf of Mexico waters. The submarine lava spread out along the ocean bottom, forming mushroom-shaped deposits of basalt. The surface of the basalt was highly fractured and weathered by seawater, giving it porosity and permeability. Sediments then covered the volcanoes that are now buried in the Texas coastal plain. Oil later migrated up into the weathered basalt that is called *serpentine* by the drillers. Fourteen of these fields, such as the Lyton Springs field (Fig. 11–10), have been found.

Granite wash is a potential reservoir rock formed by the weathering of granite. Granite is composed of large, well-sorted, sand-sized mineral grains and weathers to form well-sorted sandstone that can be very thick. After the granite and granite wash have been buried in the subsurface, oil and gas can form in source rocks at a lower elevation. The oil and gas can then migrate up and into the granite wash (Fig. 11–11). Granite wash reservoir rocks are common in the subsurface of southern and western Oklahoma and the Panhandle of Texas. The Elk City field of Oklahoma produces from granite wash where it has been uplifted by an anticline (Fig. 11–12).

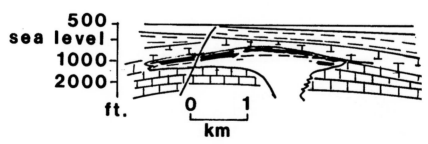

Fig. 11-10 *Cross section of Lyton Springs field, Texas (modified from Collingwood and Rettgen, 1926)*

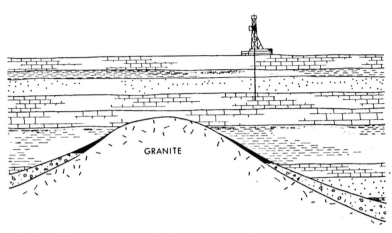

Fig. 11-11   *Migration of gas and oil up into a granite wash reservoir*

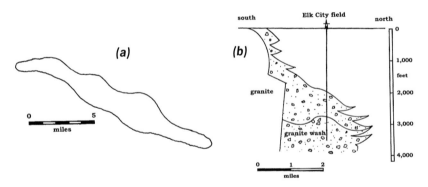

Fig. 11-12   *Elk City field, Oklahoma (a) map and (b) cross section (modified from Lyday, 1990)*

The Cretaceous age Austin Chalk is productive in Texas and Louisiana (Fig. 11–13a). The Giddings and Pearsall fields of Texas were first discovered in the 1930s, and the trend now extends into Louisiana. In the Giddings field of south-central Texas (Fig. 11–13b), the Austin Chalk occurs at a

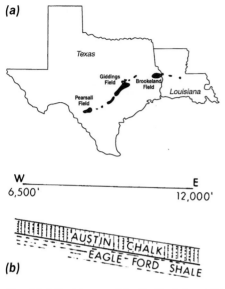

Fig. 11-13 *(a) map of Austin Chalk trend (b) east-west cross section of Giddings Field, Texas (modified from Galloway, Ewing, Garrett, Tyler and Bebout, 1983)*

depth of 6500 to 12,000 ft (2000 to 3660 m) and is underlain by the Eagle Ford Shale that is the source rock for the oil along with organic matter in the chalk itself. Vertical fractures in the chalk allowed the gas and oil to migrate up into the chalk. Without the fractures, permeability in the Austin Chalk is less than 1md.

The Niobrara chalk and Bakken shale in Colorado and Kansas also produce from fractures. Oil seeps in southeastern Colorado near Canon City lead to the discovery of the Florence field in 1876. It produces from fractures in the Pierre Shale.

Saturation

In the oil or gas reservoir, the oil or gas always shares the pore spaces with water (Fig. 11–14). The relative amount of the oil/gas and water sharing the pores of the reservoir will vary from reservoir to reservoir and is called *saturation*. It is expressed as a percent and always adds up to 100%. Saturation is why most oil wells pump not only oil, but also water called oilfield brine. *Oilfield brine* is very salty water that shared the pores with the oil.

The fluid that occupies the outside of the pore and is in contact with the rock surface is called the *wetting fluid.* Sandstones usually have oil in the center of the pore, and water is on the outside of the pore in contact with the sand grains. Because of this, most sandstones are *water wet* (water coats the sand grains). In contrast, limestones and dolomites are usually *oil wet*

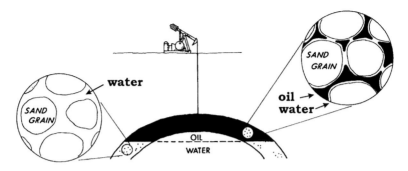

Fig. 11-14 *Magnified view of oil and water saturation in the pores of a sandstone oil reservoir. Note the 100% water saturation in the pores below the oil-water contact.*

(oil coats the rock surfaces). The percentage oil recovery tends to be greater in sandstone reservoirs than in limestone reservoirs. This is because the fluid in the center of the pore will flow easier than the fluid on the outside of the pore that is being held to the rock surface by surface tension.

Oil Shales

Oil shales are sedimentary rocks rich in a type of organic matter called *kerogen.* When the oil shale is heated to about 660°F (350°C), kerogen is transformed into crude oil called *shale oil.* Oil shales are organic-rich source rocks that are old enough but have never been buried deep enough for heat to transform the organic matter into oil. They are immature source rocks. The inorganic sediments are commonly clay, fine-grained quartz and calcite, and salts.

Very large oil shale deposits are located in Australia, Brazil, Russia, and the United States. High-grade oil shales yield more than 25 bbls (4 m^3) of oil per ton of shale. Some are located in the Eocene Age Green River Formation of the Rocky Mountains. These oil shales are actually calcareous muds and salts that were deposited in a lake and algae was a major source of the organic matter. In the Piceance basin of northwestern Colorado

(Fig. 8–5), the Green River Formation is estimated to hold about 600 billion bbls (95 billion m³) of oil in high-grade oil shale. Large oil shale deposits are also located in the Green River and Washakie basins of Wyoming and the Uinta basin of Utah. Because the oil shale crops out, it can be surface-mined. Another production method is to dig a large cavern in the oil shale. The oil shale on the sides of the cavern is set on fire for heat to generate oil in the subsurface. High production cost is a major factor preventing the commercial development of oil shales.

Tar Sands

Tar sands are composed of very heavy oil mixed with sands. The oil is too viscous to be produced by conventional methods. Very large tar sand deposits occur in northern Alberta and Venezuela. It is thought that the tar originated as good quality crude oil that was degraded on the surface during the geological past into tar.

Fig. 11-15 *Location map of Athabaska tar sands, Alberta*

The Athabaska tar sands in Alberta (Fig. 11–15, contain 800 billion bbls (125 billion m³) of oil. The tar sands occur along an angular unconformity (Fig. 11–16). They are thought to have formed from the same high-quality oil located in the Devonian reef fields of Alberta (Fig. 7–8). The oil, however, seeped on the surface during the Cretaceous time and was degraded into tar. The surface deposits are being mined with large steam shovels and treated with steam to remove sand from the tar. The syncrude produced from the Canadian tar sands is commercial.

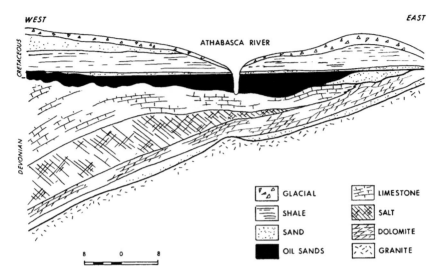

Fig. 11-16   *Cross section of Athabaska tar sands, Alberta (modified from Jardine, 1974)*

twelve

PETROLEUM TRAPS

Description

Petroleum traps were originally filled with water. The gas and oil flows into the trap later to replace the water. Because both gas and oil are both lighter than water, the trap is filled from the top downward. The trap can be filled down to a level, called the *spill point*, at which it cannot hold any more (Fig. 12–1). It is the highest point on the rim of an anticline or dome. The vertical distance from the crest of the reservoir rock down to the spill point is the *closure* of the trap. Closure is the maximum vertical amount of gas and oil that the trap can theoretically hold. The closure of a trap is a measure of the potential size of the field. Traps must have *four-sided closure* to be effective. The trap must go down or be sealed on all four sides to hold petroleum.

The *field* is the surface area directly above one or more producing reservoirs on the same trap such as an anticline (Fig. 12–2). The field is commonly given a geographic name such as a town, hill or creek. A *reservoir* is a subsurface zone that produces oil and gas but does not communicate with other reservoirs. Fluids cannot flow from one reservoir to another.

The oil or gas in a single reservoir has the same characteristics, but can be very different between reservoirs in the same field. The Wilmington oil field of California produces from several separate reservoirs. The oil is the same in each reservoir but varies between reservoirs in the field from 12 °API to 34 °API and is both sweet and sour. Reservoirs are usually named after the reservoir rock such as the Bartlesville Sandstone reservoir.

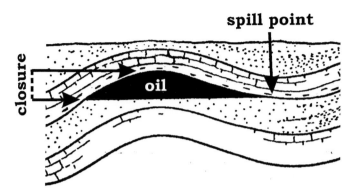

Fig. 12-1 *Description of trap*

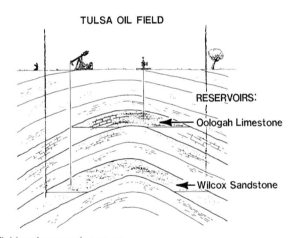

Fig. 12-2 *Field and reservoir names*

Two types of petroleum traps are structural and stratigraphic. *Structural traps* are formed by deformation of the reservoir rock, such as a fold or fault. *Primary stratigraphic traps* are formed by deposition of the reservoir rock, such as a river channel sandstone or limestone reef that is often completely encased in shale. A *secondary stratigraphic trap* is formed by an angular unconformity. A *combination trap* has both structural and stratigraphic elements.

Structural Traps

Anticlines and Domes

A characteristic of a dome or anticline is the potential for multiple producing zones. The Santa Fe Springs field near Los Angeles, California, has 25 producing zones (Fig. 10–5). Each is a sandstone reservoir separated by shale.

Most anticlines and domes are asymmetrical (Fig. 12–3), with a steep and a gentle side. The crest of an asymmetrical structure will migrate with depth. A well that is drilled on structure at the crest of a fold on the surface could be too far off structure at reservoir rock depth to encounter petroleum.

The Salt Creek field in the Powder River basin of Wyoming is formed by a large, asymmetrical anticline. The elliptical-shaped anticline is steepest on the west (where the contours on the structural contour map are closest together) and gentlest on the east (Fig. 12–4). The field has five producing zones, all sandstones, of which the Cretaceous Age Second Wall Creek Sandstone is the most important. A large oil seep was located over the trap, leading to its discovery in 1907. The field will produce 766 million bbls (122 million m³) of oil and 750 Bcf (21 billion m³) of gas.

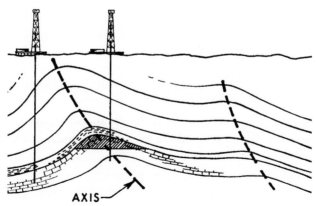

Fig. 12-3 *Asymmetric anticline*

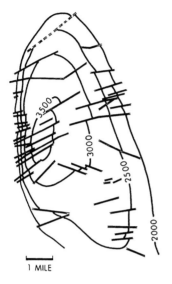

1 MILE

Fig. 12-4 *Structure contour map on reservoir rock, Salt Creek field, Wyoming (depth in ft.) (modified from Barlow and Haun, 1970)*

As the reservoir rocks are deformed into an anticline or dome trap, they are often cut by faults. The faults can sometimes be barriers to fluid flow and are called *sealing faults.* Sealing faults will divide the structure into individual producing compartments or pools (Fig. 12–5). Petroleum production from one side of the sealing fault will not affect production on the other side.

Two observations can determine if a fault is a sealing fault and has divided the structure into separate producing compartments. Oil-water and gas-oil contacts in reservoirs are level. If the contacts are at different elevations in the same reservoir rock on opposite sides of the fault, the fault is a sealing fault. If there are different fluid pressures at the same elevation on opposite side of the

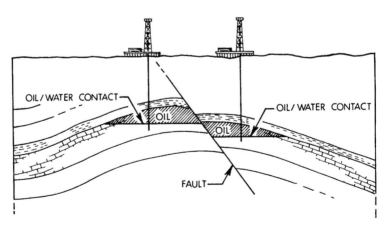

Fig. 12-5 *Sealing fault cutting an anticline*

fault, the fault is a sealing fault. Many faults in deltas and coastal and offshore areas are sealing faults because the shales are soft. When the fault moves, the soft shale is smeared along the fault plane to form a clay or shale smear that acts as a seal.

The Wilmington field, just to the southeast of Los Angeles is the largest oil field in California. The trap is a large anticline, 11 miles (18 km) long and 3 miles (5 km) wide (Fig. 12–6). The anticline is overlain by an angular unconformity at 2000 ft (600 m) depth and flat sedimentary rocks above that (Fig. 12–7). There are seven sandstone-producing zones along with the fractured metamorphic basement that also produces. Miocene to Pliocene Age turbidity currents deposited the sandstones.

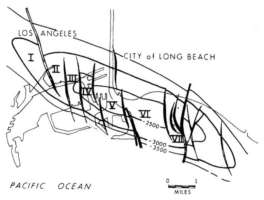

Fig. 12-6 Structure contour map on a sandstone reservoir rock, Wilmington oil field, California (depth in ft.) (modified from Mayuga, 1970)

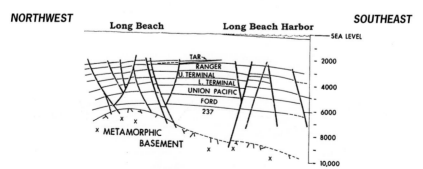

Fig. 12-7 Cross section along length of Wilmington oil field, California showing sandstone reservoir rocks (depth in ft.) (modified from Mayuga, 1970)

The anticline is cut by seven sealing faults that separate it into seven producing compartments. The four southwestern compartments under Long Beach Harbor were never fully developed until the completion of four artificial islands for drilling and production in 1965. The islands were developed by a Texaco, Humble (Exxon), Union of California, Mobile, and Shell consortium called THUMS. The Wilmington field will ultimately produced 2.75 billion bbls (450 million m^3) of primarily heavy sour oil and one Tcf (28 million m^3) of gas.

Growth Faults and Rollover Anticlines

In an area where large volumes of loose sediments are rapidly deposited, such as deltas and coastal plains, a unique type of fault and anticline forms (Fig. 12–8). A *growth* or *down-to-the-basin fault* moves as the sediments are being deposited. The fault is always parallel to and located just inland from the shoreline. The weight of the sediments being deposited along the shoreline pulls the basin side of the fault down. A growth fault is similar to a giant slump. It is called a growth fault because it moves as the sediments are being deposited. This is in contrast to other faults such as normal, reverse, and strike-slip faults that occur in sedimentary rocks millions of years old. It is also called a down-to-the-basin fault, because the basin side is moving down. Four unique aspects of the growth fault distinguish it from faults in solid rocks.

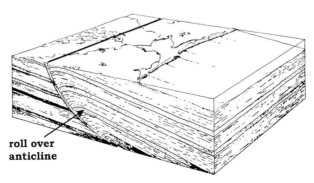

roll over
anticline

Fig. 12-8 Growth fault with a rollover anticline

First, it has a curved fault plane that is concave toward the basin. The fault becomes less steep with depth. Faults in solid rocks tend to have linear fault planes. The growth fault, however, occurs in loose sediments. As the sediments are buried, weight compacts the sediments causing the steep, near-surface fault plane to become flatter with depth.

Second, the growth fault has thicker sediment layers on the basin side of the fault. Because the fault is moving as the sediments are being deposited, a topographic low area forms on the basin side of the fault. This accommodates thicker layers on the down side than on the up side (land) of the fault.

Third, the deeper the sediments, the larger the fault displacement. This is because the deeper sediments are older and have experienced more fault movement.

Fourth, a *rollover anticline*, a large, potential gas and oil trap, commonly occurs on the basin side of the growth fault (Fig. 12–8). This anticline is caused by the curved fault plane, which is almost horizontal with depth (Fig. 12–9a). As the growth fault moves, a gap forms between the near-surface sediments on either side of the fault (Fig. 12–9b). The relatively loose sediments roll over into the gap to form the anticline (Fig. 12–9c). Rollover anticlines are prolific petroleum traps along the Gulf of Mexico coastal plain and in the Mississippi and Niger River deltas.

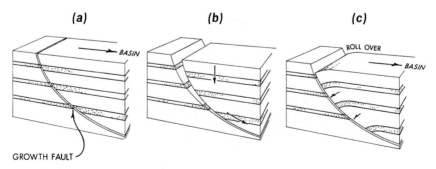

Fig. 12-9 *Formation of a rollover anticline*

Growth faults will become inactive and buried in the subsurface as the shoreline progrades out into the basin. Ancient growth faults are found both inland and offshore (Fig. 12–10). Offshore growth faults were active during times of lower sea levels.

The Vicksburg oil and gas field trend along the south Texas coastal plain is along a buried, inactive growth fault (Fig. 12–11). Rollover anticlines form a series of gas and oil fields that parallel the Vicksburg fault on the Gulf of Mexico side. The trend will produce more than 3 billion bbls (0.5 billion m^3) of oil and 20 Tcf (0.5 trillion m^3) of gas. Exploration and production started in 1934 with the discovery of the Tom O'Conner field.

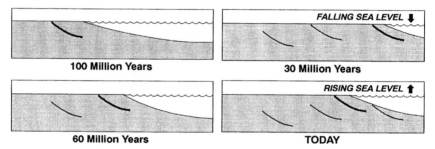

Fig. 12-10   *Cross sections showing the migration of growth faults with prograding shorelines and rising and falling sea levels. The thick line is the active growth fault and the thin lines are inactive growth faults.*

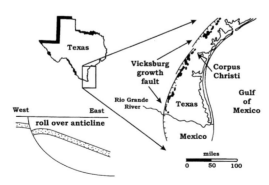

Fig. 12-11   *Map and cross section of Vicksburg fault trend, Texas Gulf Coast (modified from Stanley, 1970)*

174

The trap for the Tom O'Connor field is a rollover anticline, 10 by 3 miles (16 by 5 km) in size (Fig. 12–12) on the down-thrown (Gulf of Mexico) side of the Vicksburg fault. The field has several Oligocene to Pliocene age sandstone reservoirs. Because the coastal plain sands are relatively young and haven't been buried very deep, porosities average 31% and permeabilities range up to 6500 millidarcies. The Tom O'Conner field will ultimately produce more than 500 million bbls (80 million m^3) of oil and 1 Tcf (28 million m^3) of gas.

The Tuscaloosa trend of wet gas fields in Louisiana is similar. The traps are rollover anticlines on the Cretaceous age Tuscaloosa Sandstone at depths of 16,000 to 22,000 ft (5000 to 6700 m).

Rollover anticlines on growth faults are often cut by smaller faults called *secondary faults* (Fig. 12–13) that displace the reservoir rocks. These secondary faults often are sealing faults that divide the field into numerous smaller reservoirs.

A cross section of the Niger River Delta, Nigeria (Fig. 12–14) shows the underlying Akata Formation, the black shale source rock. It was deposited in deep waters off the delta but has been covered by younger delta sediments. The reservoir rocks are in the overlying Agbada Formation that was deposited in shallow water and on the delta. The formation consists primarily of shale but it contains numerous distributary and river channel and beach sandstone reservoir rocks.

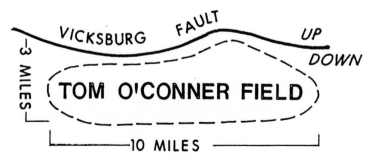

Fig. 12-12 Map of Tom O'Conner field, Texas (modified from Mills, 1970)

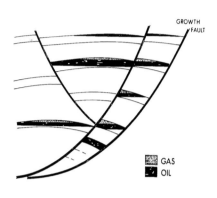

Fig. 12-13 *Cross section showing secondary faults on a rollover anticline*

The traps are rollover anticlines that are often cut by secondary faults that are sealing faults. The only active growth faults are the ones located just inland from the present shoreline. The growth faults that are buried and located further inland were active when the shoreline was inland. The inactive growth faults located offshore were active when sea level was lower (Fig. 6–22).

On the top of the delta is the Benin Formation. It consists of river channel sands and marsh and swamp deposits that are still being deposited today on the subaerial top of the delta. There are about 100 fields, each with more than 50 million bbls (8 million m³) of recoverable oil in the delta. However, only 12 fields have more than 500 million bbls (80 million m³) of recoverable oil, and there is no field that has one billion bbls (160 million m³) of recoverable oil.

Hibernia, the largest oil field in eastern Canada, is a rollover anticline trap located offshore from Newfoundland (Fig. 12–15a). The Murre fault, a growth fault, was active when sea level was lower, and the area was exposed.

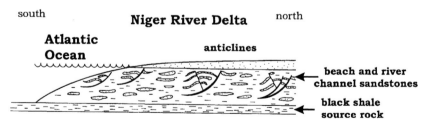

Fig. 12-14 *North-south cross section of the Niger River Delta, Nigeria*

The rollover anticline is cut by several secondary faults (Fig. 12–15b). The reservoir rocks are the Cretaceous age Avalon (7200 ft or 2200 m deep) and Hibernia (12,200 ft or 3700 m deep) sandstones. The Avalon Sandstone averages 20% porosity and 220 md permeability whereas the Hibernia sandstone averages 16% porosity and 700 md permeability. The field will produce 615 million bbls (100 million m³) of 32 to 35 °API, sweet oil.

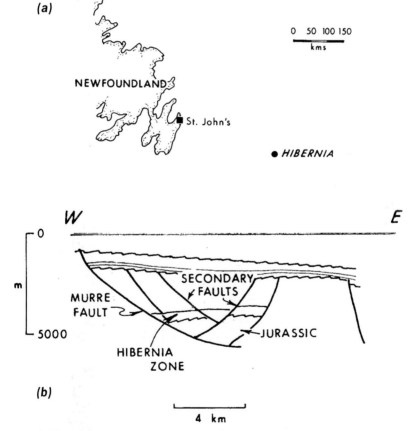

Fig. 12-15 *Hibernia oil field, offshore eastern Canada (a) map and (b) cross section (modified from MacKay and Tankard, 1990)*

Drag Folds

Drag folds are formed by friction generated along a fault plane when a fault moves. Friction causes the beds on either side of the fault to be dragged up on one side and down on the other side of the fault (Fig. 12–16).

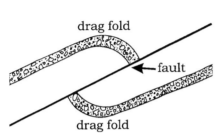

Fig. 12-16 Drag folds along a fault

Fig. 12-17 Map of Rocky Mountain or Western Overthrust Belt

Most mountain ranges on land were formed by compressional forces and display compressional features such as folds, reverse faults, and thrust faults. The thrust faults occur in zones called *overthrust* or *disturbed belts*. The most drilled of these overthrust belts is the Rocky Mountain or Western overthrust or disturbed belt of the United States and Canada (Fig. 12–17). Thrust faulting occurred from the Cretaceous through Eocene time during the formation of the mountains.

A west to east cross section shows the deformation of the earth's crust (Fig. 12–18). Several of the thrust faults moved tens of miles horizontally. In Wyoming and Utah, the 1000 ft (300 m) thick Jurassic age Nugget Sandstone, has been deformed into large subsurface drag folds along the thrust faults. These are the targets for Rocky Mountain overthrust drilling.

The Painter Reservoir field, Wyoming, discovered in 1977, is a typical overthrust field. A large-

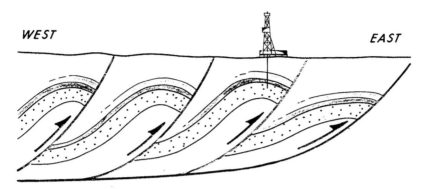

Fig. 12-18 *East-west cross section of overthrust belt*

scale, east to west cross section (Fig. 12–19), shows the numerous thrust faults and the drag fold trap in the Nugget Sandstone at 10,000 ft (3000 m) below the surface. A smaller scale, east to west cross section of the field (Fig. 12–20) shows more faulting and the complex deformation. The oil and gas pay zone is more than 770 ft (235 m) thick.

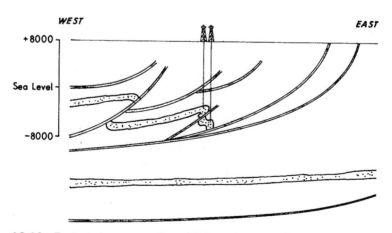

Fig. 12-19 *East-west cross section of Painter Reservoir field, Wyoming (modified from Lamb, 1980)*

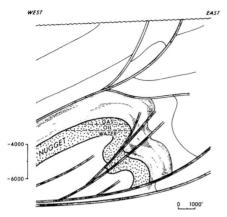

Fig. 12-20 *Small scale, east-west cross section of Painter Reservoir field, Wyoming (modified from Lamb, 1980)*

All the drag fold fields are located just to the west of a thrust fault (Fig. 12–21). The largest oil field is the Anschutz Ranch East field that will produce 180 million bbls (29 million m³) of oil and 4 Tcf (110 million m³) of gas. The largest gas field is Whitney Canyon field that will produce 5.9 Tcf (167 m³) of sour gas and 115 million bbls (18 million m³) of oil.

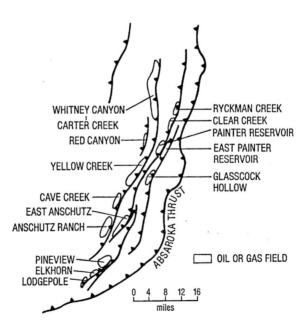

Fig. 12-21 *Map of oil and gas fields in the Western Overthrust Belt of Wyoming and Utah (modified from Sieverding and Royse, 1990*

Production on the same disturbed belt in Alberta includes Turner Valley and Jumping Pond. The Canadian production is primarily sour wet gas from Mississippian age limestones.

The location of drag folds in the subsurface cannot be predicted from the rock outcrops on the surface because of the intense deformation. It was not until the 1970s that improvements in seismic acquisition and processing opened exploration and drilling in overthrust belts.

Stratigraphic Traps

Secondary Stratigraphic Traps – Angular Unconformities

Angular unconformities can form giant gas and oil traps when a reservoir rock is terminated under an angular unconformity and overlain by a seal. The two largest oil fields on the North American continent, the East Texas field with 5 billion bbls (0.8 billion m^3) of recoverable oil and Prudhoe Bay field with 13 billion bbls (2.1 billion m^3) of recoverable oil are in angular unconformity traps.

For centuries, the Eskimos had known and used the numerous oil seeps on the tundra along the Arctic coast of Alaska for fuel. Very little exploration for oil occurred on the North Slope of Alaska until the 1960s, because the Arctic Ocean is frozen most of the year, and tankers cannot get into that area. If oil were discovered, the only way to get the oil to market would be to lay an 800-mile (1290 km) pipeline across the state of Alaska to a southern port (Valdez). The Trans Alaska pipeline would cost billions of dollars. Because of this, only a super giant oil field with more than a billion barrels of recoverable oil would be commercial.

During the early 1960s, a seismic survey was run over the North Slope, and the angular unconformity trap was revealed (Fig. 12–22). The obvious traps, however, were the anticlines in the foothills of the Brooks Range. Six exploratory wells were drilled into the anticlines during the early 1960s. All were dry holes. Atlantic Richfield then drilled a wildcat well on the angular unconformity during the winter of 1967–68 when the ground was frozen and could support a drilling rig. They put a 5% probability of success on a commercial discovery. The nearest well to it was a dry hole located 60 miles (97 km) away. At a depth of 8200 ft (2500 m), they discovered the Prudhoe Bay reservoir.

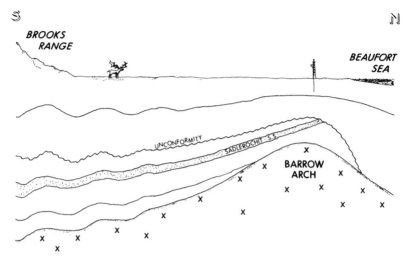

Fig. 12-22 *North-south cross section of Prudhoe Bay oil field, Alaska (modified from Morgridge and Smith, 1972)*

Thirteen billion bbls (2.1 billion m³) of oil will be produced from the Permian-Triassic age Sadlerochit Sandstone, the primary reservoir that is about 500 ft (150 m) thick, along with several smaller reservoirs including some limestones. The sandstone was deposited as horizontal layers of river and delta sands when shallow seas covered the area. Deposition of other sediment layers covered the sandstone and buried it in the subsurface. Later, an uplift occurred along the northern coast of Alaska, the Barrow arch, raising the Sadlerochit Sandstone (Fig. 12–22). This exposed the sandstone to erosion that removed it from the top of the arch. Seas later invaded the area, covering the angular unconformity with shale and other sediments, burying it in the subsurface. The oil and gas later formed in a black shale source rock. It migrated up along the unconformity until it met and filled the Sadlerochit Sandstone below the unconformity. Shale on the unconformity is the caprock.

There were 25 billion bbls (4 billion m³) of 27 °API gravity, sour (1.04% S) crude oil in place with an enormous free gas cap (27 Tcf or 760 million m³) above the oil (Fig. 12–23). Because of the size of the free gas cap that covers the top of the trap, even leases far to the sides of the trap produce oil.

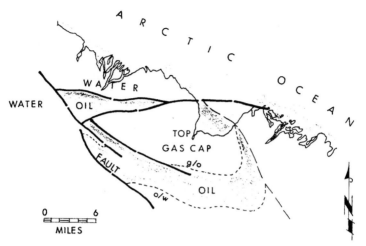

Fig. 12-23 *Map of Prudhoe Bay oil field, Alaska (modified from Jameson, Brockett and McIntosh, 1980)*

Not all angular unconformities, however, are productive. Mukluk, the world's most expensive dry hole, was drilled in 1983 on an angular unconformity trap that was very similar to Prudhoe Bay. The trap was located by seismic exploration not far from Prudhoe Bay in Alaska's Beaufort Sea. The potential reservoir rock was the same Sadlerochit Sandstone, and the caprock was shale. The leases sold for 1.5 billion dollars. A gravel island had to be constructed in the Beaufort Sea to drill a well. The dry hole cost $120 million. Why the oil was not there may never be known. Perhaps the shale caprock was not effective and the oil leaked out or perhaps the trap was not in position before the oil migrated through the area.

Primary Stratigraphic Traps

Reefs, beaches, river channels and incised valley fills, and updip pinch outs of sandstones form primary stratigraphic traps.

Reefs. Reefs are prolific gas and oil traps in North America. Permian age reefs in West Texas and New Mexico (Figs. 7–2, 7–3, and 7–4),

Devonian age reefs of Alberta (Figs. 7–8 and 7–9) and Cretaceous age reefs of Mexico (Fig. 7–13) form giant oil fields.

Petroleum production can come not only from the reef but also from a compaction anticline overlying the reef (Fig. 12–24). The *compaction anticline* forms in porous sediments, such as sands and shales, deposited on a hard rock mound or ridge, such as a limestone reef or bedrock hill. The sediments are deposited thicker to the sides of the reef than directly over the top. When the sediments are buried deeper, the weight of the overlying sediments compacts the loose sediments. The reef, composed of resistant limestone, compacts less. Because more compaction occurs in the thicker sediments along the flanks of the reef, a broad anticline forms in the sediments over the reef. Any reservoir rocks in the sedimentary rocks overlying the reef can trap petroleum.

Horseshoe atoll, a buried Pennsylvanian-Permian age reef, is located in northeastern Midland basin, part of the Permian basins (Fig. 7–2). The circular reef (Fig. 12–25) covers a large area. High points along the reef form oil fields.

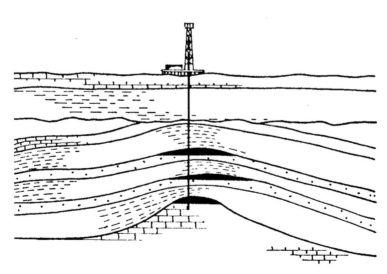

Fig. 12-24 *Compaction anticline*

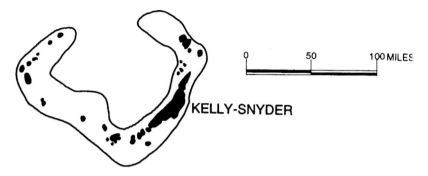

Fig. 12-25 *Map of Horseshoe Atoll, Texas showing oil fields (modified from Stafford, 1959)*

The largest field on Horseshoe atoll is the Kelly-Snyder oil field, which will ultimately produce 1.7 billion bbls (270 million m³) of oil and is located at a depth of 5000 ft (1500 m). A cross section through the field (Fig. 12–26) shows that the major production comes from the reef that has pores enhanced by solution. Production also comes from the compaction anticline in thin sandstone layers in the overlying shale of the Canyon Formation.

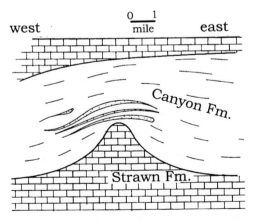

Fig. 12-26 *Cross section of Snyder field, Texas showing reef and overlying compaction anticline (modified from Stafford, 1957)*

The Leduc oil and gas field of Alberta (Fig. 12–27) produces from both a dolomitized reef (Leduc Formation, the D-3 zone) and the younger, overlying dolomite layers in the Nisku Formation, the D-2 zone. The Nisku Formation contains shales, sandstones and dolomites that were compacted over the reef. The field will produce more than 200 million bbls (32 million m³) of oil. The D-2 pay averages 63 ft (19 m) whereas the D-3 pay averages 35 ft (11 m). Leduc was discovered in 1947 when Imperial Oil Company drilled a seismic anomaly that was interpreted as an anticline. The discovery of Leduc started the oil boom in Alberta with drilling for Devonian age reefs.

The Cotton Valley Lime pinnacle reef trend of east Texas (Fig. 12–28) is a gas play. The Jurassic age reefs are 400 to 500 ft (120 to 150 m) high and 15,000 ft (5000 m) deep. The reservoir quality part of the reef, the sweet spot, covers only 20 to 80 acres and is 1000 to 2000 ft (300 to 600 m) on a side. Reef reserves range from 0.8 to 105 Bcf (0.02 to 3 billion m³) of sour gas with CO_2. The first pinnacle reef discovery was made in 1980 but because of the small drilling target at such great depth, it was not until 3-D seismic exploration became common in the 1990s that the trend was exploited.

Combination Traps

Combination traps have both structural and stratigraphic trapping elements such as the Hugoton-Panhandle Field (Fig.8–9)

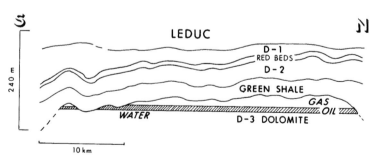

Fig. 12-27   *Cross section of Leduc oil field, Alberta (modified from Baugh, 1951)*

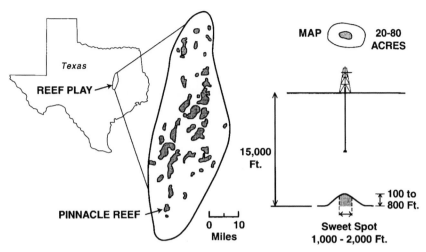

Fig. 12-28 *Cotton Valley Lime pinnacle reef trend, east Texas map and cross section*

Bald-headed structures

When an anticline or dome is formed (Fig.12–29a) the crest of the structure is exposed to erosion. Most or all of the potential reservoir rocks can be removed from the top of the structure (Fig. 12–29b). Seas later cover the area, and sediments are deposited, burying the eroded structure in the subsurface (Fig. 12–29c). When the petroleum migrates up the reservoir rocks, it is trapped below the angular unconformity. Because the crest of the structure is barren but the flanks are productive, it is called a *bald-headed structure* or *anticline.*

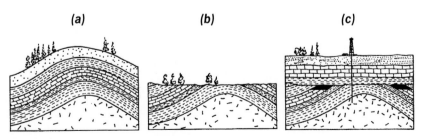

Fig. 12-29 *Formation of a bald-headed anticline*

The Oklahoma City field, just south of Oklahoma City is almost a bald-headed structure. A large north-south, strike-slip fault—the Nemaha fault—caused folding of the sedimentary rocks into a steep dome west of the fault (Fig. 12–30). Most of the potential reservoir rocks were eroded off the top of the structure, leaving an angular unconformity and only the Arbuckle Limestone as a reservoir rock on structure over the granite basement rock. Seas covered the area and deposited more sedimentary rocks. The unconformity and structure were uplifted again, forming a broad dome on the surface just southeast of downtown Oklahoma City. Gas and oil then migrated up the reservoir rocks and was trapped in 29 reservoir rocks under the angular unconformity and in the Arbuckle Limestone on the top of the dome.

In 1928, Indian Territory Illuminating Oil Company mapped the subtle dome on the surface (Fig. 12–31) and drilled the first well on structure. The well came in as a gusher, flowing 6500 bbls (1000 m³) of oil per day from the Ordovician age Arbuckle dolomite at 6600 ft (2000 m). Although highest on the structure (Figs. 12–30 and 12–32), the Arbuckle dolomite has produced only about 18 million bbls (3 million m³) of oil and the discovery well went to water by the end of 1928. Drilling out from the center of the structure located the more prolific pay zones.

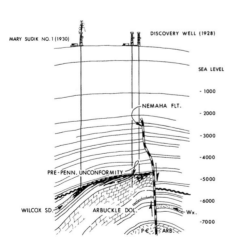

Fig. 12-30 *Cross section of Oklahoma City oil field, Oklahoma (depth in ft.) (modified from Gatewood, 1970)*

In 1930, Indian Territory Illuminating Oil Company drilled the No. 1 Mary Sudik far to the south of the Arbuckle dolomite discovery well. It was the first well into the Ordovician age Wilcox sand, a very well sorted sandstone with 20 to 30% porosity. The well blew out at the rate of 200 MMcf (6 million m³) of gas and 20,000 bbls (3200 m³) of oil per day. Because the Wilcox

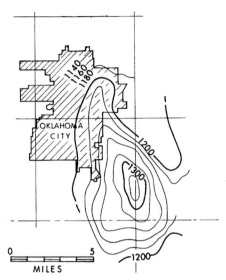

Fig. 12-31 *12–31 Structure contour map on Garber Sandstone showing the dome southeast of Oklahoma City (contours in ft.) (modified from Gatewood, 1970)*

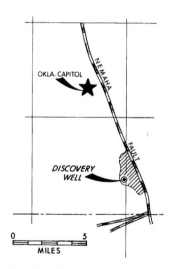

Fig. 12-32 *Map of Arbuckle production, Oklahoma City oil field, Oklahoma (modified from Gatewood, 1970)*

sand was loosely cemented, large quantities of loose sand blew up the well with the gas and oil, inhibiting the blowout preventers (BOP) from being thrown to control the well. Norman, Oklahoma, to the south, and then Oklahoma City to the north, were covered with oil for 11 days. It was called the Wild Mary Sudik Well. The well was finally capped by lowering and clamping a valve to the top of it.

The Wilcox sand is the most prolific reservoir rock on the structure (Figs. 12–30 and 12–33) producing more than 250 million bbls (40 million m³) of oil. The Oklahoma City field has 755 million bbls (120 million m³) of recoverable oil. Both the Oklahoma City and East Texas oil fields were developed with unrestricted production in the early 1930s. The market was glutted with crude oil, and the price fell to 16 cents a barrel.

Salt Domes

A *salt dome* is a large mass of salt, often miles across, rising from the subsurface through

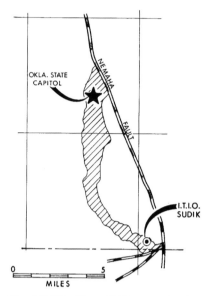

Fig. 12-33 *Map of Wilcox production,
Oklahoma City oil field, Oklahoma, (modified
from Gatewood, 1970)*

overlying sedimentary rocks to
form a plug-shaped structure
(Fig. 12–34). Salt, composed
primarily of halite, is a solid
that flows slowly as a viscous
liquid under pressure. A salt
layer is formed by the evaporation of water. When sands and
shales are deposited on the salt
layer, the weight of the overlying sediments presses down on
the salt layer. The salt lifts up a
weak area in the overlying sedimentary rocks. As the salt rises,
it uplifts and pierces overlying
sedimentary rocks to form a
piercement dome. Because the
salt is lighter in density than the
surrounding sediments, buoyancy also helps the salt rise.

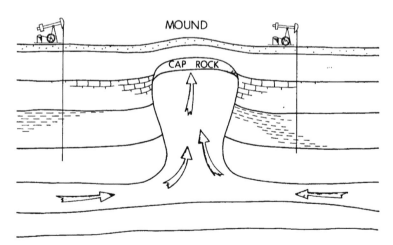

Fig. 12-34 *Cross section of a salt dome*

The salt is composed primarily of halite that is highly soluble. Large amounts of salt are dissolved as the rising salt dome comes in contact with water in the overlying sediments. However, 1 to 5% of the salt is insoluble anhydrite. As the salt dissolves, an insoluble layer called the *caprock* builds up on the top of the dome (Fig. 12–34). This caprock ranges from 100 to 1000 ft (30 to 300 m) thick. Some of the anhydrite is altered by bacterial and chemical reactions to gypsum, limestone, dolomite, and sulfur. The caprock is often highly fractured and has vugular pores.

Many subsurface salt domes have mounds, one to two miles (1.6 to 3.2 kms) in diameter, on the surface above the rising dome. The highest of these—Avery Island—Louisiana, rises 150 ft (50 m) and the salt is only 16 ft (5 m) below the surface.

There are two areas for drilling on salt domes (Fig. 12–35). Above the salt dome, any shallow reservoir rocks such as sandstones are domed. Near the top of the salt dome, uplift has caused the overlying sedimentary rocks to be faulted with normal faults that sometimes form grabens. These form fault traps. Because the caprock is often fractured and porous, it can be productive reservoir rock. Deep, along the flanks of the salt dome, reservoir rocks that were uplifted and pierced formed traps against the impermeable salt dome.

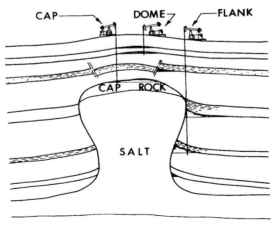

Fig. 12-35 *Salt dome gas and oil traps*

Individual salt domes often have numerous petroleum traps. Bay Marchand is a salt dome in the shallow waters of the Gulf of Mexico south of New Orleans (Fig. 12–36). This salt dome has more than 125 separate producing reservoirs that have been discovered and will eventually yield 615 million bbls (98 million m^3) of oil. Bay Marchand is part of a 27 mile (43 km) long salt ridge that also includes the Caillou Island and Timbalier Bay salt domes.

Some salt domes form overhangs of salt on the top of the dome. Upbent sandstones under the salt overhangs form prolific oil and gas reservoirs. The Barbers Hill salt dome of Texas (Fig. 12–37) has salt overhangs on all four sides. It will produce more than 100 million bbls (16 million m^3) of oil from under the overhangs.

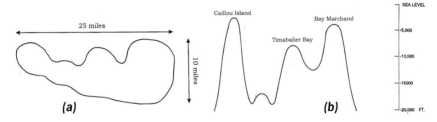

Fig. 12-36 *Bay Marchand, Timbalier Bay and Caillou Island salt domes, offshore Louisiana (a) map (b) cross section (modified from Frey and Grimes, 1970)*

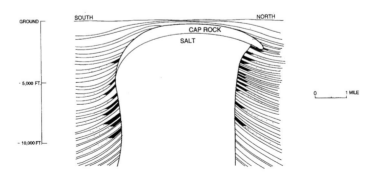

Fig. 12-37 *North-south cross section of Barbers Hill salt dome, Texas (modified from Halbouty, 1979)*

A large part of the bottom of the Gulf of Mexico is underlain by patches of the Jurassic age Louann Salt. The Louann Salt also underlies extensive areas of the Texas, Louisiana, and Mississippi coastal plains that were part of the Gulf of Mexico when the salt was deposited. This salt has formed more than 500 salt domes throughout the Gulf coastal plain and on the bottom of the Gulf of Mexico.

The Spindletop salt dome near Beaumont, Texas (Fig. 12–38), was the first salt dome successfully drilled for petroleum (1901). Up to that time, no highly productive wells had been discovered anywhere in the world. No one had ever drilled a gusher west of the Mississippi River; all the wells had to be pumped.

A large mound with gas seeps marked the surface expression of the Spindletop salt dome. A well on that mound was drilled into the Spindletop caprock at just below 1000 ft (300 m) in 1901. The caprock was very cavernous dolomite with high-pressure oil in it. The well blew out as the world's first great gusher at the rate of 100,000 bbls (16,000 m³) of green-black oil per day to a height of 100 ft (30 m) above the derrick. It came in at a rate greater than the production rate of all the other wells in the United States combined.

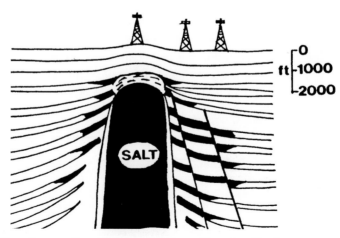

Fig. 12-38   *Cross section of Spindletop salt dome, Texas (modified from Halbouty, 1979)*

Within 9 months, 64 more wells were completed into the Spindletop caprock, each producing a gusher. Caprock production from Spindletop rapidly declined and was depleted by 1903 because of unrestricted production. It was not until 1925 that the deeper oil along the flanks of Spindletop was discovered. Spindletop eventually produced more than 49 million bbls (8 million m^3) of oil from the caprock and 82 million bbls (13 million m^3) from the flank traps. It was the world's first great oil field.

Spindletop oil resulted in the creation of more than 100 companies, including Gulf and the Texas Company (Texaco), to drill, produce, transport, refine, and/or market the oil. Crude oil had become cheap and plentiful, and gasoline refined from crude oil for internal combustion engines had become popular. Salt domes also occur in Kansas, Utah, and Michigan but are unproductive.

A thick salt layer, the Permian age Zechstein Salt, underlies the North Sea and forms salt domes. Until the giant Gronigen gas field was discovered on land in the Netherlands in 1959, no exploration was done in the North Sea. During the late 1960s, more than 200 exploratory wells were drilled in the North Sea. Some small gas fields were found in the southern, United Kingdom sector but most, however, were dry holes. Many companies were abandoning the North Sea when a well being drilled by Phillips Petroleum in 1969 encountered a 600 ft (183 m) column of oil above the Ekofisk salt dome in the Norwegian sector. It was the first major oil discovery in the North Sea. The reservoir rock, the Ekofisk Chalk, has 25–48% porosity but is permeable only because of fractures that were probably caused by the salt dome uplift. Chalk permeability is only 1 to 5 md, but with fractures it is 1 to100 md. The caprock is the overlying shale. Ultimate oil production from the Ekofisk field will be 1.7 billion bbls (270 million m^3) and 3.9 Tcf (110 million m^3) of gas.

thirteen

PETROLEUM EXPLORATION— GEOLOGICAL AND GEOCHEMICAL

A *geologist* is a scientist who studies the earth by examining rocks and interpreting their history. A *petroleum geologist* specializes in the exploration and development of petroleum reservoirs. An *exploration geologist* searches for new gas and oil fields. A *development geologist* directs the drilling of wells to exploit a field. A *petroleum geochemist* uses chemistry to explore for and develop petroleum reservoirs.

Seeps

The first commercial oil well was drilled in 1859 to a depth of 69 $^1/_2$ ft (21 m) with a cable tool drilling rig along the banks of Oil Creek in Pennsylvania (Plate 13–1). The driller was William "Uncle Billy" Smith, and the operator was Edwin L. Drake for the Seneca Oil Company. The site was chosen because of a natural oil seep, and the purpose of the drilling was to increase the flow of oil to the surface. The well initially flowed 20 barrels (3 m^3) of oil per day from the sandstone reservoir rock. In the previous year (1858), an oil pit was dug on an oil seep to a depth of 60 ft (18 m) at Oilsprings, Ontario, Canada by James M. Williams. The pit was lined with timber to prevent caving.

Plate 13-1   *Drake well (cable tool rig), Titusville, Pennsylvania*

How petroleum formed, migrated through subsurface rock, and accumulated in traps was not understood at that time. For the next 50 years, exploration wells were randomly drilled or located next to seeps; a technique that was relatively successful. The creekology theory, which promoted creeks as drilling sites, was also popular for a time. Drillers selected the drillsites while envisioning large, flowing underground rivers and subterranean crevasses filled with oil. Geologists were seldom used to select drillsites. Once an oil field was discovered, the "closeology" principle applied. The closer a proposed well was to a producing well, the better the well proposed was.

Every major petroleum-bearing basin of the world has numerous *oil seeps* or *seepages* where oil is leaking on the surface (Plate 13–2). This is because not all the oil is trapped as it migrates up from the source rock (Fig. 11–2). Even if the petroleum is trapped, the seal above the reservoir often has been fractured by folding or another process. Some of the gas and oil from the trap leaks through the fractures and onto the surface to form a seep above the *leaky trap*

(Fig. 13–1). If the trap has been filled with oil and gas down to the spill point and more migrates into the trap, some of the oil will spill out the side of the trap. In this case the trap is full, and the seep is located to the side of the trap. This is why drilling on or near oil seeps was so successful.

Plate 13-2   *Natural oil seep, Dorset coast, England (insert shows close-up of seep)*

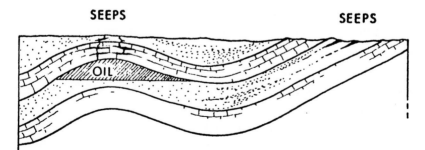

Fig. 13-1   *Oil seeps related to an oil trap*

When oil seeps on the surface, three processes degrade the crude oil to heavy oil, tar, and asphalt. These are water washing, evaporation, and bacteria. Water flowing by the oil will "wash it" by dissolving and removing some of the soluble, lighter fractions. Heating by the sun causes evaporation of the lighter fractions. Bacteria on the surface consume the lighter fractions. The effect of these processes is to increase the relative percentage the heavy fractions of the oil. The Athabaska tar sands in Alberta (Figs. 11–15 and 11–16) are thought to be a Cretaceous age oil seep.

Geological Techniques

In the early 1900s, it was finally accepted that oil accumulated in high areas of reservoir rocks such as anticlines and domes. This was known as the *anticlinal theory* that was originally suggested in the late 1800s but was not immediately accepted. Oil companies then hired geologists to map sedimentary rock layers cropping out on the surface. Bull's-eye (Fig. 5–11) and nose patterns (Fig. 5–9) were used to locate subsurface traps such as domes and anticlines. With the development of airplanes and aerial photography, surface mapping became more efficient. Vertical black-and-white aerial photographs were made into geologic maps. Spot checks of selected areas on the surface, called *ground truthing*, identify the rocks seen on the aerial photographs. Aerial photography can also be done with radar. The ground image is similar to a black-and-white photograph. Radar has the advantage that it can see through clouds and can be run at night.

Since the 1970s, satellites have been orbiting the earth at distances of several hundred miles in space. Some of these satellites photograph the surface of the earth in infrared, visible, and ultraviolet light and transmit these images back to earth. Many of these are spy satellites. However, five United States *Landsat* satellites that were launched between 1972 and 1984, and a new one was launched in 1999, are orbiting the earth at altitudes of 438 to 570 miles (705 to 917 km). They operate under the open-sky policy. The images for almost everywhere on the earth are made available at a uniform price to anyone without restrictions. The images can be digitally enhanced

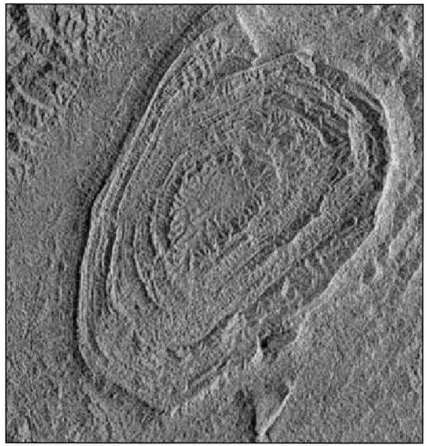

Plate 13-3   *Radar image of Kalimantan, Indonesia (note the eroded dome and fault)
(Courtesy RADARSAT International)*

by computer. They are useful in mapping remote areas and for giving a different perspective to a previously mapped area. There are two similar French satellites called *SPOT*. A Canadian satellite, *RADARSAT*, is making radar images of the earth's surface (Plate 13–3).

Geological reasoning can be used to find gas and oil. During the mid-1950s, several wells were drilled in eastern New Mexico. It was recognized from well samples that all wells drilled to the north of a line had drilled through lagoonal facies limestones deposited during a specific time in the Permian (Fig. 13–2). All the wells to the south of that line drilled through relatively deep-water reef facies limestones deposits called fore reef that were deposited at the exactly the same time. There had to be a reef located between these two facies.

In 1957, Pan American Petroleum Co. drilled a well between the two facies and discovered the Empire Abo field (Figs. 7–3 and 7–4). The first few wells missed the top of the reef and had a relatively thin pay from 20 to 60 ft (6 to 18 m). A later well, however, drilled into the reef crest and had 725 ft (221 m) of pay proving it to be a major discovery with more than 250 million bbls (40 million m^3) of recoverable oil.

In a frontier basin where relatively few wells have been drilled, a geologist starts by looking for large structural traps. First, the size and shape of the basin is determined. The stratigraphy (*i.e.*, sequence of rock layers) in the basin is established to identify potential source rocks, reservoir rocks, and seals. Structures that can be identified by field mapping and reconnaissance (large scale) seismic surveys are located.

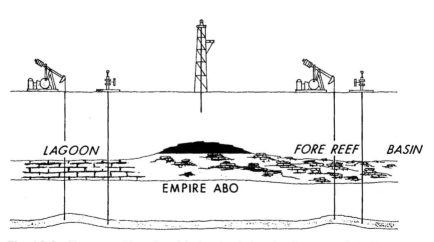

Fig. 13-2 *The recognition of reef facies that led to the discovery of the Empire Abo oil field, New Mexico (modified from LeMay, 1972)*

In mature areas that have been relatively well drilled, geologists spend most of their effort in subsurface mapping and constructing cross sections. Most of the large structural traps have been found leaving the more subtle stratigraphic traps. Subsurface maps (*i.e.*, structural and isopach) are made of potential reservoir rocks with scales ranging from the entire basin to a single field. Every time a new well is drilled in that area, more information is obtained. This new data is then plotted on the maps, and the maps are recontoured and reinterpreted. Common geological principles are applied to predict where hidden subsurface structures and facies changes might form petroleum traps.

Correlation

Constructing cross sections to find gas and oil is done by *correlation*. Correlation is the matching of rocks from one area to another. On the surface, correlation of sedimentary rocks is often started with a marker bed (Fig. 13–3). A *marker bed* is a distinctive rock layer that is easy to identify. Volcanic ash layers, thin beds of coal, limestone, or sandstone, and fossil zones are good marker beds. After correlating the marker bed, the rock layers above and below the maker bed can then be correlated on physical similarity and their position in the sequence of layers. The boundary between two rock layers is called a *contact*.

Marker beds are also used in subsurface correlation. A *key horizon*, the top or bottom of a thick, distinctive rock layer, can also be used to start the correlation. In areas complicated by faulting, facies changes, or unconfor-

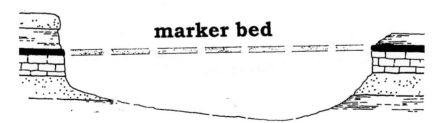

Fig. 13-3 *The use of a marker bed to start correlation*

mities, the most accurate correlation is done by matching microfossils between wells (Fig. 13–4). Wireline well logs and seismic surveys can also be used to trace rock layers in the subsurface.

A *cross section* is a vertical slice or panel of the subsurface rocks. It is made by correlating the rock layers from one well to another. When a well is drilled, a record of rocks in the well, called a *well log*, is made. Comparison of rock layers on well logs is used to correlate between wells. There are two types of cross sections, depending on how the well logs are arranged to make the correlation. Each well log must be arranged or *hung* along a common horizontal surface going through the well logs.

A *structural cross section* is made by hanging the well logs by modern sea level in each well (Fig. 13–5). The well logs are then correlated. Structures such as folds and faults and potential traps are illustrated on structural cross sections. Structural cross sections are used to find structural petroleum traps.

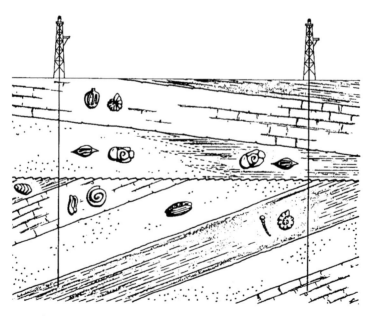

Fig. 13-4 *The use of microfossils for correlation between wells*

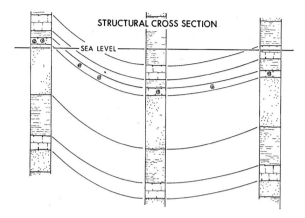

Fig. 13-5 *A structural cross section with well logs hung from modern sea level*

A *stratigraphic cross section* is made by hanging the well logs from the same marker bed in each well (Fig. 13–6). Because the marker bed was originally deposited horizontal, a stratigraphic cross section restores the rocks to their original horizontal position before they were deformed. Stratigraphic cross sections are used to illustrate the relationship between rock layers such as facies changes and to locate stratigraphic petroleum traps.

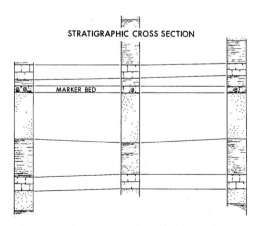

Fig. 13-6 *A stratigraphic cross section with well logs hung from a common marker bed*

A *fence diagram* is used to show how wells correlate in three dimensions (Fig. 13–7). The diagram is arranged like a map. North is at the top, south at the bottom, east to the right, and west to the left. Each well is located (*spotted*) on the map. The well log for each well is drawn vertically under the well's position. The rock layers are then correlated from one log to another. Each set of correlations forms a panel. The panels extend from well to well to form a closure. The entire diagram is called a fence diagram.

A lot of subsurface information from wells, no matter who drilled them, eventually becomes public. A company drilling an exploratory well in a new area might want to keep as must information as secret as possible by running a *tight hole*. The information can then be used by the tight hole operator to locate other wells and leasing any land that is still open (*loose acreage*).

However, state, provincial, and federal laws require that a specific suite of well logs be released to the government regulatory agency within a certain time. This time limit varies, but is six months in the states of Oklahoma and Texas. For United States offshore federal land, it is 5 or 10 years. This is immediately public information. In many countries, there is a national oil company that is a partner with any foreign oil company drilling in that country. Because the foreign oil companies have a common partner in the national oil company, well information can be obtained through that source.

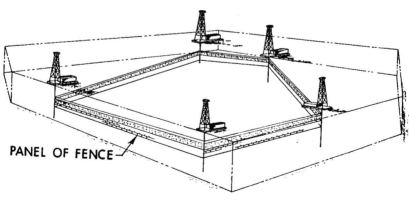

PANEL OF FENCE

Fig. 13-7 *Fence diagram*

An important source of subsurface well information is wireline well logs. *Well-log libraries* collect copies of wireline logs for regional areas. There is a well-log library in almost every oil patch city in the United States and Canada. Well-log library members can review, copy and even check out well logs and other well information from the library. The major oil companies have been digitizing well logs for their computers. A company geologist can call up logs on a workstation.

Large oil companies use *scouts* who gather information on any petroleum-related activity in their assigned region. A scout uses all kinds of "ethical" methods to find out where competitors are exploring, leasing, and drilling. Every time a well is drilled, the scout obtains as much information as possible about the well. The scout fills out a form, called a *scout ticket*. It usually includes the well name, location, operator, spud and completion dates, casing and cement data, production test data, completion information, the tops of certain zones or formations, and a chronology of the well. Because a single company scout cannot keep an eye on all the activity going on in their assigned region, scouts from different oil companies meet periodically to coordinate their efforts and exchange information in *scout checks* or *meetings*. A *czar* or *bull scout* is elected to conduct the meetings. An organization known today as the International Oil Scouts Association annually publishes the statistical information gathered by the scouts.

Commercial scouting firms publish daily, weekly, and monthly reports on regional drilling activity. Some publish *completion cards* for each well drilled in the United States and Canada. The information on the completion card includes name, location, spud date, total depth drilled, depths to the tops of formations in the wells, intervals completed, completion techniques and initial petroleum production. The source of the information is scout tickets and government regulatory agencies. For a fee, the firm will provide completion cards for a regional area and update the information periodically.

Well-cutting libraries collect well cutting for a regional area. Well cuttings can be sold to the library, and well cuttings in the library can be examined for a fee. *Well-core libraries* collect cores drilled in that state, province, or country. In Canada, all cores must be eventually submitted to a central facility in each province for storage and examination.

Each state in the United States and each province in Canada has a *geological survey* that publishes reports and maps on the petroleum geology of that area. The federal government in almost every country has a geological survey.

Geochemical Techniques

Geochemistry is the application of chemistry to the study of the earth. Traces of hydrocarbons in soil and water are often good indications of the proximity of a petroleum trap (Fig. 13–8). In an exploratory area, surface samples of waters and soils are taken. These samples are analyzed in the laboratory with instruments such as gas chromatographs for minute traces of hydrocarbons. Many subsurface petroleum reservoirs are leaky and have obvious seeps on the surface. Some traps, however, are not as leaky and have only *microseeps* on the surface that cannot be detected visually. The microseeps often occur in a pattern called a *hydrocarbon halo* that outlines the subsurface trap. Seeps are also common in the ocean. Ships towing water sampling equipment and shipboard hydrocarbon sensing devices called *sniffers* are used to detect their locations.

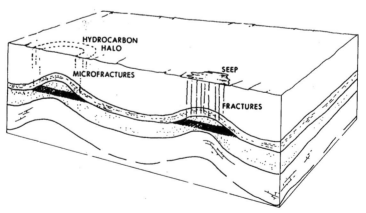

Fig. 13-8   *Geochemical exploration for microseeps*

Water samples can be taken from subsurface rocks for chemical analysis. The subsurface water salinity must be known before some wireline well-log calculations, such as oil saturation, can be accurately made. Traces of hydrocarbons in the subsurface waters of a dry hole could indicate the presence of a petroleum reservoir in the area. No traces of hydrocarbons were found in subsurface waters from wells drilled in the 1960s on the Destin anticline located offshore from the Florida panhandle. This discouraged further exploration in that area.

Vitrenite reflectance is a method used to determine the maturity of a source rock. Vitrenite is a type of plant organic matter often found in shale. The source rock sample is polished and then examined under a reflectance microscope. The percentage of light reflected from the vitrenite is dependent on the maturity of the source rock. Vitrenite reflectance can determine if oil and gas have been generated. If all the source rock samples from an unexplored basin show that hydrocarbons were never generated in that basin, further exploration would be discouraged.

Geochemistry can also be used to identify the source rock for a specific crude oil. The crude oils in traps can then be correlated with source rocks to determine the migration path for the petroleum.

Plays and Trends

A *play* is a combination of trap, reservoir rock, and seal that has been shown by previously discovered fields to contain commercial petroleum deposits in an area. An example is the Tuscaloosa trend play in Louisiana. The Cretaceous age Tuscaloosa Sandstone generally ranges from 35 to 200 ft (11 to 61 m) thick in Louisiana and is reservoir rock quality. A shale seal overlies it. In the Louisiana coastal plain, large growth (down to the basin) faults cut the Tuscaloosa Sandstone. On the basin (Gulf of Mexico) side of the growth fault, the Tuscaloosa Sandstone can form a rollover anticline. By drilling 16,000 to 22,000 ft (4900 to 6700 m) to the Tuscaloosa Sandstone rollovers, gas and condensate fields can be discovered. This is known because several fields have already been discovered.

A *trend* or *fairway* is the area along which the play has been proven, and more fields could be found. The Tuscaloosa trend extends from Texas, through Louisiana, and into Mississippi (Fig. 13–9). The trend was opened up by the discovery of the False River field in 1974.

A *prospect* is the exact location where the geological and economic conditions are favorable for drilling an exploratory well. A prospect can be presented by using prospect maps that illustrate the reasoning for selecting that drilling location. The maps include at least a structure and isopach map of the drilling target and a map of test results and fluid recoveries from wells in the area.

There are four major geological factors in the success of a particular prospect. First, there must have been a source rock that generated petroleum. Second, there must be a reservoir rock to hold the petroleum. Third, there must be a trap. This includes a reservoir rock configuration that has four-sided closure, a seal on the reservoir rock, and no breach of the trap. Fourth, the timing must be right. The trap had to be in position before the petroleum migrated through the area. An economic analysis of the prospect should include reserves and risk calculations.

Fig. 13-9   *Tuscaloosa trend for deep, wet gas in Louisiana*

fourteen

PETROLEUM EXPLORATION—GEOPHYSICAL

Geophysics is the application of physics and mathematics to the study of the earth. *Geophysicists,* who are trained in mathematics and physics, commonly use three surface methods—gravity, magnetic, and seismic—to explore the subsurface. Most of the exploration money is being spent on seismic exploration, where the most technology advances are being made.

Gravity and Magnetic Exploration

Gravity meters and magnetometers are relatively inexpensive, portable, and easy-to-use instruments. A *gravity meter* or *gravimeter* measures the acceleration of the earth's gravity at that location. A *magnetometer* measures the strength of the earth's magnetic field at that location. Both are small enough to be transported in the back of a pickup truck. A magnetometer can be operated in an airplane to conduct an *aeromagnetic survey.* The magnetometer can also operate while being towed behind a boat. The gravity meter doesn't work well in either an airplane or the ocean because of vibrations.

The gravity meter is very sensitive to the density of the rocks in the subsurface. It measures gravity in units of acceleration called *milligals.* Over a typical area of earth's crust with 5000 feet of sedimentary rocks underlain by basement rock that is very dense, the gravity measurement is predictable (Fig. 14–1). A mass of relatively light rocks such as a salt dome or porous reef can be detected by the gravity meter because of lower than normal gravity

values over it. A mass of relatively heavy rocks near the surface such as basement rock in the core of a dome or anticline can be detected by higher than
normal gravity values.

The magnetometer measures the earth's magnetic field in units called
gauss or *nanoteslas*. It is very sensitive to rocks containing a very magnetic
mineral called magnetite. If a large mass of magnetite-bearing rock (*e.g.*,
basement rock) occurs near the surface, it is detected by a larger magnetic
force than the normal, regional value (Fig. 14–2). The magnetometer is
primarily used to detect variations basement rock depth and composition. It
can be used to estimate the thickness of sedimentary rocks filling a basin and
to locate faults that displace basement rock.

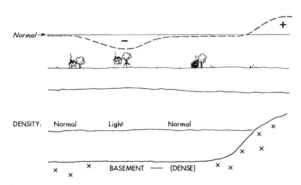

Fig. 14-1 *Gravity meter measurements over an area*

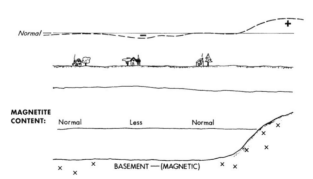

Fig. 14-2 *Magnetometer measurements over an area*

In order to explore the subsurface of an area using a gravity meter, a grid pattern of points is located on the surface. A gravity reading is made at each point. The gravity values are then plotted on a base map and contoured similarly to a topographic map. With an aeromagnetic survey, the plane flows in two sets of parallel lines that intersect at right angles. The magnetic data is also contoured. Most of the area will have "normal" gravity and magnetic measurements. Anomalies of abnormally high (maximum) or low (minimum) gravity and magnetics are noted.

A subsurface salt dome is seen as a surface anomaly of relatively low gravity and magnetics because the salt is light in density and has no magnetite mineral grains compared to the surrounding sedimentary rocks (Fig. 14–3). Many salt domes in the coastal areas of Texas and Louisiana were discovered in the 1920s by gravity meter surveys.

A subsurface reef can have a gravity anomaly that ranges from abnormally high to abnormally low. The abnormally high anomaly is caused by a dense (non-reservoir) limestone reef, and the abnormally low anomaly is caused by a porous limestone reef. Magnetics are generally not useful in locating reefs.

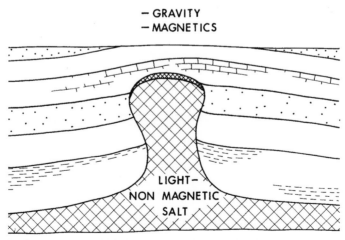

Fig. 14-3 Gravity and magnetic anomalies over a salt dome

A dome or anticline can be identified by both a high gravity and high magnetic anomaly. This is caused by dense and magnetite-bearing basement rock that is close to the surface in the center of the structure (Fig. 14–4). Ghawar, the largest conventional oil field in the world, was found by a gravity survey in Saudi Arabia in 1948. The surface of the ground there was covered by sand dunes, but there was a large gravity maximum anomaly.

A subsurface dip-slip fault can produce a sharp change in both gravity and magnetic values along a line because the basement rock is higher on one side of the fault than on the other side (Fig. 14–5).

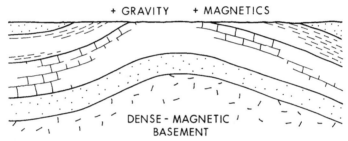

Fig. 14-4 *Gravity and magnetic anomalies over a dome or anticline*

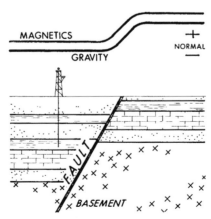

Fig. 14-5 *Gravity and magnetic anomalies over a fault*

Small features at shallow depths and large features at deep depth have similarly sized gravity and magnetic anomalies. Because of this, it is difficult to determine the size and depth of a feature causing the anomaly.

Seismic Exploration

The first oil field found by seismic exploration alone was the Seminole field of Oklahoma in 1928. The seismic data, however, was originally recorded by analog in the field on a sheet of paper. It was noisy and not very accurate. The greatest improvements in petroleum exploration in the last several decades have involved new seismic acquisition techniques and computer processing of digital seismic data.

Acquisition

The seismic method uses sound energy that is put into the earth. The energy travels down through the subsurface rocks, is reflected off subsurface rock layers, and returns to the surface to be recorded. Seismic exploration images the shape of the subsurface sedimentary rocks and locates petroleum traps. A source and a detector are used. The source emits an impulse of sound energy either at or near the surface of the ground or at the surface of the ocean. The sound energy is reflected off subsurface rock layers. Like a mirror, the maximum reflection occurs when the angle of incidence between the seismic energy and reflector is equal to the angle of reflection (Fig. 14–6). The reflected sound energy then returns to the surface to be recorded on a detector.

The detector on the surface records both the *signal,* wanted information from the subsurface rock layers, and *noise,* unwanted energy. Noise can be caused by surface traffic, wind, surface and air waves, and subsurface reflections that are not direct (*primary*) reflections from a subsurface rock layer. A high signal/noise ratio is desired. A *noise survey,* which is a small seismic survey, can be run first to determine the nature of noise in that area and plan the optimum seismic program to reduce noise.

On land, the most common seismic sources are explosives and Vibroseis. Dynamite was the first seismic source but is not as common today. Explosives are used where the surface is covered with loose sediments,

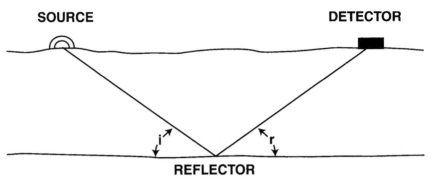

Fig. 14-6 *A seismic reflection with the angle of incidence (i) equal to the angle of reflection (r)*

swamps, or marshes. To use explosives, a small, truck-mounted drilling rig often accompanies the seismic crew to drill a *shot hole*, usually 60 to 100 ft (18 to 30 m) deep, to below the soil. The explosives are planted in solid rock on the bottom of the hole. *Primacord*, a length of explosive cord, can be planted in a trench about 1 ft (0.3 m) deep or suspended in air as a seismic source. Explosives as a seismic source are expensive.

About 70% of the seismic exploration run on land today is done by *Vibroseis*, a technique developed by Conoco. In Vibroseis, a *vibrator truck* (Plate 14–1) with hydraulic motors mounted on the bed of a truck and a plate called a *pad* or *baseplate* located below the motors is used. The vibrator truck drives to the shot point and lowers the pad until most of the weight of the truck is on the pad. The hydraulic motors use the weight of the truck to shake the ground for a time (*sweep length*) that is often 7 to 20 seconds. A range of frequencies, called the *sweep*, is imparted into the subsurface. Vibroseis is very portable and can be run in populated areas. Other less common, land seismic sources include weight drop, gas gun, land air gun, and guns such as a shotgun.

The location of the seismic source is called the *shot point.* To reduce the noise generated from the source, several linked explosive shots or vibrators can be used simultaneously in a *shot point array.* The seismic energy

Plate 14-1 *Vibrator truck*

commonly has a useable frequency of 8 to 120 Hertz (Hz) or cycles per second. The human ear can hear 20 to 20,000 Hz.

At sea, a common seismic source is an air gun. The *air gun* is a metal cylinder that is several feet long. It is towed in the water at a depth of 20 to 30 ft (6 to 9 m) behind the ship. On the ship are air compressors. High-pressure air at 2000 psi (140 kg/cm^2) is pumped through a flexible, hollow tube into the air gun in the water. On electronic command, ports are opened on the air gun. An expanding, high-pressure air bubble in the water provides a seismic source that is not harmful to marine life.

Air guns are described by the capacity of their air chamber, such as 200 in.^{3} (3000 cm^3). Air guns of different sizes (tuned air gun array) are often fired at the same time to cancel any noise from the source such as the air bubble expanding and contracting after the first impulse. The air gun is also used in some applications in swamps and marshes. Other seismic sources used at sea include water gun, sleeve gun and sparker.

A *seismic contractor* who owns and operates the seismic equipment runs the seismic survey. The seismic contractor can run the seismic survey under contract with an exploration company. A *spec survey* can also be run by a seismic contractor. A limited number of exploration companies then pays for and views the non-exclusive seismic records. In another method, several exploration companies can share the cost and results of a seismic survey run by a seismic contractor in a *group shoot*.

Before seismic exploration is run on private land in countries such as the United States and Canada, a *permit man* obtains permission from the surface rights owners of the land. A fee per shot hole or seismic line mile is paid, and damage fees are negotiated. A *survey crew* then cuts a path through the trees and brush (if necessary), accurately locates and flags the shot points and geophone locations, and records them in the survey log book. Members of the seismic crew called *jug hustlers* lay the cable and arrange and plant the geophones.

At sea, permitting is not necessary. The ship's crew does the navigation while the seismic crew runs the seismic equipment. Surveying at sea is done by global positioning using navigational satellites.

The seismic energy travels down through the subsurface rocks (Fig. 14–7). Each time the sound impulse strikes the top of a subsurface layer, part of the sound is reflected back to the surface as an echo. The rest of the sound impulse travels deeper into the subsurface to bounce off deeper and deeper layers or to dissipate. The returning echoes are recorded on land by vibration detectors called *geophones* or *jugs*. They detect vertical ground motion and translate it into electrical voltage.

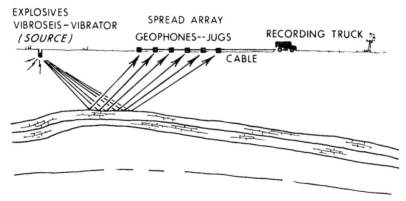

Fig. 14-7 *The seismic method on land*

The geophone often has a spike on the bottom so it can be planted in the ground (Plate 14–2). One to dozens of geophones are connected to form a *group* that records as a single unit called a *channel.* By using several geophones in a group, noise is reduced. The geophones in a group are arranged in a line, several parallel lines, a star, a rectangle, or another geometric pattern. The groups of geophones are deployed in a larger geometric pattern called the *spread.* The large number of geophone groups is used to cover a large area of the subsurface with each seismic shot (Fig. 14–7). A reading of 96 channel or trace data means 96 geophone groups were used for each shot point. Forty-eight to 96 geophone groups, each located 55 to 110 ft (17 to 34 m) apart, are commonly used. The geophones are all connected to the lead cable that goes to the recording truck or doghouse. The data can also be transmitted digitally by a radio telemetry system that uses radio signals to make the connection. The *recording truck (doghouse)* contains the equipment such as a high-quality tape recorder used to record and store the seismic data digitally on magnetic tape and run field checks on the quality of the data.

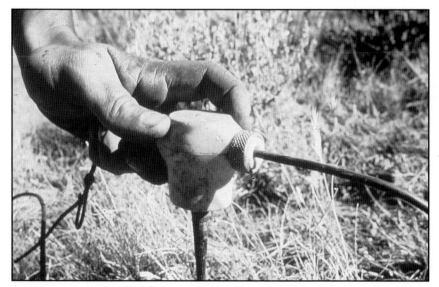

Plate 14-2   *Geophone (courtesy American Petroleum Institute)*

A common spread consists of a long main cable stretched out in a line several miles long. Shorter cables at specific intervals connect the individual geophone groups located at equal distances with the main cable. This is called a *linear spread* in which the geophone groups are arranged in a line. A *split spread* with the source in the middle of the linear spread is commonly used on land (Fig. 14–8a).

The *roll along technique* is often used to move the geophones. After each seismic shot, a portion of the geophone cable is detached from one end of the linear spread and moved to the other end. The shot point is then moved an equal distance in the same direction.

At sea, the source is towed in the water behind a boat (Fig. 14–9) that steams at about five knots (5 nautical miles per hour). The seismic energy is powerful enough for much of the sound to penetrate the ocean bottom. The returning reflections are recorded on vibration detectors, called *hydrophones*, contained in a long, plastic tube (the *streamer*) that is towed behind the boat. Wires run from the hydrophones through the streamer to the doghouse on the ship where the recording equipment is located.

The streamer is filled with a clear liquid such as kerosene to be neutrally buoyant. It is strung out for up to 5 miles (8 kms) in a straight line behind the boat. Several hundred hydrophones are used in evenly spaced groups of

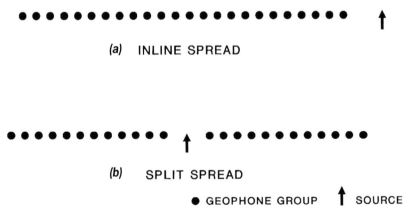

(a) INLINE SPREAD

(b) SPLIT SPREAD

● GEOPHONE GROUP ↑ SOURCE

Fig. 14-8   *Linear spread of geophone groups (a) split spread (b) inline or end on spread*

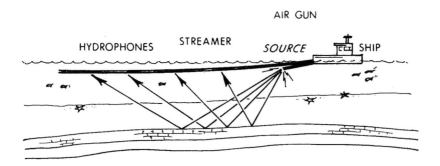

Fig. 14-9   *The seismic method at sea*

25 to 40 hydrophones. There are two, three, or four hydrophone groups in a cable section with connections at each end of the section. Several sections are joined to form the streamer of 96 to 240 groups. Devices called *birds* or *depth controllers* are used to keep the streamer at a depth of 20 to 50 ft (6 to 16 m). A buoy with a light and positioning equipment is located on the end of the streamer.

Offshore seismic is acquired by an inline spread with the source at the end of a linear spread (Fig. 14–8b). To cover a large area, the seismic ship can often tow 12 parallel streamers and 4 source arrays (Plate 14–3). In a variation called *seafloor seismic* the hydrophone streamer is located on the ocean bottom. The seismic source ship steams parallel to the ocean bottom cable or cables. This is used where there are obstructions such as production platforms, or in shallow water or areas of limited access.

Common-depth-point (*CDP*) or *common-mid-point* (*CMP*) stacking is a process used to improve the signal/noise ratio by reinforcing actual reflections and minimizing random noise. It involves recording reflections for each subsurface point from different source and detector distances (*offsets*) and combining (*stacking*) the reflections (*traces*). The number of times that each subsurface point is recorded is called the *fold*. It is the number of reflections (traces) that are combined in stacking to produce one stacked reflection. A 48 fold or 4800% stack uses 48 reflections (traces) off the same subsurface point at different offset distances to form one stacked reflection (trace).

Seismic exploration is most expensive on land, especially in rugged terrain. It is less expensive and of better quality at sea.

Plate 14-3 *Seismic ships towing air gun arrays and streamers (courtesy Western Geophysical Division of Baker Hughes)*

Seismic Record

A *seismic* or *record section* on which the seismic data is displayed is similar to a vertical cross section of the earth (Fig. 14–10). The original vertical scale is in seconds. Zero seconds is always at or near the surface of the ground or exactly at the surface of the ocean. *Time lines,* usually in 1/100s of a second, run horizontally across the section. The 1/10th second lines are heavier, and the full second lines are heaviest.

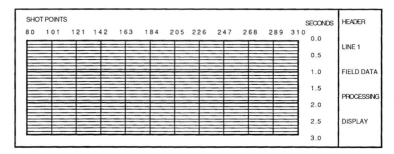

Fig. 14-10 *Seismic record format*

Across the top of the record are the shot point locations. On the side of the record is the header. The *header* displays information such as the seismic line number and how the data was acquired, processed and displayed. A *shot point base map* (Fig. 14–11) accurately shows the location of the seismic lines and individual shot points.

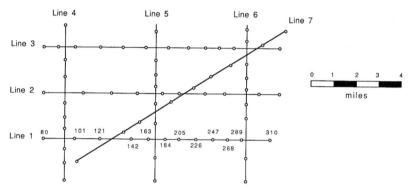

Fig. 14-11   *Shot point base map*

Depth to a seismic reflector on a seismic record is measured in *milliseconds* (1/1000th of a second). The seismic energy travels down, is reflected off a subsurface rock layer, and returns to the surface. Because it travels twice the depth (down and up), time on a seismic record is *two-way travel time.* The deeper the reflecting layer, the longer it takes the echo to return to the surface.

Seismic data can be recorded three ways. A *variable area wiggle trace* (Fig. 14–12a) uses vertical lines with wiggles to the left or right to record seismic energy. Wiggles to the right are reflections from the subsurface (the geophone detected an upward motion) and are usually shaded black. Wiggles to the left (the geophone detected a downward motion) are left blank. A *variable density display* (Fig. 14–12b) uses shades of gray to represent seismic energy amplitude. The darker the shade: the stronger the reflection.

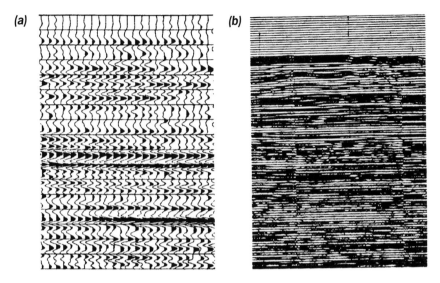

Fig. 14-12 *(a) variable area wiggle trace (b) variable density display*

Colored seismic displays have become common (Plate 14–4). The human eye can distinguish many different colors and see more information on a colored display. Seismic interpreters can identify far more subtle trends on a colored display than with shades of gray. In one method, peaks and troughs of reflections are colored blue or red on a white background (Plate 14-4a). In another method, a larger spectrum of colors is used. An example would be cyan-blue-white-red-yellow with cyan and yellow the maximum peak and trough amplitudes.

Interpretation

Each reflection that can be traced across a seismic section is called a *seismic horizon*. Layered rocks on a seismic record are sedimentary rocks (Fig. 14–13). The area with no good, continuous reflections below the layered sedimentary rocks is basement. Basement has short, discontinuous, and disordered reflections. Salt domes and reefs, however, do not show layered reflectors.

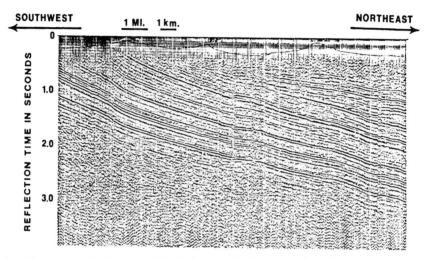

Fig. 14-13 *A seismic record in the Wind River basin of Wyoming. Well-layered sedimentary rocks are seen dipping down to the northeast. Unlayered basement rock is located below the sedimentary rocks. (courtesy of Conoco)*

Any deformation such as tilting, faulting, or folding is apparent on a seismic record (Fig. 14–14). The primary purpose of seismic exploration is to determine the structure of the subsurface rocks. Reefs are identified on a seismic record as a mound without internal layering (Fig. 14–15). Salt domes are seen as an unlayered plug (Fig. 14–16). Salt dome edges are defined by uplifted and terminated sedimentary rocks.

Contoured maps of the subsurface can be made using seismic sections. A map of depth in milliseconds to a seismic horizon is called a *structure-contour map*. It is very similar to a structural map made from well data. An *isotime, isochron,* or *time interval map* uses contours to show the time interval in milliseconds between two seismic horizons (Fig. 14–17). It is similar to an isopach map made from well data. If the seismic velocities through the rocks are known, the structure-contour and isotime maps can be converted into structural and isopach maps.

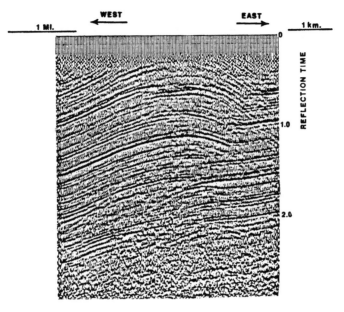

Fig. 14-14 *A seismic record in the Big Horn basin of Wyoming. A drag fold occurs on a curving thrust fault. This is the Elk Basin oil field. (courtesy of Conoco)*

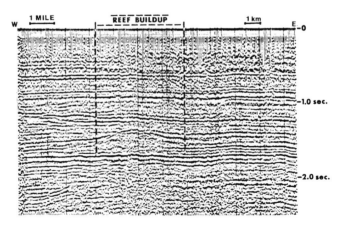

Fig. 14-15 *A seismic record in the Midland basin, Texas. A reef, part of Horseshoe Atoll, is shown. (courtesy of Conoco)*

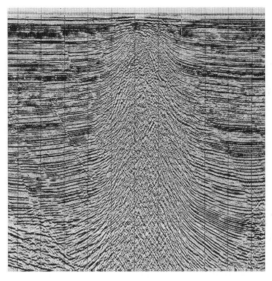

Fig. 14-16 *A seismic record in the Gulf of Mexico south of Galveston, Texas. Note the salt dome. (courtesy of TGS Calibre Geophysical Company and GECO/Prakla)*

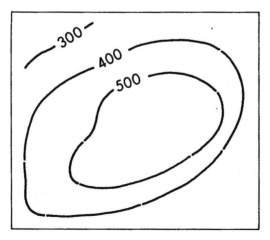

Fig. 14-17 *An isochron map contoured in milliseconds (courtesy of TGS Calibre Geophysical Company and GECO/Prakla)*

225

The amplitude of the seismic echo off the top of a surface depends primarily on the contrast in *acoustic impedance* (sound velocity times density) between the upper and lower rock layers that form the surface. The greater the contrast, the larger the reflection. The percent of seismic energy reflected is called the *reflection coefficient.* Typical sedimentary layers have a reflection coefficient of 2 to 4%.

Because gas has a very slow velocity, the slowest velocity sedimentary rock is an unconsolidated gas sand. If overlain by a seal, the acoustic impedance contrast will produce an echo of about 16% of the seismic energy, called a *bright spot.* It is seen as an intense reflector on the seismic profile (Fig. 14–18).

Bright spots have been used very successfully to locate gas reservoirs and free gas caps on saturated oil fields. Not all bright spots, however, are commercial deposits of natural gas. A *dim spot* where the reflection amplitude becomes less, occurs over some reefs.

If a sonic log is used to compute the velocity, and a density log is used to compute the density of each rock layer in a well that is located on a

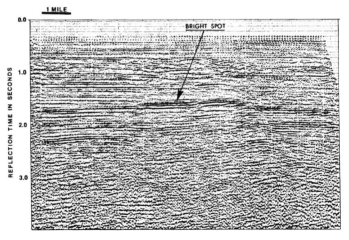

Fig. 14-18 *A seismic record in the Gulf Coast over a gas field. Note the bright spot on the gas sand. (courtesy of Conoco)*

seismic line or its vicinity, the acoustic impedance of each surface can be computed. A *synthetic seismogram*, an artificial, computer-generated seismic record, can be made from the computed reflections caused by differences in acoustic impedances. It is compared to the actual seismic record. The composition of the various rock layers on the seismic record can be identified. The reflection response of seismic energy to various rock layer configurations can be modeled.

Two types of *direct hydrocarbon* indicators on a seismic record are bright spots and flat spots. A *flat spot* is a level, flat seismic reflector in a petroleum trap formed by rock layers that are not flat such as an anticline (Plate 14–4a). The flat spot is a reflection off a gas-oil.

Processing

Digital recording of seismic data in the field and computer processing of the seismic data has greatly improved the accuracy and usefulness of seismic exploration. A correction (*statics*) is made on the seismic data for elevation changes and the thickness and velocity of the near-surface, loose sediments called the *weathering layer* or *low-velocity zone.*

As the seismic energy travel through the subsurface rocks, the relatively sharp impulse of seismic energy tends to become spread out, and some portions of the energy is lost. *Deconvolution* is a computer process that compresses and restores the recorded subsurface reflections so that they are similar to the original seismic energy impulse. This makes the reflections sharper and reduces some of the noise.

A seismic section is accurate only over flat, horizontal rock layers. Dipping rock layers have a different path for the seismic energy from source to detector than horizontal rock layers in the same position. Because of this, dipping rock layers do not appear on the seismic record in their actual positions. They are shifted to a down dip position and appear flatter than they are. This effect causes anticlines to look larger, and synclines look smaller than they actually are. It causes the rock layers in a deep, steeply dipping syncline to cross forming a bow tie (Fig. 14–19a). Rock layers sharply terminated against a fault, appear to cross with rock layers on the other side of the fault (Fig. 14–19b). A computer process called *migration* moves the dipping rock layers into a more accurate position on the seismic record.

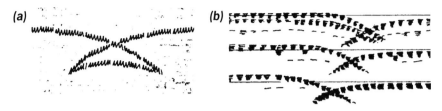

Fig. 14-19 *Unmigrated seismic events (a) a bow tie as a result of a deep, steep syncline (b) crossing events due to a fault*

Many basins such as the Gulf of Mexico and the North Sea have extensive salt layers. Passage of seismic energy through the salt blurs the seismic image of any potential petroleum structures below the salt. A computer processing technique called *prestack migration* of subsalt seismic data results in a clearer seismic image of the deeper structures but involves significantly more computer time. The first subsalt discovery in the Gulf of Mexico was Mahogany in 1993. The trap is an anticline with a sandstone reservoir located at a depth of 18,500 ft (4500 m). The Louann Salt is located above it at a depth of 15,000 ft (4500 m).

Each seismic line is run to intersect another seismic line (*tie in*) so that the reflections can be correlated from one record to another. If the reflections from two intersecting seismic records do not correlate, it is called a *mis-tie*.

A typical seismic record shows the structure of the subsurface rocks and identifies sedimentary rocks by their characteristic layering. It does not, however, identify the individual sedimentary rocks layers, such as San Andres Limestone or even the rock type. A seismic record is more valuable when the individual sedimentary rock layers have been identified, and potential reservoir rocks and seals can be traced. To do this, the seismic line is often run through a well (*tied in*) that has been already drilled. The well logs from that well then provide the basis for identifying subsurface rock layers on the seismic record. If no well is available, a *stratigraphic test well* or *strat test* is drilled on the seismic line. The primary purpose of the well is to collect subsurface samples and run wireline well logs. This identifies the ages and composition of reflections on the seismic profile.

In *reprocessing*, new methods of computer processing are applied to old digital data that were recorded on magnetic tapes. Because new fields can be found by reprocessing old data, the seismic data is not given away. Any information kept secret is called *proprietary*. *Seismic brokers* are used to sell and buy seismic data.

Time-to-Depth Conversion

Seismic data is recorded in seconds (*time domain*), and a well log is recorded in feet or meters (*depth domain*). Because of this, the vertical scales on both are different and cannot be directly compared. If the seismic velocity though each rock layer is known, a *time-to-depth conversion* can be made on the seismic data.

Two ways to do this is by checkshot survey or vertical seismic profiling. In a *checkshot survey*, a geophone is lowered down the well. The seismic source (*e.g.*, dynamite, air gun, or Vibroseis) is then detonated on the surface. The geophone is then raised up the well a distance of 200, 500, or 1,000 ft (60, 150 or 300 m), and another measurement is made. This is repeated until the geophone is on the surface. *Vertical seismic profiling* (*VSP*) is the same as a checkshot survey except the geophone interval is shorter (50 to 100 ft or 15 to 30 m).

Amplitude Versus Offset

Amplitude versus offset (*AVO*) is an analysis of seismic data to locate gas reservoirs and help identify the composition of the rock layers. *Offset* is the distance between the seismic source and the receiver. The amplitude of a reflection usually decreases with increasing offset distance. Gas reservoirs and different sedimentary rocks such as sandstones, limestones, and shales have different reflection amplitudes versus offsets. Some increase, and others decrease with offset.

3-D Seismic Exploration

In the 1980s and '90s, 3-D seismic exploration (Fig. 14–20) that gives a three-dimensional seismic image of the subsurface was developed. On land, 3-D seismic data is often acquired by *swath shooting* with receiver cables laid out in parallel lines, and the shot points run in a perpendicular direction. In the ocean, 3-D seismic exploration is often run with *line shooting* in closely spaced, parallel lines from a single ship towing several arrays of air guns and streamers.

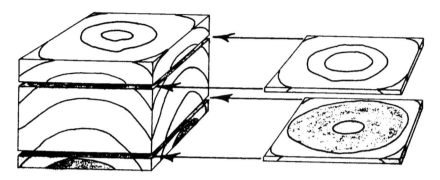

Fig. 14-20 *3-D seismic image with two time slices*

There are many different patterns of sources and geophones that can be used for 3-D seismic acquisition. In *undershooting*, the sources and geophones are not even located on the land being surveyed. The source is located on one side, and the geophones are located on the other side of the land.

The 3-D seismic survey is divided into horizontal squares called *bins*. All reflections whose midpoints fall within a particular bin are combined for common-mid-point (CMP) stacking. The CMP fold is the number of midpoints in each bin. Bins are commonly 55 by 55 ft., 110 by 110 ft., 20 by 20 m, or 30 by 30 m.

After computer processing, a 3-D view of the subsurface is produced. Rock layers are migrated more accurately, and details are shown better than on a 2-D seismic image. A *cube display* is very common (Plate 14–4b). The cube can be made transparent so that only the highest amplitude reflectors are shown. The 3-D seismic image on a computer monitor can be rotated and viewed from different directions. A *time* or *horizontal slice* (Fig. 14–20 and Plate 14–4c) of the subsurface is a flat seismic picture made at a specific depth in time (milliseconds). Various reflectors that intersect the slice are shown. A single seismic reflection can be displayed as a *horizon slice*, and a fault surface can be shown as a *fault slice*.

Special rooms called *visualization centers* (and several other names) are used to display the 3-D seismic images in three dimensions (Plate 14–5). In one type of 3-D seismic visualization room, there are screens on the walls. The viewers wear stereoscopic glasses and sit in chairs. A computer operator projects the seismic image on the screens and can move the image (Plate 14–5a). In another type, there are screen on three walls and the floor. The viewer uses stereoscopic glasses. The viewer is immersed in the 3-D seismic image and can walk through the subsurface (Plate 14–5b). As the viewer turns his head and moves, the subsurface perspective moves with the viewer

Three-D (3-D) seismic exploration is expensive because of acquisition costs and computer processing. A 3-D seismic survey can have 500 gigabytes of information. However, more 3-D seismic exploration, both on land and in the ocean, is being run today than 2-D seismic exploration. It saves money by decreasing the percentage of exploration dry holes. It also saves money during developmental drilling by accurately imaging and defining the subsurface reservoir. The optimum number of developmental wells can then be drilled into the best locations to drain the reservoir efficiently.

4-D and 4-C Seismic Exploration

Four-D (4D) or *time lapse seismic exploration* uses several 3-D seismic surveys over the same producing reservoir at various time intervals such as two years to trace the flow of fluids though the reservoir. As a reservoir is drained, the temperature, pressure, and composition of the fluids change. Gas bubbles out of the oil, and water replaces gas and oil as it is being produced. Time slices of the reservoir are compared, and changes in the seismic response such as amplitude can document the drainage. Undrained pockets of oil can be located and wells drilled to drain them.

Four-C (4-C) or *multicomponent seismic exploration* records both compressional and shear waves that are given off by a seismic source. A *compressional wave (P-wave)* is how sound travels through the air. Particles through which the compressional wave is traveling move closer together and then farther apart (Fig 14–21a). A *shear wave (S-wave)* is like a wave on the surface of the ocean. The particles move up and down (Fig. 14–21b). Shear waves are slower than compressional waves and cannot pass through a liquid or gas.

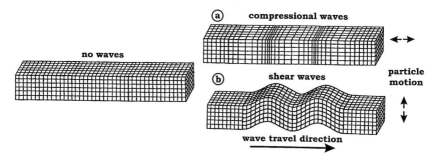

Fig. 14-21 *Seismic waves a) compressional waves, b) shear waves*

The conventional seismic method records only the compressional waves with a one-component geophone. The 4-C seismic method records both the compressional wave (one component) and also uses three geophones that are perpendicular to each other to record the shear wave (three components).

Four-C seismic exploration is used to locate and determine the orientation of subsurface fractures and determine the composition of the sedimentary rock layers and their fluids through which the seismic energy passes. Compressional waves are distorted by gas in sedimentary rocks, but shear waves are not affected. Seismic shear wave recording produces a more accurate picture of any sedimentary rocks that contain gas.

PLATE 14-4

14-4a Anticline showing a flat spot on the oil-water contact (courtesy Geophysical Service Inc.)

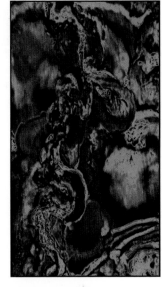

14-4c Horizontal slice from a 3-D seismic survey showing a subsurface, meandering submarine channel on a submarine fan, offshore Niger River Delta (courtesy Veritias).

14-4b Cube display of 3-D seismic data (courtesy Bureau of Economic Geology, The University of Texas at Austin)

PLATE 14-5:
THREE-D VISUALIZATION CENTERS

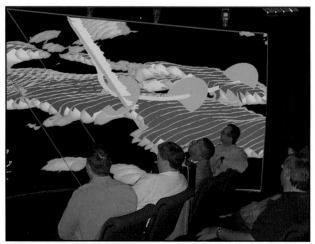

14-5a   *(Courtesy ARCO and Silicon Graphics)*

14-5b   *(Courtesy Landmark Graphics, a Halliburton Company)*

fifteen

DRILLING PRELIMINARIES

A geologist, geophysicist, and/or engineer will locate the *wellsite* or *drillsite*, a location to drill and possibly find commercial oil and gas. A drilling target or targets are identified. A *drilling target* is a potential reservoir such as the Bartlesville Sandstone. Depth to the drilling target is estimated.

The American Petroleum Institute reported that in the United States in 1998, the average exploratory oil well on land was drilled to 5080 ft (1549 m) and cost an average of $87/ft, and the average exploratory gas well on land was drilled to 6220 ft (1896 m) and cost an average of $99/ft. In 1999, there were 42,709 wells drilled for gas and oil in the world. Of those, 28% (12,023) were drilled in the United States, 27% (11,531) were drilled in Canada, with 224 wells (0.5%) drilled in Saudi Arabia, 134 in Iran, and 99 in Kuwait. The deepest well drilled for petroleum was the Bertha Rogers No. 1 to a depth of 31,441 ft (9,583 m) in Oklahoma during 1974. It was a dry hole.

Land and Leasing

In the United States and parts of Canada, private land is called *fee land* and has two, separate ownerships, a surface rights and a mineral rights owner. The *surface rights owner* can build a house, ranch, or farm on the land. The *mineral rights owner* can explore and drill for gas and oil and owns and can produce the gas and oil. The surface and mineral rights owner are not necessarily the same person. The mineral rights can have been separated from the surface rights and sold. In every other country of the world, the federal government owns the mineral rights to the land. About one-third of the mineral rights on land in the United States is owned by the federal or state government in public land. In Canada, land owned by the federal or provincial government is called *crown land.*

It is the job of a *landman* to identify and locate the mineral rights owner of fee land. This is done by searching through the county or parish courthouse records or by using a commercial title company. Commercial land ownership maps that are frequently updated are used to determine the status of leases, the names of lessees, and the identity of surface rights and mineral rights owners.

A *title opinion* can be obtained from an attorney who determines the mineral rights ownership history of the land. The landman then approaches the mineral rights owner and attempts to persuade the owner to sign a lease. A *lease* is a legal document that grants the right to explore, drill and produce oil and gas during the life of the lease. The mineral rights owner (the *lessor*) receives a bonus and a royalty, both negotiated. A *bonus* is money for signing the lease. Typical bonuses today are $25 to $50 per acre and can be up to $250, depending upon how promising the land is. A *royalty* is a promised fraction of the gross revenue from the oil and gas that will be produced, free and clear of any production costs. A standard royalty used to be 1/8th but now is commonly either $3/16^{th}$ or a $1/5^{th}$. The exploration company, the *lessee*, is granted the right to explore and drill on that land during the time of the lease. If petroleum is found, the company has the right to produce it for the life of the field. Leases are printed on standard forms.

Leases have a definite time period for exploration, called the *primary term.* Between one to five years is common. If commercial petroleum in paying quantities is not established and production started by the end of the primary term, the lease expires. If commercial production is established, the lease is automatically extended into the *secondary term* to cover future production. Two types of leases are paid up and delay rental. In the *delay rental lease*, a specific sum (*delay rental*) must be paid to the lessor if drilling has not commenced by the end of each year during the primary term to maintain the lease. If not, the lease expires. The *paid-up lease* does not require delay rentals.

Leases are real property. They can be bought, sold, and traded. One company can transfer (*farm out*) a lease for a royalty (such as $1/4$th) or some other consideration to another company that will drill the lease. Leases accepted from another company are *farmed in*.

Producing properties have both royalty and working interest owners. The *royalty interest* receives a fraction of the gross oil and gas revenue, free and clear of the cost of production. A *landowner's royalty* is granted to the mineral rights owner when the lease is signed. Other royalties might go to the landman, geologist, or promoter who put the deal together. This is called an *overriding royalty* that was created from the working interest. The *net revenue interest* is the percentage such as 87.5% that is left after all royalties have been deducted. The *working interest owners* receive the remaining portion of oil and gas revenue after the royalty interest owners have taken their share and production expenses have been paid. Working interest owners are responsible for all drilling and production costs. After a discovery is made, the operating company files division orders on that well or unit. *Division orders* establish the distribution of oil and gas revenues to the royalty and working interest owners.

Foreign Contracts

An oil company that operates in many countries is called a *multinational* or *international company*. A company that is owned by a federal government and operates only in that country is called a *national* or *host company*. There are three phases of oil and gas operations. The *exploration phase* includes geological, geochemical, and geophysical exploration and drilling of exploration wells. The *exploitation* phase involves the development of newly discovered fields. The *production* phase occurs during oil and gas production.

A contract by a foreign government gives an international oil company the right to explore for gas and oil in a specific area. If commercial amounts of gas or oil are found, the international oil company has the right to develop the field and produce the gas or oil. The contract states how the international oil company will be reimbursed for their expenses and how the gas or oil will be shared. If no commercial amounts of gas or oil are found, the international oil company cannot recover any costs.

Several types of contracts (concession agreement, production sharing agreement, service contract and production contract) can be negotiated between a multinational company and a foreign government. The oldest is a *concession agreement* (also known as license/concession agreement or a tax and royalty agreement). A multinational company is granted an exclusive concession and bears the entire cost and risk of exploration, exploitation and production. The host country is paid bonuses, taxes, and royalties on production. In a variation of this contract, the multinational company still bears the costs and risks of exploration but the host country will share the cost and risk of exploitation.

A *production-sharing contract* is common today. The multinational contractor explores, exploits and produces the gas and oil, bearing the entire cost and risk. If commercial amounts of oil or gas are found, an agreed share of the gross oil and gas production, called *cost oil*, goes to the contractor to recover the costs of exploration, drilling and production. After costs have been recovered, the remaining oil, called *profit oil*, is split by an agreed formula between the multinational company and the host government or company.

A *service contract* provides a contractor with a fee for specific services such as exploration or production. A *production contract* involves a contractor taking over an existing or underdeveloped field and improving production. The contractor is paid a portion of the increased production.

Authority for Expenditure

Before a well is drilled, an *authority* or *authorization for expenditure* (*AFE*) is completed (Fig. 15–1). This form estimates the cost of drilling and completing the well, both as a dry hole and a producer. Costs such as drilling intangibles, completion intangibles, and equipment are listed. *Intangibles* are salaries, services, and equipment that cannot be salvaged after the well is drilled. The AFE includes the cost of the drilling rig, mud, logging, testing, cementing, casing, well stimulation, prime movers, pumps, tubing, separator, and other services and expenses.

The AFE is used to economically evaluate the well before it is drilled. The operator and any other financial contributors to the well approve the AFE. The operator then uses the AFE as a guideline for expenditures.

AFE
(Authority For Expenditure)

	Dry Hole	Completed Well
Drilling Intangibles		
Site Preparation		
Drilling Footage		
Cement and Surface Casing		
Logging		
Mud		
.		
.		
Completion Intangibles		
Cement		
Perforating		
Frac or Acid		
.		
.		
Equipment		
Casing		
Tubing		
Pump Jack		
Separator		
Tanks		
.		
.		
Total Well Cost	$	$

Fig. 15-1 *Authority for expenditure (AFE)*

Drilling Contracts

Drilling rigs are usually owned and operated by drilling contractors. The exploration company signs a drilling contract with the drilling contractor to drill the well to a specific depth or horizon (drilling target) at a specific location (drillsite). The drilling contract includes the *spud date* (when the well is to be started), hole diameters, how much the well can deviate from vertical, drilling mud to be used, logging and testing to be done, casing sizes and depth of each casing string, cementing, how the well is to be completed, the drill collars to be used, and the subsurface formations to be drilled. The drilling contract also contains rates for when the well is not being drilled, such as during standby or logging operations.

There are three common types of drilling contracts. A *footage drilling contract* is very common on land. It is based on a cost per foot to drill down to the contract depth. A *daywork contract* is common offshore and is based on a cost per day to drill down to contract depth. A *turnkey contract* has an exact cost to drill down to the contract depth. A *combination contract* has a footage rate to a certain depth and a daywork rate below that. Standard drilling contracts by the American Petroleum Institute (API) and the American Association of Drilling Contractors (AAODC) are commonly used.

Service and supply companies are contracted. A *service company* performs a specific service such as wireline logging or mud engineering. A *supply company* will furnish equipment such as casing.

Joint Operating Agreements and Support Agreements

There are several ways to develop an area with a limited budget, to encourage a well to be drilled or to reduce the financial impact of possibly drilling a dry hole or holes. A company can enter into a joint operating agreement with one or more other companies. A *joint operating agreement* (*JOA*) can be for drilling a single well or the development of a larger *working interest area.* The joint operating agreement defines the rights and duties of each party including each party's share of expenses. An *operator* who is in charge of the day-to-day operations is identified. The well or wells are drilled. The joint operating agreement defines how the production is to be shared by the parties.

A *support agreement* or *contribution agreement* can be used to encourage and support drilling a well. There are three types. In a *dry hole agreement,* a party agrees to make a cash contribution if the well being drilled by another party is a dry hole. In return that party receives the geological and drilling information from that well whether or not the well is a dry hole. In a *bottom-hole contribution agreement*, a party agrees to make a cash contribution to the party drilling of a well to a certain depth in return for geological and drilling information on that well. In an *acreage contribution agreement*, a party contributes leases or interests to another party who is drilling a well in that area in return for geological and drilling information on that well.

Site Preparation

To *stake a well*, a surveyor accurately determines the well location and elevation. A *plat* (map) of the site is prepared and registered with the appropriate government agency. A bulldozer can be used to grade an access road to the site and make a turnaround. The bulldozer then clears and levels the site. Boards might be laid if the ground is wet. A matting, often made of 3 x 12 in. (7.5 x 30 cm) timbers, can be spread on the surface to support the rig and improve drainage. A gravel pad is used on permafrost ground to prevent melting. A large pit, the *reserve pit*, is dug and lined with plastic next to the drilling rig. It will hold unneeded drilling mud, cuttings, and other materials from the well.

If it is going to be deep well, a rectangular pit (*cellar*) can be dug (Fig. 15–2) and lined with boards or cement. The cellar provides space below the drilling platform for the blowout preventers. Provisions are made for a water supply at the drilling site by drilling a water well or by laying a water pipeline.

If the well is shallow (to 3000 ft or 1000 m), the entire drilling rig comes on the back of a truck or trailer. This is a *truck-mounted* or *portable rig*. If the well is deeper, the rig comes in modules on the back of several tractor-trailers. The modules are specifically designed to be trucked and be quickly fitted together at the drillsite with large pins secured with cotter pins (Plate 15–1).

The deeper the well, the larger and stronger the rig has to be to support the drillpipe on the derrick floor as the pipe is being pulled out of the well. Each drilling rig is rated for maximum depth.

In very remote areas, a *helirig* is used. It is made of specially designed modules that are transported by helicopter.

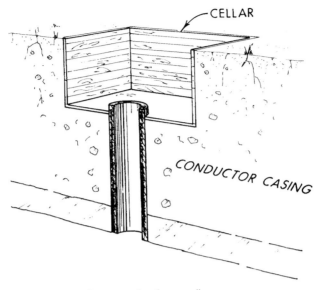

Fig. 15-2 Cellar and conductor casing for a well

Plate 15-1 Pin with cotter pin used to secure drilling rig modules

A rig is assembled during *rig up* and disassembled during *rig down*. The start of drilling a well is called *spudding in*. To spud a medium or deep well usually begins with a small, truck-mounted rig that drills a large-diameter but shallow hole (20 to 100 ft or 7 to 30 m) called the *conductor hole*. Large-diameter pipe (20 in. or 50 cm), called *conductor casing* or *pipe*, is then lowered and cemented into the conductor hole (Fig. 15–2). In soft ground, the conductor casing can be pile-driven without drilling. The conductor casing stabilizes the top of the well and provides an attachment for the blowout preventers in areas where shallow gas could be encountered.

Types of Wells

A well drilled to discover new oil or gas reserves (Plate 15–2) is called a *wildcat* or *controlled exploratory well*. The exploratory well can be drilled to 1) test a trap that has never produced (*new-field wildcat*), 2) test a reservoir that has never produced in a known field that is deeper or shallower than the producing reservoir(s), or 3) extend the known limits of a producing reservoir in a field. A *rank wildcat* is drilled at least 2 miles (3 kms) away from any known production. If the well does discover a new field, it is called the *discovery well* for that field.

As soon as possible after a discovery, the size of the field must be determined. If this is private fee land, it must be determined which leases need to be drilled to maintain the leases and which can be abandoned. If this is an offshore field or in a remote area or foreign country, the size of the field needs to be established to compute the amount of oil and gas that can be produced. This will determine if the field is large enough to economically justify further development. Field size is determined by *step out, delineation,* or *appraisal wells* that are drilled to the sides of the discovery well. If the oil-water or gas-water contact can be located on all four sides of the discovery well, the area of the field can be determined.

Wells drilled in the known extent of the field are called *developmental wells*. Wells drilled between producing wells in an established field to increase the production rate are called *infill wells*.

Plate 15-2 *Wildcat well being drilled by a rotary drilling rig (courtesy Parker Drilling)*

Government Regulations

Before the 1930s in the United States, there were no regulations as to how close the wells could be located or how fast the oil could be produced. Today governments prevent the exploitation of a field with excessive drilling and production rates. Each well in a field is given a *drilling and spacing unit* (*DSU*), a square (sometimes a rectangle) of a certain area on which only one well can be drilled and completed (Fig. 15–3). Typical for an oil well is 40 or 80 acres. Oil viscosity and reservoir permeability are two important factors in well spacing. Higher viscosity oils and lower permeability reservoirs need smaller spacing for efficient drainage. Gas wells drain a larger area, and DSUs of 640 acres are common. Usually the well does not have to located in the center of the DSU but cannot be located on the edge.

In some countries production is limited by an allowable to prevent excessive production rates. An *allowable* is the maximum amount of oil and gas production that is permitted from a single well, lease, or field during a specific unit of time such as a month.

Cable Tool Rigs

When the first commercial oil well was drilled at Titusville, Pennsylvania, in 1859, a cable tool rig was used (Plate 13–1). Cable tool rigs had been in use for hundreds of years before that to drill for fresh water or for brines that were evaporated for salt.

A *cable tool rig* is relatively simple (Fig. 15–4). The hoisting system consists of a tower with four legs called the *derrick* that was originally wooden and 72 to 87 ft (22 to 27 m) high. An engine, originally a steam engine,

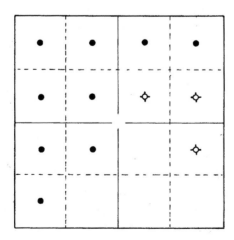

Fig. 15-3 *Drilling and spacing units*

causes a wooden *walking beam* to pivot up and down on the *Sampson post.* The *bit*, a solid, steel rod about 4 ft (1 1/3 m) long with a chisel point on it, is suspended down the well from the opposite end of the Sampson post by a rope or cable. As the walking beam pivots, it causes the rope and bit to rise and fall. The bit pounds the well down by pulverizing the rock. The rope or cable is wound around a reel called a *bull wheel.* The rope or cable goes up over a single wheel (*crown block*) at the top of the derrick and then down to a temper screw on the walking beam and down the well to the bit. As the well is pounded deeper, more rope or cable is let out by turning the *temper screw* on the end of the walking beam.

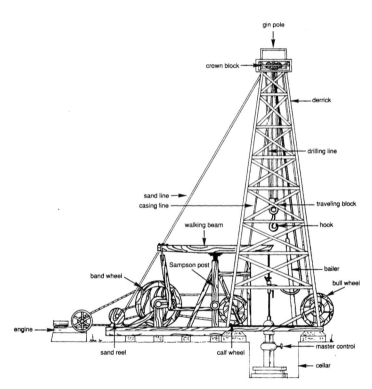

Fig. 15-4   *Cable tool drilling rig (Hyne, 1991)*

After drilling 3 to 8 ft. (1 to $2^1/_2$ m), the bottom of the well becomes clogged with rock chips. The bit is then raised, and a *bailer* is lowered into the well on a sand line to remove the rock chips and water. After the bailer is raised and emptied, the bit is lowered into the well to pound deeper.

Heavy casing is run down the well from wire rope wound around the *calf wheel.* The wire rope runs through a multiple sheave crown block at the top of the derrick. The casing (large diameter pipe) is used in the well to keep water from filling the well and to prevent the sides from caving in. Lighter equipment, such as the bailer, is run in the well on a *sand line* from the *sand reel.*

Cable tool drilling is very slow with 25 ft. ($7^1/_2$ m) per day being an average and 60 ft. (20 m) being very good. It does not effectively control subsurface pressures, and blowouts were common during cable tool operations. However, all fields that were discovered during the 1800s were drilled by cable tool rigs. A cable tool rig in New York drilled a well to a depth of 11,145 ft. (3397 m) in 1953.

The rotary drilling rig that replaced the cable tool drilling rig was introduced in various areas throughout the world during the period of 1895 to 1930. The greatest advantage of the rotary drilling rig is that it could drill the well considerably faster (several hundred to several thousand feet per day). There were, however, some major problems to be worked out with the early rotary drilling rigs, and they were not immediately accepted. In 1950, there was an equal number of active cable tool and rotary drilling rigs in the United States. Today, almost all well are drilled with rotary drilling rigs.

sixteen

DRILLING A WELL—THE MECHANICS

Today, almost all wells are drilled with rotary drilling rigs (Plate 16–1). The *rotary drilling rig* rotates a long length of steel pipe with a bit on the end of it to cut the hole called the *wellbore*. The rotary rig consists of four major systems. These include the power, hoisting, rotating, and circulating systems (Fig. 16–1).

Power

The *prime movers* are diesel engines that supply power to the rig and are usually located on the ground in back of the rig. Diesel fuel is stored in tanks near the engines. Most of the power goes to the hoisting and circulating systems. Some also goes to the rotating system, rig lights and other motors.

Depending on the size of the rig and the drilling depth, there are one, two, or four engines. The engines are rated by their horsepower and fuel consumption. They commonly develop 1000 to 3000 horsepower (hp). Power from the diesel engines is transmitted to the rig mechanically by a system of pulleys, belts, shafts, gears, and chains called a *compounder*.

On some new and on most marine rigs, a *silicon controlled rectifier* (*SCR*) is used in a *diesel-electrical system*. The diesel engines are coupled to an alternating current (AC) generator that supplies AC electrical power by cables to the rig where it is converted to direct current power in the SCR to drive the equipment.

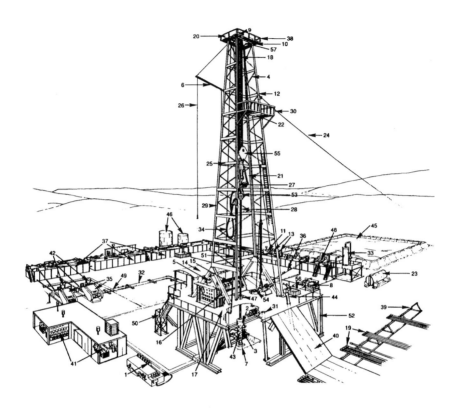

1	accumulators	15	drawworks	29	leg	43	ram blowout preventers
2	annular blowout preventer	16	driller's console	30	monkey board	44	rat hole
3	blowout preventer stack	17	drill (derrick) floor	31	mouse hole	45	reserve pit
4	brace	18	drilling line	32	mud discharge line	46	reserve tanks
5	cathead	19	drill pipe	33	mud-gas separator	47	rotary table
6	catline boom	20	duck's nest	34	mud (rotary) hose	48	shale shaker
7	cellar	21	elevators	35	mud pumps (hogs)	49	shock hose
8	choke manifold	22	fingers	36	mud return line	50	stairways
9	crown block	23	fuel tank	37	mud tanks (pits)	51	standpipe
10	crown platform (crow's nest)	24	Geronimo line	38	pigpen	52	substructure
11	degasser	25	girt	39	pipe rack	53	swivel
12	derrick (mast)	26	hoisting line	40	pipe ramp	54	tongs
13	desanders and desilters	27	hook	41	prime movers	55	traveling block
14	dog house	28	kelly	42	pulsation dampeners	56	trip tank 57 water table

Fig. 16-1 Rotary drilling rig (Hyne, 1991)

Plate 16-1   *Rotary drilling rig (courtesy Parker Drilling)*

Hoisting System

The hoisting system is used to raise and lower and to suspend equipment in the well (Fig. 16–2). The *derrick* or *mast* is the steel tower directly above the well that supports the crown block at the top and provides support for the drillpipe to be stacked vertically as it is pulled from the well. If the tower comes on a tractor-trailer and is jacked up as a unit, it is a mast. On a *cantilevered mast rig*, the mast is assembled horizontally (Fig. 16–3) and then pivoted up to a vertical position using the traveling block and drawworks. If the tower is erected on the site, it is a derrick. A derrick that can hold only two lengths of drillpipe vertically is a *double derrick*. Most, however, are *treble derricks* that can hold three lengths of drillpipe. They are 80 to 187 ft ($24^1/_2$ to 57 m) tall.

Derricks and masts have four vertical *legs* made of structural steel. The horizontal structural members between the legs are called *girts* (Fig. 16–4). The diagonal members are *braces*. An inverted, V-shaped opening in the front of the derrick or mast called the *V-door* allows drillpipe and casing to be pulled up the pipe ramp onto the drill floor.

Derricks are rated for maximum drillpipe load and have capacities of 86,000 to 1,392,000 lb (39,000 to 631,000 kg). Derricks and masts are also rated for wind load and can commonly withstand 100- to 130-miles per hour (160 to 208 km/hr) winds. The base of the mast or derrick is a flat, steel surface called the *drill* or *derrick floor* where most of the drilling activity

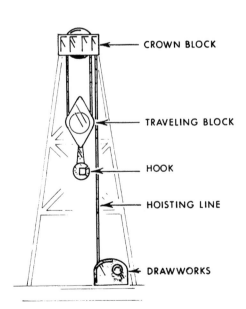

Fig. 16-2 *Hoisting system*

CROWN BLOCK

TRAVELING BLOCK

HOOK

HOISTING LINE

DRAWWORKS

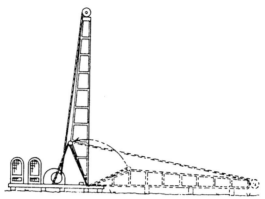

Fig. 16-3 Cantilevered mast rig

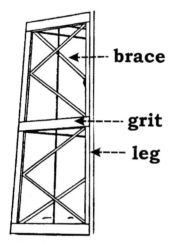

Fig. 16-4 Part of derrick showing vertical legs, horizontal grits, and diagonal braces

occurs. Two *substructures* made of a steel framework 10 to 30 ft (3 to 10 m) high are used to raise the derrick floor above the ground (Plate 16-1). This is to provide space for wellhead equipment below the drill floor such as the blowout preventers.

The *drilling* or *hoisting* line is made of braided steel wire about 1 1/8 in. (3 cm) in diameter. The line consists of several strands of braided steel wire wound around a fiber or steel core. The hoisting line is described by the type of core, number of strands around the core, and the individual wires per strand. There are several ways to wrap the strands that are wound around the core.

The hoisting line is spooled around a revolving reel on a horizontal shaft in a steel frame called the *drawworks* on the drill floor. The prime movers drive the drawworks to wind and unwind the drilling line. There are several speeds and both forward and reverse on the reel. The driller controls the drawworks from a *brake*, a hand lever, on the drill floor. Drawworks are often rated by input horsepower that ranges from 500 to 3000 hp. Small spools called *catheads* are attached to a *catshaft* that runs horizontally through the drawworks. They are used to pull lines such as the jerk or spinning line.

On the drilling rig, there are two sets of wheels on horizontal shafts in steel frames called *blocks*. The drilling line from the drawworks goes over a wheel in the *crown block* that is fixed at the top of the derrick. It then goes down to the *traveling block* that is suspended in the derrick. The drilling line goes back and forth through wheels in the crown and traveling block 4 to 12 times. The end of the drilling line is fixed to a *deadline anchor* located under the drill floor substructure. After a certain amount of usage, the drilling line is moved 30 ft. (9 m) through the anchor to prevent wear on any particular spot along the line. Below the traveling block is a *hook* for attaching equipment. As the drilling line is reeled in or out of the drawworks, the traveling block rises and falls in the derrick to raise and lower equipment in the well.

Rotating System

The rotating system is used to cut the hole (Fig. 16–5). The turning drillpipe, bit and related equipment is called the *drillstring*. Suspended from the hook directly below the traveling block is the *swivel*. The swivel allows the drillstring below it to rotate on bearings in the swivel while the weight of the pipe is suspended from the derrick.

Below the swivel is located a very strong, four- or six-sided, high-grade molybdenum steel pipe called the *kelly* that is 40 or 54 ft (12 or $16^{1}/2$ m) long (Plate 16–2). The kelly has sides to enable it to be gripped and turned by the rotary table. The kelly turns all the pipe below it to drill the hole.

The *rotary table* is a circular table in the derrick floor that is turned clockwise (*to the right*) by the prime movers. If it were turned in the opposite direction, the drillpipe would unscrew. The kelly goes through a fitting called the *kelly bushing* (Plate 16-2 and Fig. 16–6a), which fits onto to the *master bushing* (Fig. 16–6b) in the rotary table. Rollers in the kelly bushing

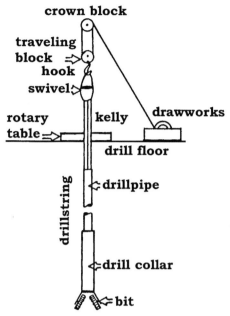

Fig. 16-5 *Rotating system*

Plate 16-2 *Kelly, kelly bushing, and rotary table on drill floor*

allow the kelly to slide down through the kelly bushing as the well is drilled deeper. The rotary table, master bushing, kelly bushing and kelly rotate as a unit. *Turning to the right* and *making hole* are terms for drilling a well.

Below the kelly is the *drillpipe* (Fig. 16–7). The round, heat-treated alloy steel drillpipe ranges from 18 to 45 ft ($5^1/_2$ to 14 m) long but is commonly 30 ft (10 m). The pipe ranges in outer diameter from 2 7/8 to $5^1/_2$ in (7 to 14 cm) and is threaded with male (*pin*) connections on each end. The larger diameter portion on one end is the *tool joint* that has been screwed and welded onto the drillpipe. It has a female (*box*) connection. The threaded ends are tapered for easy connection and to keep the drillpipe screwed.

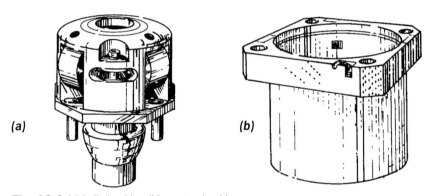

(a) *(b)*

Fig. 16-6 *(a) kelly bushing (b) master bushing*

Fig. 16-7 *Joint of drillpipe*

Pipe dope, a lubricant such as grease, is applied to the threads of each joint of pipe as it is being added to the drillstring. The drillpipe wall is thicker with an *upset* where it is threaded on each end to strengthen that area. Most upsets are internal in that they decrease the inner diameter of the pipe.

API specifications of drillpipe include three length ranges and five grades for strength. Drillpipe is also described by nominal weight per foot, inner diameter, collapse resistance, internal yield strength and pipe body yield strength. Each section of drillpipe is called a *joint*. The drillpipe is reused after each well is drilled and is graded for wear. There are five API drillpipe wear grades. After the drillpipe is worn, it is replaced with new pipe.

The kelly must always be located on top of the drillpipe. After drilling the well down 30 ft (10 m), another joint of drillpipe must be added to the drillstring to make it longer in a process called *making a connection*. The next joint of drillpipe, however, must be added to the bottom of the kelly to keep the kelly at the top of the drillstring.

The pipe is raised from the well by *pipe elevators* that are attached to the bottom of the traveling block and are designed to clamp onto the pipe. The *tongs* and *spinning wrench* are large clamp and wrench devices that are suspended from cables above the drill floor (Plate 16–3). They are used to screw (*make up*) and unscrew (*break out*) pipe. A steel wedge with handles (Fig. 16–8), the *slips*, can be placed in the rotary table bowl to hold the pipe with teeth and prevent the drillpipe from falling down the hole.

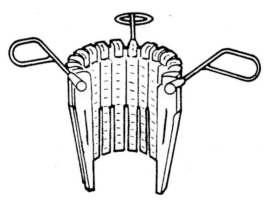

Fig. 16-8 *Slips*

Plate 16-3 *Tongs and spinning wrench being used on drillpipe (courtesy American
Petroleum Institute and Chevron)*

The next joint of drillpipe used to make a connection (Fig. 16–9a) is kept in a hole in the drill floor called the *mouse hole*. The drillstring is raised until the entire kelly is above the rotary table. The slips are then put into the rotary table bowl, and the kelly is unscrewed from the top of the drillstring (Fig. 16–9b). The kelly is then swung over to the mouse hole and screwed into the next joint of pipe. The drillpipe is then raised out of the mouse hole and swung over to the rotary table where it is screwed into the drillstring (Fig. 16–9c). The slips are then removed from the bowl in the rotary table. The well is drilled another 30 ft (10 m) deeper, and another connection is made again. A *spinning chain* used to be used to wrap around a joint of pipe and pulled to start screwing two joints together. This, however, is dangerous and is being eliminated.

Drillpipe is stored horizontally on the *pipe rack* located on the ground next to the front of the rig. Individual joints are dragged up to the drill floor along the pipe ramp and through an opening in the derrick, the *V-door*.

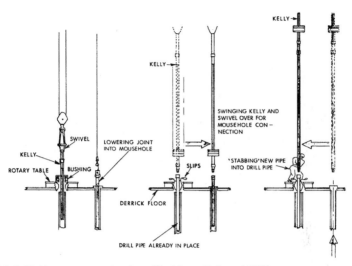

Fig. 16-9 *Making a connection (modified from Baker, 1979)*

    The section of the drillstring below the drillpipe is called the *bottomhole assembly*. Below the drillpipe are thicker-walled, heavier, stronger pipes called *drill collars* (Fig. 16–10). Drill collars are made of heat-treated alloy steel and are 31 ft. (9 $^1/_4$ m) long. Box and pin connections are cut into each end. Drill collars are designed to put weight on the bottom of the drillstring to drill straight down and prevent the drillpipe from kinking and breaking. Two to ten joints of drill collars are often used.

    Heavyweight drillpipe is intermediate in strength and weight between drillpipe and drill collars. It has the same outer diameter but a smaller inner diameter than drillpipe and comes in 30 $^1/_2$ ft (9 $^1/_4$ m) joints. Heavyweight drillpipe is often run between the drillpipe and drill collars to minimize stress between the two and prevent drillstring failure in that area.

Fig. 16-10 *Joint of drill collar*

Fig. 16-11 *Stabilizer*

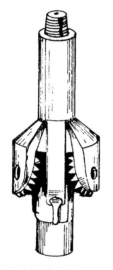

Fig. 16-12 *Hole opener*

Smaller sections of pipe called *subs* can be run between and below the drill collars to do various functions. A *stabilizer* is a common sub (Fig. 16–11) that uses blades to contact the well walls. It is designed to keep the drillstring central in the well.

A *vibration dampener* or *shock sub* uses rubber, springs, or compressed gas to absorb vibrations from the bit. It is usually run just above the bit. A *bit sub* is used to make the connection between a bit and the drill collar or sub above it. A *crossover sub* is used to make a connection between two different sizes of pipe or two different thread types. A *hole opener* (Fig. 16–12) uses roller cones to enlarge the wellbore. A *reamer* has three or six tungsten steel rollers along its sides and is often run above the bit to provide a gauge hole. A *gauge hole* is a hole with a specific minimum diameter.

The bit has a pin connection on the top that screws into the bit sub on the bottom of the drillstring. The most common bit used is a *rotary cone bit* with three rotating cones, called a *tricone bit* (Fig. 16–13). The body of the tricone bit is made of three legs of heat-treated steel alloy that have been welded together. Each leg has a jet and channel for drilling mud to flow though it. Each cone is mounted on a bearing pin that protrudes from the leg. The cones rotate on sealed and self-lubricating bearings and rollers in a race. Some surfaces on the bit are hard-faced with tungsten carbide to increase resistance to abrasion wear.

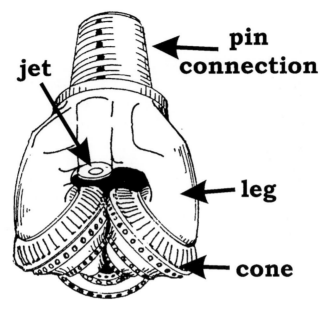

Fig. 16-13 *Tricone drill bit*

As the bit is turned on the bottom of the drillstring, the cones rotate. Teeth or buttons on the cones are designed to either flake or crush the rock on the bottom of the well. The rock chips that are formed are called *well cuttings* or *cuttings.*

There are hundreds of different tricone bits that are classified as either milled teeth or insert bits. The *milled teeth* or *steel tooth* type of tricone bit has teeth machined out of the solid cones that are designed to flake the rock (Plate 16–4). The teeth of adjacent cones fit between each other to clean out cuttings. This bit is used for relatively soft (long, widely spaced teeth) and medium (short, closely spaced teeth) hardness rocks.

The *insert* or *button* type of tricone bit has holes drilled into the solid cones (Plate 16–5). Buttons of tungsten carbide stick out of the holes and are designed to crush the rock. It is used to drill relatively hard rocks.

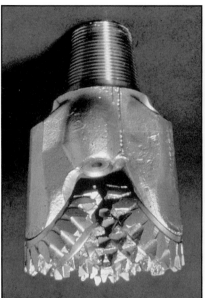
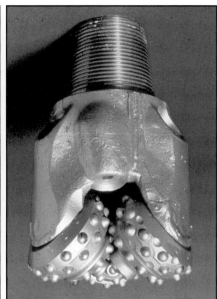

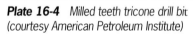

Plate 16-4 *Milled teeth tricone drill bit*
(courtesy American Petroleum Institute)   **Plate 16-5** *Insert tricone drill bit*

A *diamond bit* is made of solid steel with no moving parts (Fig. 16–14). Hundreds of industrial diamonds are attached in geometric patterns on the bottom and sides of the bit. Grooves in the face of the bit (*watercourses*) allow drilling mud to flow out of the center of the bit, across the diamonds, and to the side of the bit. Diamond bits are used for drilling very hard rocks.

A *polycrystalline diamond compact* (*PDC*) bit is also a solid metal bit with no moving parts (Fig. 16–15). It has metal cutters on the bottom that are faced with blanks containing man-made industrial diamonds designed to shear away the rocks to produce cuttings. These are the most expensive bits. A PDC bit, however, can often be used for several hundred hours and drill more footage in a well than with other types of bits.

Bits commonly come in diameters of 3 $^3/_4$ to 26 in. (9 $^1/_2$ to 66 cm). The International Association of Drilling Contractors identifies bits with three numbers such as 334. These numbers are based on 1) cutting structure such as milled teeth or inserts, 2) hardness of formation to be drilled, and 3) mechanical design.

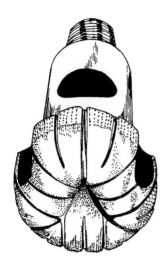

Fig. 16-14 *Diamond bit*

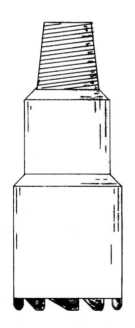

Fig. 16-15 *Polycrystalline diamond compact (PDC) bit*

The bit is turned at a rate of 50 to 100 rpm. It is generally turned slightly faster at shallow depths and slower at deeper depths. Not all the weight of the drillstring is let down on the bit because that would crush it. The larger the bit, the more weight is applied. About 3000 to 10,000 psi per inch of bit diameter is used.

A tricone bit wears out after 8 to 200 hours of rotation, with 24 to 48 hours being common. A worn bit can be detected from the change in noise that the drillpipe makes on the derrick floor and by a decrease in rate of penetration. The bit is changed by *making a trip* during which the drillstring is pulled from the well, the bit changed, and the drillstring is put back in the well (Fig. 16–16).

First, the kelly is raised above the rotary table. The slips are put in the rotary table bowl, and the kelly is unscrewed from the top of the drillstring. The kelly is then put in a hole in the drill floor called the *rat hole*. A member of the crew called the *derrickman*, climbs up and onto a small platform called the *monkeyboard* located 90 ft. (27 m) above the drill floor near the top of the derrick. The drillpipe is then pulled from the well in a process called *tripping out*, unscrewed and stacked in the derrick. The pipe is usually pulled and unscrewed three joints (a *tribble*)

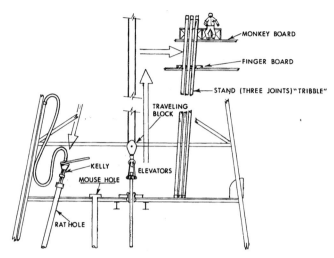

Fig. 16-16 *Changing a bit (modified from Baker, 1979)*

at a time, and is called a *stand.* A stand could also be two or four joints on some rigs. The derrickman guides the stands between the fingers in a *finger board* located under the monkeyboard. The bottom of the stand rests on the drill floor. The bit is then changed, and the pipe run back into the hole in a process called *tripping in.* This process takes rig time and the deeper the well, the longer making a trip take.

To unscrew a bit, a *bit breaker* (Fig. 16–17) is placed in the rotary table to grip the bit. The rotary table is then turned to screw and unscrew the bit from the drillstring.

Circulating System

The circulating system pumps drilling mud in and out of the hole. Drilling mud is stored in steel *mud tanks* on the ground beside the rig (Plate 16–6). The drilling mud is kept mixed in the tanks by rotating paddles on a shaft called a *mud agitator* or by a high pressure jet in a *mud gun.*

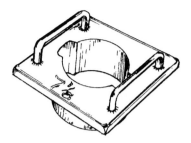

Fig. 16-17 *Bit breaker*

Large pumps driven by the prime movers, called *mud hogs,* use pistons in cylinders to pump the drilling mud from the mud tank. The mud pumps are either duplex or triplex. A *duplex pump* uses two double-acting pistons in cylinders that drive the mud on both the forward and backward strokes. A *triplex pump* uses three single-acting pistons in cylinders that drive the mud only during the forward stroke.

The mud flows from the pumps, through a long, rubber tube, the *mud hose,* and into the swivel. The drilling mud then flows down through the hollow, rotating drillstring and jets out through the holes in the drilling bit on the bottom of the well. The holes, called *nozzles* or *jets,* are located between each pair of cones (Fig. 16–13). On a diamond bit, the drilling mud flows through the bit into grooves, called *watercourses,* on the face of the bit, and across the diamonds (Fig. 16–14). The drilling mud picks the rock chips (cuttings) off the bottom of the well and flows up the well in the space (*annulus*) between the rotating drillstring and well walls.

Plate 16-6 *Mud tanks and reserve pit*

At the top of the well, the mud flows through the blowout preventers (BOPs), along the *mud return line* and on to a series of vibrating screens made of woven screen cloth in a steel frame called the *shale shaker.* The shale shaker is located on the mud tanks and is designed to separate the coarser well cuttings from the drilling mud. It can be either single or double deck. The *double deck shaker* has a coarser screen above a finer screen. The screens are tilted 10° from horizontal to cause the cuttings to vibrate down the screen and into the reserve pit. The mud then flows through other solids-control devices such as cone-shaped *desanders* and *desilters* (Plate 16–7) that centrifuge the mud to remove finer particles. The mud then flows back into the mud tanks to be recirculated down the well.

The mud tanks are 6 ft (2 m) high, up to 8 ft (2 $^1/_2$ m) wide and are usually 26 ft (8 m) long. They have two, three, four, or more compartments. A common mud tank configuration has the *shaker tank* receiving the drilling mud from the well after the cuttings have been removed. The drilling mud flows from the shaker tank to the *reserve tank* and then to the *suction tank.* Drilling mud from the suction tank goes to the mud hogs. On the shaker tank is a *mud-gas separator* that removes any subsurface gas that was dissolved in the drilling mud.

Plate 16-7 *Desanders and desilters*

Adjacent to the mud tanks but away from the rig is a large earthen pit called the *reserve pit* (Plate 16–6). It holds discarded mud for reuse and the cuttings from the shale shakers.

Drilling mud is a mixture of clay with either water (*water-based drilling mud*), oil (*oil-based drilling mud*), a mixture of oil and water (*emulsion mud*) or a synthetic organic matter and water mixture (*synthetic-based drilling mud*). The water is usually fresh but can be saline. Oil-based drilling mud is made from diesel, mineral, or synthetic oil and brine. It has excellent bit lubricating properties and does not affect the formations being drilled. It is, however, expensive, hard to dispose of after drilling, and can be flammable. The emulsion mud with 8 to 12% oil in water has advantages of both. The synthetic-based drilling mud has the oil-based drilling mud advantages and is relatively easy to dispose.

A common drilling mud is made of fresh water and bentonite. *Bentonite* is a type of clay that forms a colloid and will stay suspended in the water a long time after agitation has stopped. Drilling mud viscosity can be increased by adding more bentonite (*mud up*) or decreased by adding water (*water back*). Chemicals mixed with the mud for various effects are called *additives*. A mass of additives put into drilling mud to remedy a situation is called a *pill*. Heavier drilling mud used to exert more pressure in the well is made by mixing in high-density substances called *weighting material* such as barite ($BaSO_4$) or galena (PbS).

Other mud additives include:
- alkalinity or pH control agents
- bactericides
- defoamers
- emulsifiers
- flocculants
- filtrate reducers
- foaming agents
- shale control agents
- surface active agents
- thinners
- lost circulation material

The clay and additives are brought onto the drill site in dry sacks and are stored in the *mud house*. They can be added to the mud in the mud tanks though a *hopper*, a funnel-shaped device.

Drilling mud is described by weight. Fresh water weighs 8.3 pounds per gallon. Typical bentonite drilling mud weighs 9 to 10 pounds per gallon. A heavy drilling mud designed to exert a greater pressure on the bottom of the well can weight 15 to 20 pounds per gallon.

The density, viscosity, and other properties of the drilling mud are frequently checked during drilling by a *mud man* or *drilling fluids engineer*. The mud man is often a service company employee.

A *Marsh funnel* (Fig. 16–18) is used to determine mud viscosity (the resistance to flow). The time the mud takes to flow through the funnel in seconds is calibrated to mud viscosity. In the laboratory, an instrument called a *viscometer* is used to determine mud viscosity and gel strength. *Gel strength* is the ability of the mud to suspend solids. A *filtration test* is made by passing the mud through a filter in a filter press. It measures the thickness and consistency of the mud cake on the filter and the amount of liquid (filtrate) that passed through the filter. The pH (alkalinity) and solid content of the mud are also measured.

Fig. 16-18 *Marsh funnel*

Circulating drilling mud in a well serves several purposes. The mud removes well cuttings from the bottom of the well to allow drilling to continue. Without removing the cuttings, drilling would have to stop every few feet to remove the cuttings that clog up the bottom of the well as has to be done on a cable tool drilling rig. As the mud flows across the bit, it cleans the cuttings from the teeth. The drilling mud also cools and lubricates the bit. In very soft sediments, the jetting action of the drilling mud squirting out of the bit helps cut the well.

The drilling mud also controls pressures in the well and prevents blowouts. At the bottom of the well, there are two fluid pressures. Pressure on fluids in the pores of the rock (reservoir or fluid pressure) tries to force the fluids to flow through the rock and into the well (Fig. 16–19). Pressure exerted by the weight of the drilling mud filling the well tries to force the drilling mud into the surrounding rocks. If the pressure on the fluid in the subsurface rock is greater than the pressure of the drilling mud (*underbalance*), water, gas, or oil will flow out of the rock into the well. This can cause the sides of the well to cave in (*sluff in*), trapping the equipment. In extreme cases, it can cause a blowout where fluids flow uncontrolled and often violently onto the surface.

In order to control subsurface fluid pressure, the weight of the drilling mud is adjusted to exert a greater pressure on the bottom of the well than the pressure on the fluid in the rocks (*overbalance*). Some of the drilling mud is then forced into the surrounding rocks during drilling. The rocks act as a filter, and the solid mud particles are plastered to the sides of the well to form a *filter* or *mud cake* as the fluids enter the rock. The filter cake is very hard. It stabilizes the sides of the well and prevents subsurface fluids from flowing into the well.

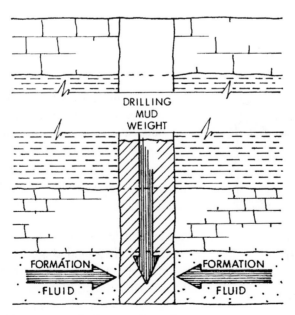

Fig. 16-19 *Pressures on the bottom of a well*

After a well has been drilled, the drilling mud is disposed of and not reused. If it is fresh water drilling mud, it can be spread on the adjacent land to fertilize the crops (*land farming*). Salt water, oil-based, and emulsion drilling mud, however, usually have to be trucked away to a disposal site. On an offshore drilling rig, a barge can be used to bring the mud ashore to a disposal site.

The *kelly cock* (Fig. 16–20), a valve, is used either above or below the kelly. It allows drilling mud to be circulated down the drillstring but can be closed by hand with a hexagonal wrench to prevent fluids from flowing up the drillstring. The kelly cock is closed during making a connection to prevent mud from spilling out of the kelly.

Blowout preventers (*BOPs*) are used to close off the top of the well. They are bolted (*nippled up*) to the top of the well and are located below the derrick floor (Fig. 16–21). Arranged vertically in the *blowout preventer stack* are a series of rams and a preventer.

Blind rams are two large blocks of steel that can close over the well. Because they have flat surfaces, they can be thrown only when the drillstring is not in the well. *Pipe rams* are two large blocks of steel that are designed to close around pipe in the well. They have inserts cut into the surfaces to fit around a specific size pipe. *Variable-bore rams* can be thrown around a range of pipe sizes. Blind, pipe, and variable-bore rams have rubber coatings on their metal closing surfaces to prevent pipe damage and improve the seal. *Shear rams* are designed to cut across the pipe to close the well quickly. They are used primarily on offshore wells. The *annular preventer* is made of synthetic rubber with steel ribs in a doughnut-shape that fills a steel body. It is compressed by pistons to fit around any size and shaped equipment in the well. It goes in a cylinder on the top of the BOP stack and is the first one closed. If it is not effective, the rams are then thrown.

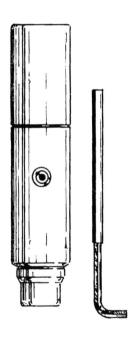

Fig. 16-20 *Kelly cock with wrench*

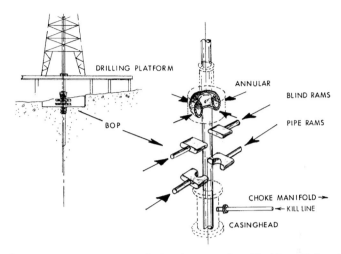

Fig. 16-21 *Blowout preventer stack showing rams (modified from Baker, 1979)*

A typical blowout preventer stack has an annular preventer at the top with one or more rams in line below it. Between the rams and the annular preventer are drilling spools. A *drilling spool* is a spool-shaped fitting that permits attachment of kill and choke lines to the stack.

The power to activate the blowout preventers is stored pneumatically in cylinders called the *accumulator* that are mounted on skids next to the rig. The cylinders contain hydraulic fluid and nitrogen gas. A charging pump always keeps pressure in the cylinders so that the blowout preventers can be thrown even if the rig's prime movers are down. There is a *blowout preventer panel* on the drill floor and another one a safe distance away from the rig. Handles on the panel are used to thrown individual rams or the annular preventer.

Blowout preventer stacks are designed to API standards. The API describes blowout preventers by 1) working pressure, 2) inside bore diameter, and 3) type of rams and annual preventer. *Working pressure* is the maximum pressure that equipment is designed to operate under. As the well is drilled deeper, blowout preventers with higher working pressures are installed to replace those with lower working pressures. Working pressures range from 2000 to 15,000 psi (140 to 1050 kg/cm^2). The *substructure*, a steel framework, raises the elevation of the derrick floor to make space for the BOP stack (Plate 16-1). A cellar (Fig. 15–2) can also be used to provide space for the BOP stack.

The *choke manifold* is a series of lines, automatic valves, gauges, and chokes on the ground next to the drilling rig. It is connected to the blowout preventer stack outlet by a *choke line*. The choke manifold is used to direct flow from the well to the reserve pit, burning pit, mud tank, or mud conditioning equipment. It can be used to relieve pressure buildup in a well after the BOP stack has been thrown and to circulate heavier drilling mud into the well through a *kill line*. It is operated from a control panel on the drill floor.

Drilling Operations

Operating a drilling rig is very expensive. Except for small, shallow rigs, drilling occurs 24 hours a day. Three 8-hour shifts of workers usually operate the rig. Each shift is called a *tour* (pronounced *tower*). The *graveyard* or *morning tour* is from midnight to 8 a.m. The *day tour* is from 8 a.m. to 4 p.m. The *evening tour* is from 4 p.m. to midnight. In remote areas and on offshore rigs, two 12-hour tours are used each day.

The *drilling contractor* is the company that owns and operates the rig. The *tool pusher* is a drilling company employee who is similar in authority to the captain of a ship. The tool pusher supervises the drilling operations and usually lives at the drill site 24 hours a day. Each morning the tool pusher compiles the results of the past 24 hours of drilling into a *daily drilling* or *morning report*. The report is transmitted back to the *drilling superintendent* at the drilling contractor's office and to the operator. The report includes the depth of the well at 6 a.m., footage drilled during the last 24 hours, rig time spent on different activities, supplies used, and other drilling and geological data. If the drilling contract specifies it, the tool pusher can make sure the drilling supplies are ordered and delivered to the rig on time.

The *operator* is the company that organizes and finances the drilling, selects the drill site and contracts with the drilling contractor to drill the well. The operator has a *company representative* or *company man* at the drill site who works with the tool pusher to make sure the well is being drilled to specifications. The company man can also be responsible for the daily drilling report. If the drilling contract specifies it, an employee of the operator called the *materials man* can be responsible for calculating the amount and ordering and supervising the timely delivery of supplies to the rig.

A *driller* is in charge of each tour. The driller operates the drilling machinery from a drilling console on the derrick floor and gives orders to the crew (Plate 16–8). The *drilling console* contains instruments and controls for the rotary table, drawworks, mud pumps and chain drives. Dials on the console include a weight indicator, mud pump pressure gauge, rotary tachometer, rotary torque indicator, pump stroke indicator, and rate of penetration recorder. The *weight indicator* shows the weight of the drillstring suspended from the hook on the bottom of the traveling block. Too much weight could cause the derrick to collapse. The *rotary tachometer* shows the rotary table speed. The *rotary torque indicator* measures stress on the drillpipe, and is used to prevent pipe twist-off. The *rate of penetration indicator* shows how fast the drill is penetrating the rocks, usually recorded in

Plate 16-8   *Driller and drilling console (courtesy American Petroleum Institute)*

minutes per foot or meter. The driller stands in front of the drilling console with one hand on the drawworks *brake*, a control lever used to raise and lower equipment and apply weight to the bit. After each tour, the driller fills out a *tour report* describing the activities during that tour. Tour reports are used to make the daily drilling reports.

The *derrickman* is second in command of each tour. The derrickman can also monitor the drilling mud and circulating equipment. On the derrick floor, there are two to four *roughnecks* or *rotary helpers*, depending on the size of the rig. They handle and maintain the drilling equipment. On a large rig, one person called the *motorman* is responsible for the prime movers.

seventeen

DRILLING PROBLEMS

Risk

There are two outcomes after drilling and testing a well. The well could be a *dry hole* that did not encounter commercial amount of gas and oil and is plugged and abandoned. The well could also encounter commercial amounts of gas and oil and be completed as a *producer*. *Risk* or *success rate* of drilling is defined as the number of successful wells completed as producers divided by the total number of wells drilled. It is expressed as a percentage or decimal. Another way of expressing risk is *success ratio*, the number of wells drilled divided by those completed as producers and then multiplying by 100.

The American Petroleum Institute reported that the success rate for new field wildcats during 2000 in the United States was 33% and for all types of exploratory wells was 39%. Nine percent of the all developmental wells drilled in the United States in 1999 were plugged and abandoned because of engineering (mechanical) problems.

Subsurface Conditions

Both temperature and pressure increase with depth. The temperature increase is called the *geothermal gradient*. It averages 2° F/100 ft (3.6°C/100 m) for the earth but varies between 0.5° to 5°F/100 ft (1° to 9°C/100 m). In a well, the geothermal gradient can be measured by a *temperature bomb* that is lowered to the bottom of the well to record the temperature or by a *temperature log* that continuously records temperatures as it is being raised in the well. An average geothermal gradient for a sedimentary basin is about 1.4°F/100 ft (2.5°C/100 m).

The average surface temperature in Oklahoma is 55°F (13°C). Oil in a reservoir at 10,000 ft (3000 m) in the Anadarko basin of Oklahoma should have a temperature of about 195°F or 90°C (Fig. 17–1). When oil and gas come up a well, they are hot.

There are two separate pressures in the subsurface. The pressure on the rock is called *earth* or *lithostatic pressure*. It increases at an average rate of 100 psi/100 ft (23 kg/cm^2/100 m). The pressure on the fluids in the pores of the rock is *reservoir, formation* or *fluid pressure*. It depends on the density of the overlying water but averages 45 psi/100 ft (10 kg/cm^2/100 m). The pressure on oil in reservoir at 10,000 ft (3000 m) should be about 4500 psi (316 kg/cm^2/100 m) (Fig. 17–2).

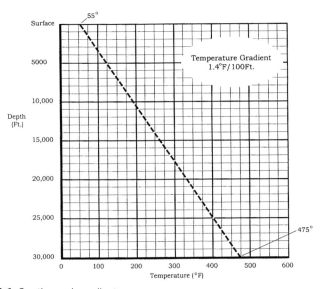

Fig. 17-1 *Geothermal gradient*

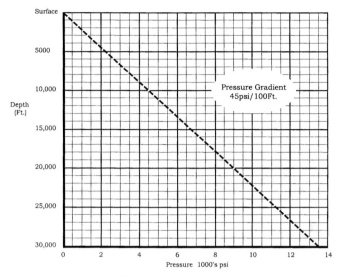

Fig. 17-2 Fluid pressure gradient

The *normal* or *hydrostatic pressure* on the fluids in the rock is caused by the weight of the overlying water in the pores of the rocks. During drilling, the pressure of the drilling mud is usually slightly higher (overbalance) than hydrostatic pressure to keep the fluids back in the rocks.

Problems While Drilling

Fishing

A common drilling problem is that something breaks in or falls down the well during drilling. For example, the drillstring twists off and falls to the bottom. A cone can break off the tricone bit or a tool such as a pipe wrench falls from the rig floor into the well. This is called a *fish* or *junk* and cannot be drilled with a normal drill bit. Drilling is suspended and special tools called *fishing tools* are leased from a service company to grapple for the fish in a process called *fishing*.

To retrieve pipe in a well, either a spear, or overshoot is used. It is screwed into the bottom of a *fishing string* composed of drillpipe and run into the well. The *spear* is designed to fit into and grip the inside of the pipe (Fig. 17–3a) whereas the *overshoot* fits around and grips the outside of the pipe (Fig. 17–3b). A *washover pipe* or *washpipe* consists of a section of large diameter pipe (casing) with a cutting edge (Fig. 17–3c). The cutting edge grinds (*dresses*) the surface of a fish smooth. Drilling mud is pumped through the washover pipe to clear debris from around the fish to prepare it for another fishing tool.

A *tapered mill reamer* (Fig. 17 –3d) is rotated to open collapsed casing and mill irregular-shaped fish. A *junk mill* (Fig. 17–3e) is run on a fishing string and rotated to grind a fish into small pieces. The pieces can be then retrieved by a *junk* or *boot basket* (Fig. 17–3f) that is run on a fishing string to just above the bottom of the well. Drilling mud is then pumped down either down the center or along the outside of the fishing string. Turbulence picks up the metal pieces and they fall into a container or basket. A *wireline spear* (Fig. 17–3g) uses barbs to hook a broken wireline. Both *permanent* and *electric magnets* (Fig. 17–3h) can retrieve magnetic fish. An *impression block* (Fig. 17–3i) is used to

| (a) | (b) | (c) | (d) | (e) | (f) | (g) | (h) | (i) |

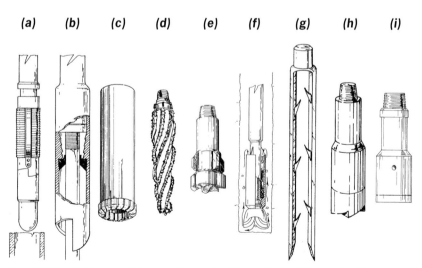

Fig. 17-3   *Fishing tools (a) spear (b) overshoot (c) washpipe (d) tapered mill reamer (e) junk mill (f) boot basket (g) wireline spear (h) fishing magnet (i) impression block*

determine the nature of a fish in the well to select the correct fishing tool. It is a weight with soft lead or wax on the bottom. The tool is run in the well on a wireline or tubing string to give an impression of the fish.

A *jar* is often used in the fishing string above the tool. The jar is a section of pipe that either mechanically or hydraulically imparts a sharp upward or downward jolt to the tool on command (Fig. 17–4). Explosives can be used to blow up the junk. The pieces are then retrieved with a magnet or junk basket.

Fishing takes time (often days) while the drilling rig is not drilling. The operator, however, is still paying for the rig during fishing. Many drilling contractors sell *fishing insurance*. For a fee, the operator is not financially responsible for fishing operations.

Fig. 17-4 *Jar*

Stuck Pipe

The drillstring can become stuck in a well due to either mechanical problems or differential wall pipe sticking. This is called *stuck pipe*. During *differential wall pipe sticking*, the drillpipe adheres to the well walls due to suction. The driller first tries to free the pipe by sudden jarring. The impact can be provided by a jar in the drillstring. A lubricant called a *spotting fluid*, often a mixture of diesel or mineral oil and a surfactant, can be applied along the well walls. The drilling mud can also be made lighter to decrease the suction.

Mechanical pipe sticking is often caused by a dogleg in the well. A *dogleg* is any deviation in the well greater than 3° per 100 ft. (30 m). Doglegs are caused by dipping hard rock layers or a change in the weight on the bit during drilling. A dogleg can result in *keyseating*, the formation of a wellbore cross section in the form of a key hole (Fig. 17–5). It is caused by the drillpipe abrading a groove in the side of the well that is smaller that the hole drilled by the bit. Larger diameter drill collars cannot pass through the keyseat. The well has to be enlarged by reamers.

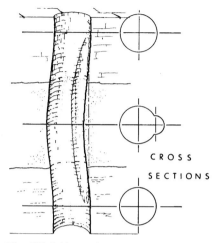

CROSS
SECTIONS

Fig. 17-5 Keyseat

Ledges are hard rock layers that ring the wellbore and can cause pipe sticking. They are formed when drilling through alternating layers of hard and soft rocks. The soft rock washes out above and below the hard rock layers to form ledges.

A *stuck-point indicator tool* or *stuck-pipe log* can be run to determine exactly where the pipe is stuck. A *back off operation* is performed as a last resort. The stuck pipe can be cut with either a string shot or a chemical cutter. A *string shot* uses an explosive cord that is detonated one joint above the stuck point while the pipe is being unscrewed. A *chemical cutter* (Fig. 17–6) is run on a wireline and activated by an electrical signal. It uses a chemical propellant, a hot, corrosive fluid that jets out of the cutter under high pressure to slice through the pipe. After the pipe is cut, a washpipe (Fig. 17–3c) is then run on a fishing string to wash around the stuck pipe and detach it from the well wall.

Wall sticking can be prevented by using *spiral-grooved drill collars* (Fig. 17–7). The three grooves, located 120° apart, decrease the area of pipe in contact with the well walls but have little effect on the weight and strength of the pipe.

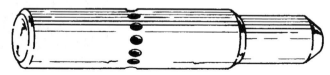

Fig. 17-6 Chemical cutter

Fig. 17-7 *Spiral-grooved drill collar*

Sloughing Shale

Sloughing shale is soft shale along the wellbore that adsorbs water from the drilling mud. It expands out into the well and falls to the bottom of the well in large balls that are not easily removed by the drilling mud. Chemicals such as potassium salts added to the drilling mud or oil-base drilling muds are used to inhibit sloughing shales.

Lost Circulation

If a very porous, cavernous or highly fractured zone is encountered while drilling, an excessive amount of drilling mud is lost to that zone during *lost circulation*. The zone is called a *thief* or *lost circulation zone*. A pill of *lost circulation additive* or *control agent* can be mixed with the drilling mud and pumped down the well to clog up the lost circulation zone.

Lost circulation additives are fibers, flakes, granular masses, or mixtures. They include ground pecan hulls, redwood and cedar shavings, hay, pig hairs, shredded leather, mica flakes, laminated plastic, cellophane, sugar cane hulls, ground pecan nut hulls, ground coal, ground tires, and asbestos. After the lost circulation zone has been drilled, it can be isolated by running and cementing a string of protection casing into the well.

Formation Damage

When drilling a well with overbalance, part of the drilling mud liquid with some fines called *mud filtrate* is forced into any permeable rock adjacent to the wellbore. The mud filtrate can decrease or destroy the permeability of a reservoir rock near the wellbore (*formation damage* or *skin damage*). Formation damage in a well can be treated by well stimulation such as acidizing (wash job) or hydraulic fracturing.

Formation damage can be prevented by drilling the formation using *brine* (very salty water) or an oil-base, emulsion or synthetic-base drilling mud. It can also be avoided by drilling with a light-weight drilling mud that exerts less pressure than formation pressure (*underbalance drilling*). The well can be drilled faster using underbalanced drilling but it will not prevent fluids from flowing out of the rocks and into the well. To maintain pressure control during underbalanced drilling, a *rotating control head* is used on the rotating table. It has a rotating inner seal assembly that fits around the kelly in a stationary outer housing. Underbalanced drilling is usually done only during part of the entire drilling operation. The well has to be killed by filling with heavier drilling mud before tripping out when drilling with underbalance.

Corrosive Gasses

In some areas, *corrosive gasses* such as carbon dioxide (CO_2) and hydrogen sulfide (H_2S) can flow out of the rocks and into the well as it is being drilled. These gasses can weaken the steel drillstring and cause *hydrogen sulfide embrittlement*. To prevent corrosion, a drillstring made of more resistant and expensive steel can be used, and chemicals can be added to the drilling mud.

Abnormal High Pressure

Unexpected abnormal high pressure in the subsurface can cause a *blowout*, an uncontrolled flow of fluids up the well. Natural gas flowing out the well can catch fire (Plate 17–1), causing the loss of the drilling rig. *Abnormal high pressure* is fluid pressure that is higher than expected hydrostatic pressure for that depth (Fig. 17–2). The drilling mud pressure may not be able to contain the formation fluids. Fluids flow out of the subsurface rocks into the well in what is called a *kick*. As the water, gas, or oil flows into the well, it mixes with the drilling mud, causing it to become even lighter and exert less pressure on the bottom of the well. The diluted drilling mud is called *gas cut*, *water cut*, and *oil cut*.

A kick and possible blowout can be detected by several different methods during drilling. As subsurface fluids flow into the well during a kick, more fluids will be flowing out of the top of the well than are being pumped

Plate 17-1 *Blowout on an offshore drilling rig (courtesy American Petroleum Institute)*

into the well. The sudden increase of fluid flow out of the well and the rise of fluid level in the mud pit are detected by an instrument called a pit-volume totalizer. The *pit-volume totalizer* uses floats in the mud tanks to continuously monitor and record drilling mud volume. It sounds an alarm if the mud volume decreases due to lost circulation or increases due to a kick.

The drilling mud can also be continuously monitored for sudden changes in weight, temperature, or electrical resistivity that would indicate the mud is being cut by subsurface fluids. Another detection method is

based on the principle that shale should become more dense and less porous
with depth as it is compacted. The density and porosity of shale can be
determined from both well cuttings and well logs. If the shale density
increase and porosity decrease are less than predicted from computations
based on normal conditions, abnormally high pressures can be expected.
The drilling rate of penetration also increases in undercompacted shales.

Abnormal high pressures often occur in isolated reservoirs of limited
extent. As a reservoir of large, aerial extent is buried in the subsurface, it reacts
to increased overburden pressure by compacting. The reservoir compacts by
decreasing porosity and squeezing some of the fluids out of the pore spaces.
This maintains normal, hydrostatic pressure. If the reservoir is isolated and
limited in extent, such as encased in shale or cut by sealing faults, it cannot
compact because fluids cannot be expelled from the reservoir. The pressure on
the overlying rocks is then transferred to the pressure on the fluids. Abnormal
high pressure can be more than twice hydrostatic pressure (Fig. 17–8).

Fluids cannot flow uncontrolled up the center of the drillstring during
a kick because of the kelly cock (Fig. 16–20). Where fluids can flow up the
well is along the outside of the drillstring in the space called the annulus.

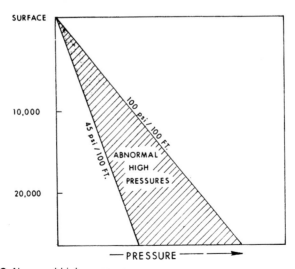

Fig. 17-8 *Abnormal high pressure*

After a kick is detected, the well is *killed* (flow from the well is stopped) by throwing the blowout preventers. Heavier drilling mud (*kill mud*) is pumped into the well through the choke manifold, kill line, and a valve in the blowout preventer stack (*kill connection*) to circulate the kick out of the well.

Two methods to control the kick are used, depending on whether the heavy (kill) mud is ready to be circulated. The kill mud is stored next to the rig. If it is already mixed and ready to use, the kick is controlled by the driller's method. If the kill mud is in dry sacks and needs to be mixed, the wait-and-weigh method is used.

In the *wait-and-weigh method*, the well is shut in as the kill mud is being prepared. The mud pumps are started, and using a slower pumping speed, the original mud and kick fluids are replaced with the heavier kill mud during one circulation. In the *driller's method*, the first circulation by the mud pumps is used to replace the original drilling mud under high pressure that was cut by kick fluids with original drilling mud without the kick fluids. The second circulation replaces the undiluted, original mud with kill mud. The abnormal high-pressure zone is then drilled, and protection casing is run and cemented into the well to isolate the zone.

About 50% of the blowouts occur during tripping out. The drillstring displaces a volume of drilling mud in the well. As the drillstring is raised, the level of drilling mud falls in the well, and the pressure is decreased on the bottom of the well. If the level of the drilling mud is not maintained in the well, overbalance is lost, and a kick could occur. A *trip tank* (a 10- to 40-barrel-volume steel tank that holds drilling mud on the drill floor) is used to keep the well filled with mud during tripping out. Also, if the drillstring is pulled from the well too fast, it could suck gas out of the formation to start the kick. *Blowout-preventer tests* are periodically run on drilling rigs to test the equipment and crew response.

eighteen

DRILLING TECHNIQUES

Straight Hole

Drilling contracts often have a clause stipulating that the well deviates no more than 3° per 100 ft (30 m) and is contained within a cone with a maximum angle of 5° (Fig. 18–1). This is called a *straight hole*. Because the bit is being turned clockwise (to the right), the wellbore tends to *walk to the right* in a clockwise corkscrew pattern as it goes down.

When rotary drilling rigs were originally introduced, the drillers often could not drill the well straight down because of dipping beds of hard rocks such as limestone (Fig. 18–2). If the bit hits a subsurface rock layer with a dip greater than 45°, the bit tends to be deflected down dip. If the hard rock layer dips less than 45°, the bit tends to be deflected updip.

A well with an excessive angle in it that has not been drilled on purpose that way is called a *crooked hole*. An area that has dipping rock layers that cause crooked holes is called *crooked hole country*. A *slick bottomhole assembly* that has no stabilizers can be used to attempt to drill a straight hole.

Directional Drilling

Modern rotary rigs can be controlled so that the well is drilled out at a predetermined angle during *directional* or *deviation drilling* and ends up in a predetermined location called the *target*. The angle in which the well goes out

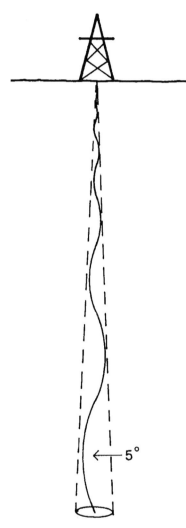

Fig. 18-1 *A straight hole*

from vertical is called the *deviation.* Two types of deviated wells are *straight-in* (Fig. 18–3a) and *S-shaped wells* (Fig. 18–3b).

Starting a straight well out at an angle is called *kicking off the well.* If the well has been cased, a hole, called a *window,* is cut in the casing with a casing mill to kick off the well. The first device used to kick off a well was a *whipstock* (Fig. 18–4), a long, steel wedge designed to bend the drillstring. The whipstock is run into the well on a drillstring and oriented by survey instruments. Weight is then applied to the drillstring to break a shear pin and separate the whipstock from the drillstring that is then pulled out. A small diameter bit is then run in the hole on a drillstring to drill a *pilot hole* out for 10 to 15 ft (3 to 5 m). The pilot hole is then surveyed. If it has the right orientation, the hole is then enlarged with a normal bit.

In relatively soft sediments, a jet bit can be used to kick off the well. A *jet bit* is a tricone bit that has one large and two small nozzles. It is run into the well and then oriented with a surveying instrument. If the orientation is correct, mud is circulated at maximum possible flow rate without rotating the drillstring. The hydraulic action of mud jetting out of the large nozzle erodes the well out at that angle. The drillstring is then pulled and the pilot hole is surveyed. If it is orientated right, the pilot hole is then drilled out with a normal tricone bit.

A modern method used to kick off a deviated well is to run a downhole assembly with a bent sub, a downhole mud motor and a diamond bit (Fig.

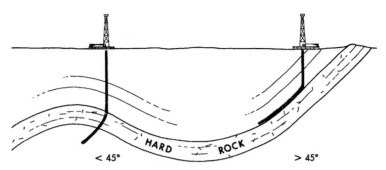

Fig. 18-2 *Cause of crooked holes*

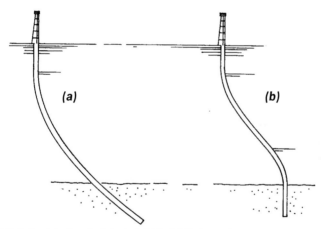

Fig. 18-3 *Deviated wells (a) straight in (b) S-shaped*

18–5). A *bent sub* is a short section of pipe with an angle of ½ to 2½° in it. Drilling mud flowing down the drillstring drives the *turbine* or *downhole mud motor*. The mud strikes either a spiral shaft or blades in a turbine motor to activate the motor. A diamond bit is usually used with a downhole mud motor.

The assembly is lowered into the well and oriented in the right direction by survey instruments. The drillstring remains stationary as the mud motor

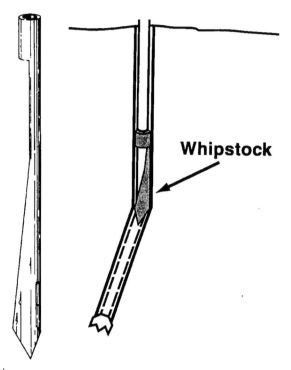

Fig. 18-4 Whipstock

is activated to drill a pilot hole. The drillstring is then pulled, and the well is surveyed. If it is oriented right, the well is then drilled out with a normal bit.

After the pilot hole has been drilled out, the deviated well can then be drilled out straight (*maintain angle*) or drilled to increase the well deviation (*build angle*) or decrease the well deviation (*drop angle*). A *steerable downhole assembly* (Fig. 18–6) is a combination of stabilizers, bent subs, downhole turbine motor, and diamond bit that can maintain, drop, or build angle. To maintain angle, the assembly is rotated similarly to normal rotary drilling in the *rotating mode* (Fig. 18–7a). To build or drop angle, the assembly is oriented in the right direction and not rotated. The downhole turbine motor is activated to drill the well out in the direction the assembly is pointing in the *sliding mode* (Fig. 18–7b) as the assembly slides along the bottom of the wellbore. The well is drilled slower in the sliding mode than in the rotating

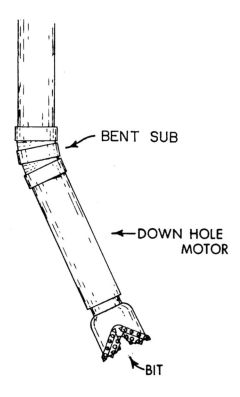

Fig. 18-5 *Downhole assembly used to kick off a deviated well. The angle on the bent sub is exaggerated.*

mode. Some steerable downhole assemblies have *adjustable bent subs* in which the angle in the bent sub can be adjusted from the surface as the assembly is in the well.

Some wells have been "accidentally" drilled to drain oil out from under adjacent leases in what is called *subsurface trespass.* Deviation drilling, however, is commonly done for legitimate reasons (Fig. 18–8).

Drilling offshore is considerably more expensive than drilling on land. An oil field in very shallow waters can often be more economically developed by deviation drilling from the beach. Offshore drilling and production plat-

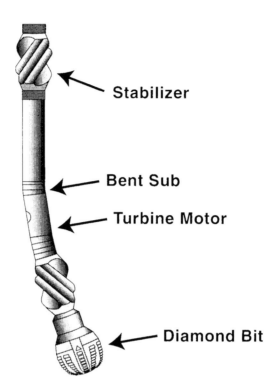

Stabilizer

Bent Sub

Turbine Motor

Diamond Bit

Fig. 18-6 *Steerable downhole assembly*

forms are very expensive. A deep-water offshore petroleum field is best developed using a large production platform with numerous, deviated wells that radiate out to the sides. Cognac, a production platform off the Mississippi River delta, has 62 deviated wells.

If a well is on fire, the *wild well* can be brought under control by two methods. The fire is first extinguished, often by detonating an explosive on top of the well to remove oxygen. Then, a valve can be lowered and attached to the wellhead and closed. Another method is to drill a *relief well* at a safe distance from the wild well. The relief well does not have to intersect the wild well in the subsurface but just come close. It drills into the abnormal high pressure zone that causes the blowout, and the pressure is relieved by producing the gas. After the pressure has been reduced, heavy drilling mud

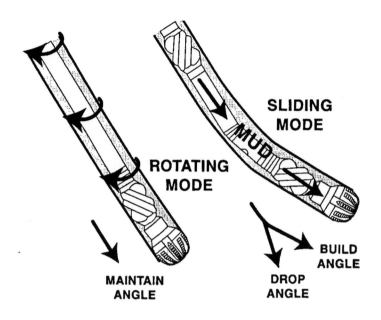

ROTATING MODE

SLIDING MODE

MUD

MAINTAIN ANGLE

DROP ANGLE

BUILD ANGLE

Fig. 18-7 *(a) rotating mode (b) sliding mode*

(*kill mud*) is then pumped from the relief well through subsurface rocks and into the wild well to control it.

If something breaks off or falls down the well and cannot be removed by fishing, the well can be drilled around the fish (*sidetracked*). A deviated well can be drilled to test several potential petroleum reservoirs rather than drill several straight holes to test each reservoir. Deviation drilling is also used to overcome a poor drilling location.

Directional surveys that measure both angle (*deviation*) and compass orientation (*azimuth*) were originally run on deviated wells when they were first being drilled (Fig. 18–9). The surveys were either single (one measurement) or multi-shot. The survey instrument contained a magnetic compass or gyroscope. The magnetic instrument was run when the downhole assembly was

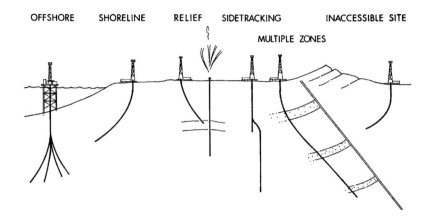

Fig. 18–8 Deviated well uses (modified from Beebe, 1961)

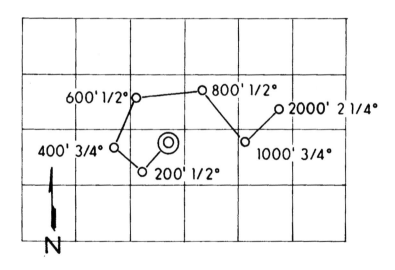

Fig. 18–9 Directional survey showing depth, deviation, and location of the wellbore

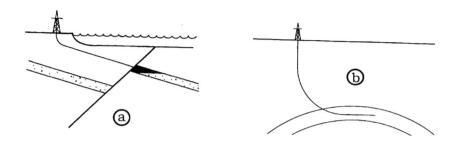

Fig. 18–10 *(a) extended reach well (b) horizontal drain well*

made of non-magnetic drill collars called *K-Monel.* The gyroscope instrument was used with magnetic drill collars or when there was magnetic iron such as casing in the area. Today a directional log is continuously made as the well is being drilled by a system called measurements-while-drilling.

An *extended reach* well is a deviated well that has one bend in the well and bottoms out several thousand feet horizontally from its surface location (Fig. 18–10a). The world record for the *horizontal reach* on an extended reach well is more than 6 miles (10 kms).

A *horizontal drain* well is a deviated well that is drilled along the pay zone parallel to the reservoir (Fig. 18–10b). The horizontal part of the well is called the *horizontal section.* A horizontal drain well contains both a geometric and geosteering section. The top (*geometric section*) is drilled to a pre-set plan as a normal straight hole. The non-vertical *geosteering section* uses real time logs (logging measurements-while-drilling) to show the driller where the drill bit is in relation to the top or bottom of the target formation. The driller then adjusts the steerable downhole assembly to continue drilling to and through the target.

Horizontal drain wells are described by their *build angle.* It is the rate of change in degrees per unit length as the well goes from vertical to horizontal such as 8°/100 ft. The wells are classified as *short radius, medium radius,* and *long radius horizontal wells,* depending on how sharp the build angle is.

Generally, a horizontal drain hole will ultimately produce three to five times more oil or 5 to 20 times more gas than a straight hole and produce it

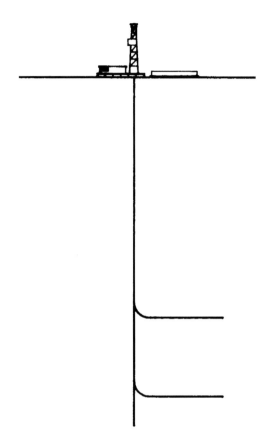

Fig. 18-11 Laterals

at higher rates. Horizontal drain wells have been most successful in drilling fractured reservoirs such as the Austin Chalk in Texas because most fractures that drain that reservoir tend to be vertical. A horizontal drain well should penetrate considerably more vertical fractures than a vertical, straight well.

Horizontal drain holes are also used in low permeability (tight) formations to increase ultimate recovery from the reservoir. They are also used to prevent coning that produces excessive gas or water from above or below the oil reservoir. Horizontal drain wells are not much more expensive to drill than a comparable, vertical well but are more difficult to log and complete.

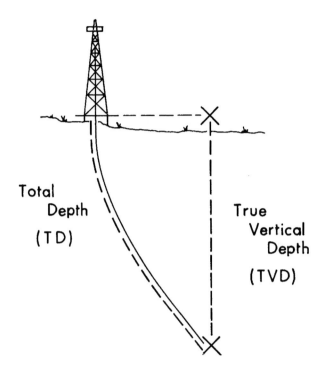

Fig. 18-12 *Depth of a Well*

Laterals (Fig. 18–11) are short, horizontal branches drilled from a well.

The depth of a well can be measured two ways (Fig. 18–12). *Total* (*TD*), *measured, logged,* or *driller's depth* is measured along the length of the well-bore. *True vertical depth* (*TVD*) is measured straight down and is less than total depth.

Air and Foam Drilling

Air drilling uses a rotary drilling rig with a skid-mounted or trailer-mounted compressor that circulates air instead of drilling mud (Plate 18–1). The rig and operations are almost exactly the same as a rotary drilling rig except without the circulating drilling mud system. Air is pumped down the drillstring and out the bit. It picks up well cuttings on the bottom of the well and returns up the annulus to blow out a *blooie line* into a pit adjacent to the rig. A *rotating head* is used on the top of the well to allow the kelly to rotate through it while maintaining a pressure seal around the kelly.

Air drilling has a faster bit penetration rate than using drilling mud and helps avoid formation damage. It, however, does not build up a filter cake along the well walls to stabilize the well. This may allow the sides to sluff in. Air drilling also cannot control formation fluids that flow into the well. Natural gas flowing into the well can cause a flammable mixture with the air.

Foam drilling is similar to air drilling but uses detergents in the air to form foam that better lifts water from the well. Soap and water are mixed and injected by a small pump into the air circulating into the well. The foam also does a better job of lifting the well cuttings than air.

Plate 18–1 Truck-mounted air drilling rig. Note trailer with air compressor

nineteen

TESTING A WELL

Completing a well usually costs more money than drilling a well. Because of this, a well must be accurately tested after it has been drilled. Will the well produce enough gas and/or oil to make it worthwhile to complete the well? Old pictures and movies show gushers. Those, however, were on cable tool drilling rigs and don't occur today. Well testing is now based on *well logs*, record of the rocks, and their fluids in the well.

Sample or Lithologic Log

A *sample* or *lithologic log*, recorded on a long strip of thick paper, is a physical description of the rocks through which the well was drilled. At the top of the sample log will be a *header* (Fig. 19–1) with information about the well such as the operator, well name, and location. A *depth strip*, usually along the left edge, lists depths in the well. In the next column, geological symbols for various rocks are used to identify the composition of each rock drilled (Fig. 9–7). The rock symbols can be colored with yellow for sandstone, blue for limestone, and gray for shale. Next to the symbols, the rock is described. Rock texture, color, grain size, sorting, cementation, porosity, oil staining, and microfossil content can be noted.

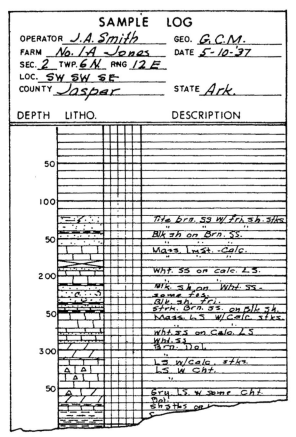

Fig. 19–1 Sample log

The source of information for the sample log is primarily from well cuttings, the small rock chips that are made by the drill bit and flushed up the well by the drilling mud. The coarser cuttings are caught on the shale shaker screen and sampled at regular intervals by the *wellsite geologists* (who is at the drilling rig 24 hours a day), the mud logger, or the driller. The well cuttings are typically sampled as a *composite sample* over each 10 ft. (3 m) of depth. The sample interval is shortened as the reservoir rock is drilled. The well cuttings are washed to remove the drilling mud and stored in cloth or paper bags. The cuttings are examined under a binocular microscope. If oil stains are present in the cuttings, they can be verified with ultraviolet light. Different °API gravity oils fluoresce with different colors.

The cuttings take time (*lag time*) to circulate from the bottom of the well to the shaker screens. Lag time is estimated by timing tracers such as corn and rice that are inserted into the circulating mud. Engineering calculations using the flow from the mud pumps, and the volume of the well can also be used to estimate lag time. As a rule of thumb, in an 8-in (20-cm) hole, it takes about 10 minutes for mud to circulate each 1000 ft. (300 m).

A *core*, a cylinder of rock drilled from the well, is the most accurate source of information about the reservoir. To take a core, drilling must stop at the top of the subsurface interval to be cored. The drillstring is then run from the hole, and the drill bit is replaced with a rotary coring bit. The drillstring is then run back into the hole.

The most common *rotary core bits* are solid metal with either industrial diamonds or tungsten steel inserts for cutting. The rotary coring bit cuts in a circle (Fig. 19–2) around the outside of the well and is hollow. The core is left to pass up into the hollow core bit and pipe. The core is retained in the *core barrel* located above the bit (Fig. 19–3). The core barrel consists of an inner and outer barrel separated by ball bearings. This allows the inner barrel to remain stationary to receive the core while the outer barrel is rotated by the drillstring to cut the core. The inner barrel has a core catcher with flexible fingers pointing upward and a check valve to retain the core.

During coring, drilling mud is circulated down the space between the inner and outer core barrel. After the core has been cut, the drillstring is raised to the surface and the rotary cone bit, core barrel and core removed. The cores are usually stored in cardboard boxes. The normal drill bit is then reattached, the drillstring is run into the well, and drilling is resumed.

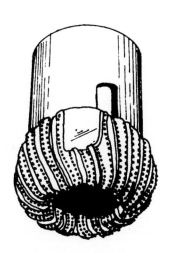

Fig. 19–2 *Rotary coring bit*

(a)   **(b)**

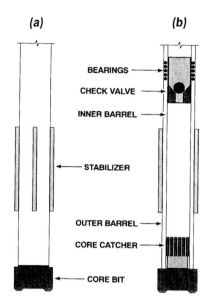

BEARINGS

CHECK VALVE

INNER BARREL

STABILIZER

OUTER BARREL

CORE CATCHER

CORE BIT

Fig. 19–3 *Rotary coring barrel (a) outside (b) cross section*

Because of the extra rig time involved, cores are expensive. The deeper the well is, the more expensive the core. This is why usually only the reservoir interval is cored.

A *full-diameter core* ranges in diameter from 1¾ to 5¼ in (4½ to 13½ cm) and can be up to 400 ft (120 m) long but is commonly 20 to 90 ft (6 to 27½ m) long. Subsurface rocks that are highly fractured, very porous, or unconsolidated are not usually retained in the core barrel. Loss of core can indicate good reservoir rock.

Oriented cores are made with reference to geographic or magnetic north by cutting a groove along the length of the core as it is being drilled. A *native state core* is enclosed in a rubber sleeve as it is being drilled to retain all the fluids in the core under reservoir conditions.

A faster and less expensive way to take samples is *sidewall coring*. After the well has been drilled, a sidewall coring device is lowered down the well until it is adjacent to the sample interval (Fig.19–4). A *percussion sidewall coring tool* commonly has 30 small core tubes called *bullets* with explosive charges behind them. These are detonated, and the bullets are shot into the sidewall to take samples. The bullets are attached to the tool by wires so that the bullets and their samples are brought to the surface with the sidewall coring tool. The percussion method can disturb the grains and alter the porosity and permeability of the sample.

During *rotary sidewall coring*, the tube is drilled into the well wall to minimize sample alteration. The rotary sidewall coring tool is lowered into position in the well, and a small bit pivots out to drill the sample. The sample then falls into the tool. The tool can be moved to take another sample. The samples are kept separate by discs in the tool. A limitation of *sidewall cores* is that they are very small, 1 in. (2½ cm) in diameter and 1¾ in. (4½ cm) long.

Many older sample logs were made by drillers with little or no training in geology. These are called *driller's logs* and display a considerable range of accuracy and usefulness. Drillers could usually distinguish between limestone, sandstone, and shale. Unfortunately, they would often use ambiguous terms such as "gumbo."

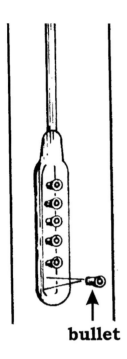

bullet

Fig. 19–4 *Percussion sidewall coring tool*

Drilling Time Log

A *drilling-time log* is a record of the rate of drill bit penetration through the rocks (*rate of penetration* or *ROP*). It is recorded in minutes per foot or meter drilled on a paper chart around a drum in a *geolograph* or *drilling recorder* on the drill floor. Because this log is recorded as the well is drilled, the drilling-time log is a real time log. Changes in the subsurface rocks are recorded instantaneously.

The rate of bit penetration depends on both drilling parameters and rock properties. Revolutions per minute, weight on bit, and bit type affect the drilling rate. If the drilling parameters are relatively constant, then rock properties will cause the dominant variations in drilling rate. With a tricone drill bit, sandstones have the fastest drilling rate, shales are intermediate, and carbonates are slowest (Fig. 19–5).

A sudden change in the drilling rate is called a *drilling break*. It occurs when the bit penetrates the top of a different rock layer. Drilling breaks on a drilling-time log are used to accurately determine the top and bottom elevations of subsurface formations. Because porous zones are less dense and easier to drill, drilling breaks can also be used to locate porous zones in a dense rock (Fig. 19–5).

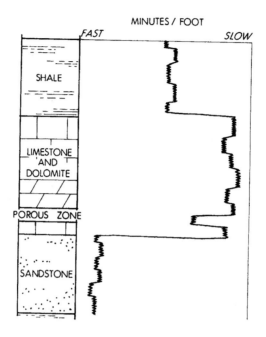

Fig. 19–5 *Drilling-time log*

Mud Log

A *mud log* is a chemical analysis of the drilling mud and well cuttings for traces of subsurface gas and oil as the well is being drilled. It is made by a service company in a *mud logging trailer* at the wellsite (Plate 19–1). The purpose of the mud log is to identify oil and/or gas bearing rocks in the subsurface. Drilling mud circulating out of the well is sampled from a gas trap in the shale shaker along with well cuttings from the shaker screens.

A mud log (Fig. 19–6) will have a header at the top with operator, well name, location, elevation, and other information. A depth strip, showing depth in the well, will run down near the middle of the log. To the left of the depth strip can be a rate of penetration and a sample log. To the right of the depth strip will be the mud log showing amounts of gas and oil detected in the drilling mud and well cuttings.

Plate 19–1 *Mud logging trailer*

Oil or gas in the drilling mud is called *oil-cut* and *gas-cut mud*. Any oil or gas above the normal expected background is called a *show*. A more detailed chemical analysis (*show evaluation*) can be made of a show. This includes a gas analysis showing the percentages of methane (C_1), ethane (C_2), propane (C_3), butane (C_4), and pentane (C_5).

Wireline Well Logs

A well log made by running an instrument down a well on a wireline is called a *wireline well log*. They were invented in France in the mid-1920s and were first applied to the oil fields in California and Oklahoma in 1930. The first log was a resistivity log that was hand-recorded, point by point, on a strip of paper. The wireline well logs helped rotary drilling become accepted. Because the early logs could be run in wells with filter cakes along their wellbore and still filled with drilling mud, they could be used to evaluate wells drilled with a rotary drilling rig.

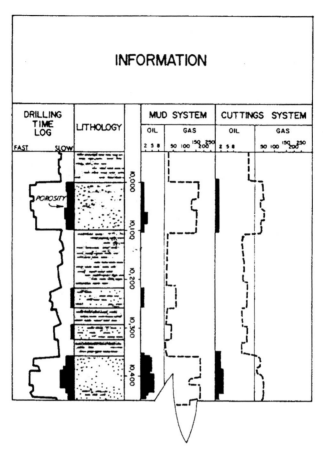

Fig. 19–6 Mud log

To make a wireline well log after the well is drilled, the hole is first cleaned by circulating drilling mud and then the drilling equipment is pulled from the well. A sonde is lowered down the well (which is still filled with drilling mud) on a logging cable (Fig. 19–7). The *logging cable* is an armored cable with steel cables surrounding conductor cables in insulation (Fig. 19–8a). It is reeled out from a drum in the back of a logging truck (Plate 19–2).

The *sonde* or tool is a cylinder (Fig. 19–8), commonly 27 to 60 ft (8 to 19 m) long and sometimes up to 90 ft (27½ m) long that is 3 to 4 in. (8 to

Plate 19–2 *Logging truck (courtesy Halliburton)*

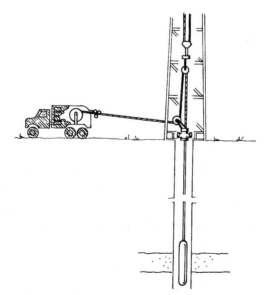

Fig. 19–7 *Making a wireline well log*

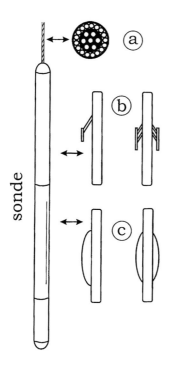

Fig. 19–8 *Sonde on wireline a)
cross section of the armored cable,
b) sondes with arm(s), and c) sondes
with bow spring(s)*

10 cm) in diameter and is filled with instruments. Several instrument packages such as formation density, neutron porosity and gamma ray can be screwed together to form the sonde. The sonde has either one expandable arm or bow spring the puts the sensors in contact with the well walls or three expandable arms or bow springs that centers the sonde in the well (Fig. 19–8b and 19–8c). As the sonde is run back up the well, it remotely senses the electrical, acoustical, and/or radioactive properties of the rocks and their fluids and sometimes the geometry of the wellbore. In a directional well with a high deviation or a horizontal drain hole, the sonde must be pushed down the well with tubing or the drillstring. One trip down and up with a sonde is called a *run.*

The data from the sonde are transmitted up the cable to instruments in the logging truck where they are recorded on a strip of paper called the *field print.* The data are also processed later, and a cleaner log (*final print*) is made. The logging data are digitized and recorded on magnetic tapes in the logging truck. The data can also be encoded and sent by radio telemetry back to the logging company office. There the data is decoded, and a server puts the log information on the Internet where it can be viewed with limited access. On an offshore drilling rig, a permanent, small cabin is used for logging. Logging a well can take from several hours to several days.

A wireline well log is commonly recorded on one of two common formats (Fig. 19–9). A *header* at the top of the log contains well information, followed by logging information. A *depth strip* or *track* runs down near the middle of both formats. The depth is measured from below kelly bushing (KB), rotary table (RT), or ground level (GL). On the left side of the depth strip on both formats is a graph called *track 1*. On the right side of the depth strip in both formats is another graph called *track 2*. On some logs, a third graph is located on the far right (*track 3*).

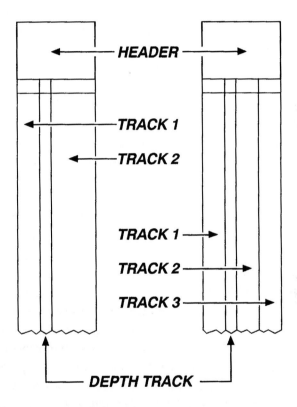

Fig. 19–9 Wireline well-log formats

At least one line in each track records some property of the rocks and their fluids in the well or the geometry of the wellbore. The caption at the top of each track (Fig. 19–10) identifies the property being recorded. Each measurement line is also labeled in the track. There can be one, two, or three different measurements in each track. If more than one property is being recorded in the same track, a different line (heavy or light, solid or dashed) is used for each. When a measurement in a track goes from one side to the other, it is called a *kick*. An accurate scale for each measurement (either linear or logarithmic) is located on the top of each track. Two common vertical scales are used. A *correlation log* uses 1 in. to 100 ft, and a *detail log* uses 5 in. to 100 ft.

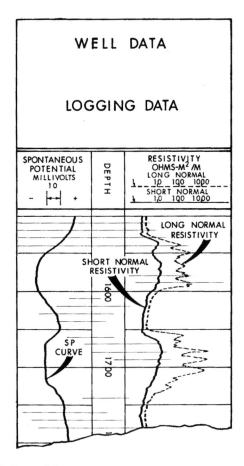

Fig. 19–10 *Wireline well log*

Most wireline well logs are *open-hole* logs, which can be run only in wells with bare rock walls. *Cased-hole logs* are less common and can be run accurately both in open holes and wells in which pipe (casing) has been cemented into the well. *Compensated* logs are measurements that have been adjusted for irregularities in the shape and roughness of the wellbore sides.

Certain measurements are usually recorded in specific tracks. Spontaneous potential, natural gamma ray, and caliper logs are usually recorded in track 1. Resistivity (electrical and induction), formation density, neutron porosity, and sonic logs are usually recorded in tracks 2 and 3.

Electrical log

The first wireline log was an *electrical log* that measured resistivity. An electrical sonde with electrodes in contact with the rocks along the wellbore was raised in the well (Fig. 19–11). An electrical current was passed through the rocks, and the *resistivity (R)* of the rocks and their fluids to that current was measured.

Two or more resistivities were often measured. *Short normal resistivity* was measured with electrodes closely spaced, 16 in. (40½ cm). *Long normal resistivity* was measured with electrodes spaced far apart, 64 in (162½ cm). Short normal resistivity was recorded in track 2 on the electrical log (Fig. 19–12) with resistivity increasing to the right. Long normal resistivity was usually recorded in track 3 with resistivity also increasing to the right.

Because of the short distance between the electrodes used in short normal resistivity, the curve sharply recorded the top and bottom of subsurface layers. It was used to accurately determine the subsurface elevation of rock layers.

An important factor in the resistivity of porous rock is the fluid in the rock. By measuring the resistivity of the rock, the fluid (water, gas, or oil) in the pores can be identified. However, when drilling with a rotary drilling rig, permeable rocks adjacent to the well have been flushed with drilling mud. Because of the pressure on the drilling mud, some of the large, solid particles in the mud are plastered against the well walls to form the mud cake, and some of the drilling mud liquid along with finer, solid particles (mud filtrate) is forced into the rock. The area adjacent to the wellbore is flushed with mud filtrate (*invade zone*) and goes from 0 to 100 in (0 to 254 cm) back from the wellbore, depending on the porosity and permeability of the rock.

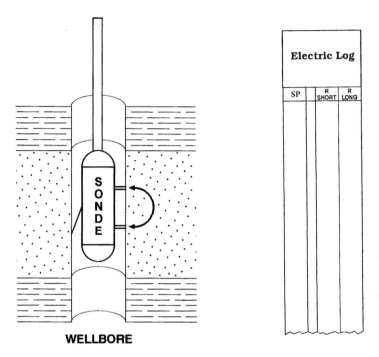

Fig. 19–11 *Making a resistivity measurement*

Fig. 19–12 *Electrical log*

Long normal resistivity puts the electrical current back behind the invade zone to measure the true resistivity of the rock and determine the natural fluids in the pores of the rock. Salt water conducts some electricity and has moderately low resistivity (Fig. 19–13). Oil and gas, however, have very high resistivity. An oil or gas reservoir will have a long normal resistivity kick to the right in track 3.

Oil and gas cannot be differentiated on a resistivity log. However, an oil-water or gas-water contact will appear as a kick on the resistivity curve (Fig 19–14). Also, if the resistivity of the salt water is known, the oil saturation in a reservoir can be calculated from a resistivity log (Fig. 19–15). The higher the oil saturation, the greater the resistivity. Electrical log resistivity has been replaced by induction and focused log resistivity measurements.

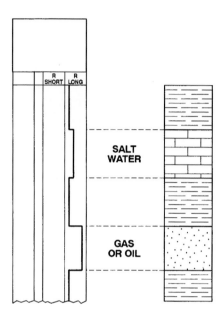

Fig. 19–13 *Resistivity responses*

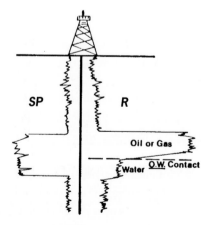

Fig. 19–14 *An oil-water contact on a resistivity curve*

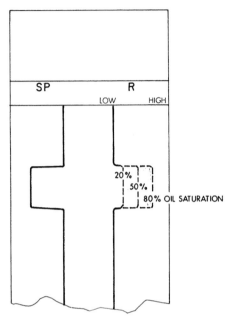

Fig. 19–15 *Resistivity responses to different oil saturations*

A common measurement made along with resistivity on an electrical log is *spontaneous* or *self potential (SP)*. It is made with an electrode that was grounded on the surface and connected to another electrode in the sonde. As the sonde is run back up the well, the electrode is in contact with the rocks along the wellbore.

When two fluids of different salinities are in contact, a potential electrical current is created. A permeable reservoir rock drilled with a rotary drilling rig using mud overbalance has an invade zone flushed with mud filtrate adjacent to the wellbore. The mud filtrate usually has a different salinity than the water in the pores of the rock. This creates a potential electrical current along the top and bottom of the reservoir rock where it is in contact with shales (Fig. 19–16).

Spontaneous potential measures the magnitude of this current to identify potential reservoir rocks in the well. It is recorded in track 1 with positive on the right and negative on the left (Fig. 19–17). The SP curve kicks

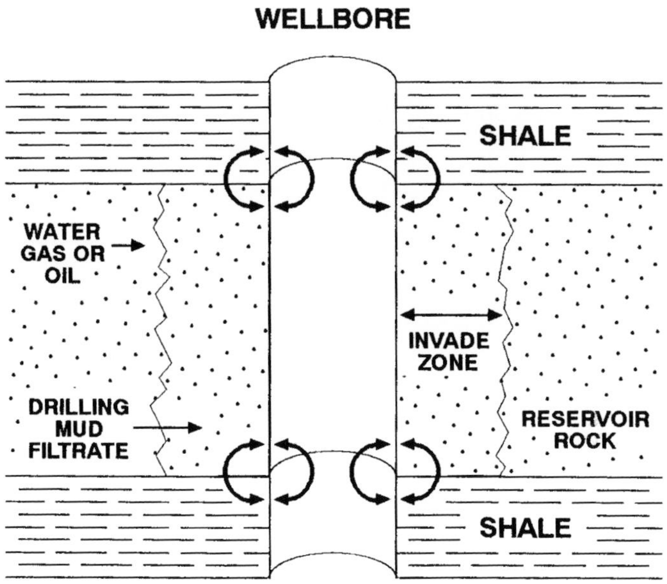

WELLBORE

Fig. 19–16 *Spontaneous potential*

to the left to identify a potential reservoir rock and to the right for non-reservoir rocks such as shale, tight sand or dense limestone. Spontaneous potential is still commonly run today.

Tight sands, dense limestones, and shales have a characteristic signature on SP and R logs (Fig. 19–18). Shales cause both curves to kick to the center of the log with the SP kicking to the right and the R kicking to the left. Tight sands and dense limestones also cause the SP to kick to the right, but they have high resistivities and cause R to kick to the right. There is no way to distinguish between a tight sand and dense limestone on these logs.

An important use for electric logs is the correlation of subsurface rocks (Fig. 19–19). A *pick* is the top or bottom of a sedimentary rock layer on a wireline well log. It is located at the deflection of a SP, R, or other measurement. Lines are drawn along the same picks to correlate between well logs. The wireline well log is often the only record of the rocks in the well available for correlation.

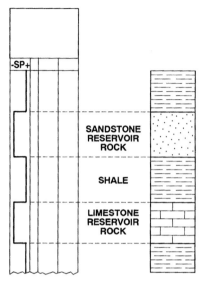

Fig. 19–17 *Spontaneous potential responses*

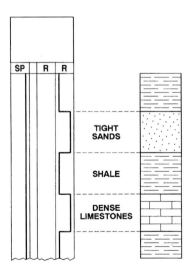

Fig. 19–18 *The responses of a tight sand, dense limestone, and shale on SP and R*

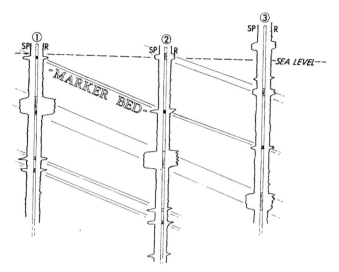

Fig. 19–19 Correlation with electrical logs

Induction and Guard Logs

Resistivity is a valuable measurement as it is the only common log that identifies the fluid in the pores of the rock. In the 1950s, wells were being drilled with oil-base mud to prevent formation damage. A normal resistivity measurement could not be run in those wells because oil does not conduct electricity.

Today, resistivity is commonly measured by an induction or a focused log. An *induction log* uses coils in the sonde to focus and induce an electrical current in the rocks adjacent to the wellbore. A *focused log*, known as a *laterolog* or *guard log*, uses guard electrodes in the sonde to focus an electrical current into the rocks. The resistivity for both types of logs is plotted in track 2 in a format similar to an electrical log (Fig. 19–20).

Deep, medium, or shallow resistivity describes how far back into the rock from the wellbore the resistivity measurement was made. A *dual induction log* and *dual laterolog* measures deep and medium resistivities. A *microresistivity log* does not penetrate very deep and measures the resistivity of the rocks along the wellbore. Sometimes *conductivity*, the inverse of resistivity, is recorded.

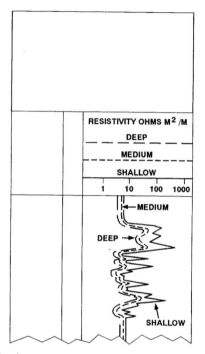

Fig. 19–20 Induction log

Natural Gamma Ray Log

A *gamma ray* or *natural gamma ray log (GR)* uses a scintillation counter to measure the natural radioactivity from potassium, thorium and uranium in the rocks along the wellbore. Of the three most common sedimentary rocks, only shale is radioactive. The gamma ray log is plotted in track 1 (Fig. 19–21) with low radioactivity to the left and high radioactivity to the right. Shales are "hot" and kick to the right. Sandstones and limestones, potential reservoir rocks, kick to the left. The amount of shale in a sandstone or limestone can be computed from its radioactivity on a gamma ray log.

A natural gamma ray log is relatively inexpensive and can be run accurately in both an open-hole and cased-hole. A *spectral gamma ray log* is a type of gamma ray log that also identifies the source of the radiation (potassium, thorium, and uranium).

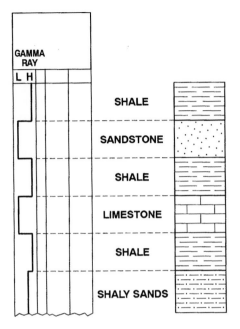

Fig. 19–21 *Gamma ray log responses*

In most logs, either an SP or gamma ray curve will be located in track 1. Both logs are used to locate potential reservoir rocks, which have characteristic kicks to the left.

Radioactive Logs

A radioactive log is made by running a radioactive source into the well. The radioactive source is stored in a compartment in the back of the logging truck. The logging engineer uses a metal pole to screw into the radioactive source and remove it from the compartment. The radioactive source is then screwed into the sonde and lowered down the well. Two types are commonly run today.

Neutron Porosity Log. The *neutron* or *neutron porosity log (NL)* is used to measure the porosity of rocks in the well. The tool has a radioactive source that bombards the rocks adjacent to the well with high-speed atomic particles (neutrons) as the tool is raised in the well. If a high-speed neutron collides with a large rock atom, the atom will bounce the high-speed neutron back with almost no loss of energy. If the high-speed neutron collides with a hydrogen atom (a very small atom), the hydrogen atom absorbs some of the neutron's energy. The neutron will bounce back as a slow-moving neutron. The slow-moving neutron can be captured by another atom in the rock, causing that atom to emit gamma rays. The more hydrogen atoms in a rock, the more slow-moving neutrons and gamma rays the rock will produce when bombarded by fast-moving neutrons. The less hydrogen atoms in a rock, the more fast-moving neutrons will bounce back as the rock is bombarded.

Hydrogen atoms occur in water, gas, or oil in the pores of a subsurface rock. Each rock is bombarded with a certain number of high-speed neutrons. Either the number of gamma rays or slow neutrons is counted. The more porous a rock, the more slow neutrons and gamma rays will be emitted and counted.

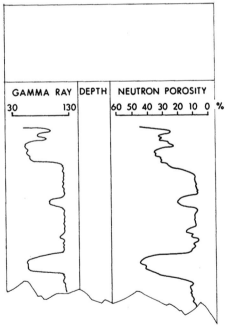

Fig. 19–22 *A neutron porosity log in track two*

The log is calibrated and recorded as percent porosity in track 2 with 0% porosity on the right side and 60% porosity on the left (Fig. 19–22). The calibration assumes there is a liquid (oil or water) in the pores. If a gas occupies the pores, the neutron porosity log will record a low porosity value. The neutron porosity log is accurate in both open-holes and cased-holes. It is usually adjusted (*compensated neutron porosity log – CNP*) for wellbore irregularities.

Formation Density or Gamma-Gamma Log. The *gamma-gamma* or *formation density log (FDL)* is another type of porosity log. A radioactive source bombards the rocks in the well with gamma rays as the tool is being raised. If a gamma ray collides with a large rock atom that has a high electron density, part of the gamma ray energy is absorbed, and the weakened gamma ray is scattered back. Gamma rays are not significantly affected by small atoms such as hydrogen. Dense rocks have more rock atoms per unit

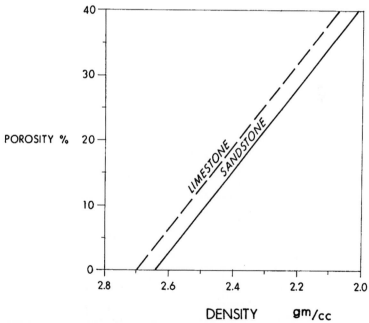

Fig. 19–23 *The relationship between rock density and porosity (modified from Gearhart-Owen)*

volume than porous rocks. The denser and less porous a rock, the more gamma ray energy will be absorbed and less scattered gamma rays will return to the detector in the logging tool. The more porous a rock, the more the gamma rays are able to return by scattering through the rock and back to the detector.

The formation density log measures the density of the subsurface rock. The porosity of the rock (sandstone or limestone) can be computed from the density (Fig. 19–23). The formation density log is recorded in track 2 in one of three ways. First, it can record bulk density in units of

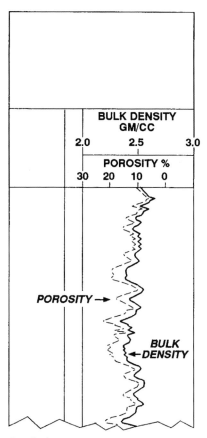

Fig. 19–24 Formation density log

grams/cubic centimeter. Second, it can record porosity that was calculated assuming a limestone and is labeled limestone matrix. There will be a percent porosity scale with 0% on the right side of the track. If the rock is a sandstone, the actual porosity is slightly lower than the porosity shown on the porosity scale. Third, both bulk density and porosity can both be displayed in track 2 (Fig. 19–24).

Gas Effect

There are two states of matter that can occupy the pores of a subsurface rock, liquid (water and oil) and gas (natural gas). The neutron porosity log is calibrated to measure porosity assuming a liquid is in the pores. It will yield an inaccurate, low porosity reading on a gas reservoir. The formation density log will give a more accurate but slightly high porosity calculation on a gas-filled rock. It is most affected by the density of the rock atoms, not the hydrogen atoms in the pores.

Natural gas is detected in subsurface reservoir rocks by running both porosity logs (formation density and neutron porosity), side by side in the well. They are plotted as porosity in track 2 on the log. If natural gas is present, the neutron porosity log will read low, and the formation density log will read high (Fig. 19–25). The divergence of the two curves is called the *gas effect*. A correction can be applied to the formation density log to calculate the accurate porosity of the natural gas reservoir rock.

Caliper Log

A *caliper log* measures the diameter of the hole. The size of the hole depends upon the size of the drill bit, the strength of the well walls, and the thickness of the filter cake. Soft rocks, such as shale and coal, break off and sluff (cave) into the well, forming a wide hole. Strong rocks such as limestones, dolomites, and well-cemented sandstones, have wellbores about the size of the drill bit. Salt layers are often dissolved by fresh-water drilling mud, forming caverns that create serious drilling problems in certain areas of west Texas.

The caliper logging tool has four arms that are expanded to touch the sides of the well. As the caliper tool is run up the well, the arms expand and

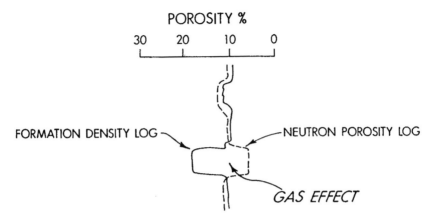

POROSITY %

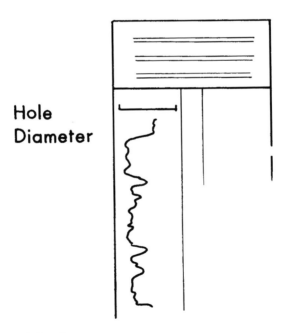

Fig. 19–25 Gas effect

**Hole
Diameter**

Fig. 19–26 Caliper log

contract to fit the well, and an electrical signal is generated to record the wellbore size. The caliper log is recorded in track 1 (Fig. 19–26). The units are inches (centimeters) of diameter with a larger diameter wellbore to the right and smaller to the left.

Caliper logs are commonly run for two reasons. First, it is necessary to know the size of the hole for future engineering calculations. If the well is going to be cased (pipe cemented to the sidewall) or plugged and abandoned, the volume of the well must be computed to order the right number of sacks of cement. Second, many of the other logs, called compensated logs, need to be calibrated for wellbore size to yield accurate results.

Also, because permeable formations will accept more drilling mud filtrate during drilling, they will have a thicker mud cake (Fig. 19–27). The caliper log can be used to locate thick filter cakes that form smaller diameter wellbores and identify permeable zones.

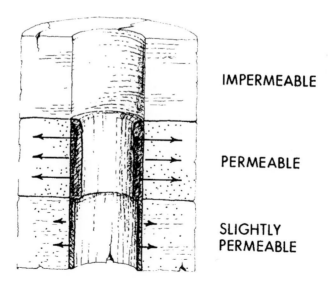

IMPERMEABLE

PERMEABLE

SLIGHTLY
PERMEABLE

Fig. 19–27 Wellbore diameters

Sonic or Acoustic Velocity Log

The *sonic (SL)* or *acoustic velocity log (AVL)* measures the sound velocity through each rock layer in the well. The logging tool has a sound transmitter at the top of the tool and two sound receivers spaced along the tool (Fig.19–28). An impulse of sound is emitted by the transmitter and is recorded on the two receivers. The time it takes the sound to travel from one receiver through the rocks to the other receiver is recorded in units of microseconds per foot. This velocity is called the *interval transit time* or Δ*t* of the rock.

Table 19–1 shows the common ranges of sound velocities through sedimentary rocks, water and natural gas. Of the common sedimentary rocks, shales has the lowest sonic velocities, sandstones have higher velocities, and limestone and dolomite have the highest. There is a wide range of sonic velocities for sedimentary rocks because sound velocity through gas and liquid is less than through solids such as rocks. The more porous a rock, the more gas or liquid it contains and the slower its sonic velocity will be.

Table 19–1

Typical sonic log velocities

	velocity (ft/second)	(m/second)	Δt(Δsecond/ft)
shale	7,000 to 17,000	2,134 to 5,182	144 to 59
sandstone	11,500 to 16,000	3,505 to 4,877	87 to 62
limestone	13,000 to 18,500	3,962 to 5,639	77 to 54
dolomite	15,000 to 20,000	4,475 to 6,096	67 to 50
natural gas	1500	456	667
water	5000	1,524	200

(modified from Gerhart-Owen)

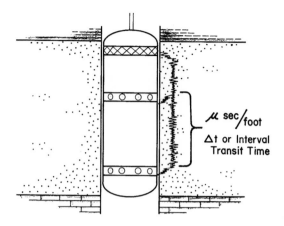

$\mu \, \text{sec}\!/\!_{\text{foot}}$

Δt or Interval
Transit Time

Fig. 19–28 *Making a sonic log measurement*

The sonic log is plotted as interval transit time in track 2 or 3 (Fig. 19–29). Fast interval transit time is on the right and slow on the left. If the composition of the rock is known, the porosity of the rock can be computed from the interval transit time of the rock (Fig. 19–30). Porosities calculated from sonic logs usually do not include any porosity formed by fractures in the rock.

Fractures greatly attenuate (decrease) the amplitude of the sonic waves through rock. A type of sonic log, the *sonic amplitude log*, measures the attenuation of the sound and detects the presence of fractures.

Dipmeter

The *dipmeter* or *dip log* is a logging tool used to determine the orientation of rock layers in a well. The dipmeter consists of four arms, each with either two closely spaced electrodes or pads of electrodes that record resistivity. The arms expand in the well to touch the sides of the well. As the dipmeter is brought up the well, the electrodes on each arm are in contact with the rock layers (Fig. 19–31). If the rock layer is dipping, different arms will contact the layer at different depths. The orientation of the dipmeter in the well is known from a gyroscope. The sequence of contacts between individual arms and each layer can be used to compute the dip of the layer. If the layer is horizontal, all arms of the dipmeter contact the layer at the same time.

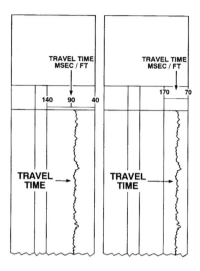

Fig. 19–29 Sonic logs

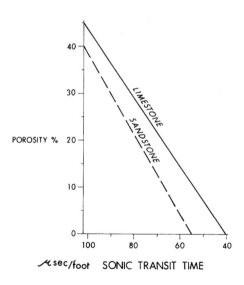

Fig. 19–30 The relationship between interval transit time and porosity (modified
from Gearhart-Owen)

Two ways to present dipmeter data are tadpole and stick plots. On a tadpole plot (Fig. 19–32), dip is plotted on the horizontal axis with zero dip on the left. Depth in the well is the vertical axis. A small circle on the log gives the depth and dip of each measurement. A small line, similar to a tadpole tail, is oriented in the compass direction of the dip with north at the top of the log.

A *stick plot* uses lines (sticks) to show the dip measurements (Fig. 19–33). Depth is recorded on the vertical axis with the well represented by a vertical line. The depth of each dip measurement is where the stick intersects the well. The angle on the stick is the dip measurement.

Nuclear Magnetic Resonance Log

A *nuclear magnetic resonance* (*NMR*) *log* uses magnetic fields to orient hydrogen atom protons in the rock. This is used to measure both porosity and pore sizes in the reservoir. It can be used to compute the formation per-

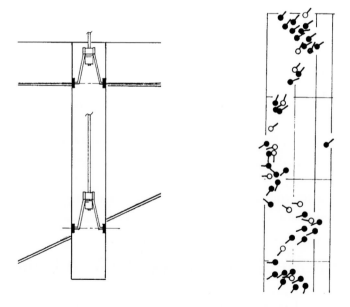

Fig. 19–31 *Making a dipmeter measurement*   **Fig. 19–32** *Tadpole plot*

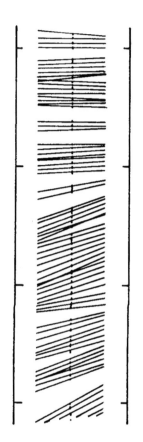

Fig. 19–33 *Stick plot*

meability, types of fluids in the formation, the volume of the fluid that will flow (movable) and the irreducible water saturation.

Wellbore Imaging Logs

Wellbore imaging logs make a picture of the rocks along the wellbore using either resistivity or ultrasonics. The resistivity tool uses a large number of small, button electrodes that are closely spaced on pads mounted on four arms. The resistivity picture of the wellbore images sedimentary rock layers including thin shale layers, unconformities, fractures (both natural and induced by drilling), and vugs. It can be used to calculate the strike and dip of rock layers and fractures.

Table 19–2 summarizes the uses and limitations of the common wireline logs.

Computer-Generated Log

A *computer-generated log* is a combination of two or more logging measurements that have been calculated by a computer on the logging truck at the wellsite or in a computing center. The wellsite computer logs are called *quick-look logs* and can show water saturation, porosity, percentage of limestone, sandstone, and shale, the presence of fractures and true vertical depth. *Computer center logs* are similar to quick-look logs but are more accurate and detailed.

Measurements and Logging-While-Drilling

Wireline well logs are run after the well has been drilled. In the 1980s, sensors for the bottom of the drillstring and a data transmitting process were

Table 19–2

The names, track recorded in, and uses of common wireline well logs

name	track	uses
Spontaneous Potential SP	1	identify potential reservoir rocks
Gamma Ray GR	1	measure natural radioactivity identify potential reservoir rocks correlate formations determine elevation of formations determine shale content of rock
Resistivity Electric and Induction R		
Short Normal or Shallow	2	determine elevation of formations
Long Normal or Deep	23	identify pore fluid
Neutron or Neuron Porosity N or NL	2	determine formation porosity
Gamma-Gamma or **Formation Density** GG or FDL	2	determine formation density and porosity
Sonic or **Acoustic Velocity** SL or AVL	2 or 3	determine sound velocity calculate porosity determine lithology
Caliper CAL	1	measure diameter of wellbore calculate wellbore-volume

developed to give a real time log as the well is being drilled called *measurements-while-drilling (MWD)* and *logging-while-drilling (LWD)*. Measurements-while-drilling measures well properties such as azimuth and deviation and logging-while-drilling measures rock and fluid properties such as short and long normal resistivity, natural gamma ray, formation density, and neutron porosity.

The sensors are located just above the drill bit on the drillstring. The power to the sensors is supplied either by a turbine driven by the circulating drilling mud or electrical batteries. The data can be transmitted to the surface by *fluid pulse telemetry*. The data are coded digitally in pressure pulses that are sent up the well through the drilling mud. They are recorded on a pressure transducer on the surface where they are decoded by a computer.

MWD is very useful in drilling deviation and horizontal drain wells. It records a *directional log* that shows the orientation of the drill bit (*tool face*), the direction in which the well is being drilled, in real time. The measurement is made with a magnetometer in the downhole tool that measures the direction of the earth's magnetic field.

Geosteering is the drilling of a horizontal drain well while continuously adjusting the direction of the bit to keep the well within the target formation. A *LWD* system is used to sense the target formation top or bottom. The MWD system shows the orientation of the bit. A steerable downhole assembly (Fig.18–6) is used to adjust the direction the well is being drilled to keep the well within a target formation that can be as thin as 7 feet (2 m).

Drillstem Test

A *drillstem test* is a temporary completion of a well. As the well is drilled and logged, it is kept filled with drilling mud. Drilling mud pressure keeps any fluids back in the pores of the rock adjacent to the wellbore. If the logs indicate a potential reservoir, a drillstem test can be run to further evaluate that reservoir. A drillstem, usually a drillstring of drillpipe, is run in the well. It has one or two packers, perforated pipe, pressure gauges and a valve assembly (Fig. 19–34). *Packers* are cylinders made of a rubber-like material that can be compressed to expand against the well walls to seal that portion of the well. The packers prevent any vertical flow of fluid in that section of the well.

If the formation is located on the bottom of the well, only one packer is used (Fig. 19–34). If the formation is located above the bottom of the well, two packers (*straddle packers*) are used (Fig. 19–35).

The packer(s) eliminate the drilling mud pressure on that formation. After the packer(s) have been seated, then water, gas, or oil can then flow out of the formation and into the well. A valve is opened on the drillstem, and the formation fluids flow into and up the drillstem. If gas is present, it will flow up the drillstem and onto the surface where it is measured and flared (burned). Sometimes oil has enough pressure to flow to the surface during a drillstem test. Usually, however, the oil fills the drillstem only to a certain height that is measured.

During the drillstem test, pressure on the fluid flowing into the drillstem is continuously measured. The valve on the drillstem is opened and closed several times, and fluid pressure build-up and drop-off are recorded on a chart (Fig. 19–36) of pressure versus time called a pressure buildup curve.

Engineers use these pressure records to calculate formation permeability, reservoir fluid pressure, and the extent of any formation damage. The test can take 20 minutes to three days. Longer tests are more accurate, but are also more expensive. The test can be run in either an open-hole or a cased well that has been perforated. It should not be run in a formation with unconsolidated sands that could collapse into the well during the test and trap the equipment.

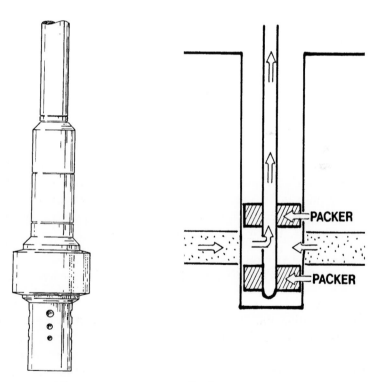

Fig. 19–34 *Drillstem tool with one packer*

Fig. 19–35 *Drillstem test with straddle packers*

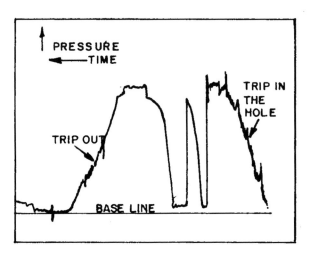

Fig. 19–36 Drillstem test record

Repeat Formation Tester

A *repeat formation tester (RFT)* is a wireline tool, several tens of feet long that is used to sample reservoir fluids and measure formation pressure versus time. It can test several levels in a well. At the zone to be tested, a backup shoe from the tool is pressed against the well wall to force a rubber pad with a valve against the opposite wall. The valve is opened, and formation fluids can flow into the tool as the pressures are measured. The pressure records are used to calculate formation permeability. Two large sample chambers, each holding several gallons, are used to obtain a sample of formation fluids at that level. This can be repeated at several locations in the well.

twenty

COMPLETING A WELL

After a well has been drilled and tested, there are two options. The well is either plugged and abandoned as a dry hole or completed as a producer by setting pipe.

Casing

A well is always cased (*set pipe*) to complete the well. *Casing* is relatively thin-walled, steel pipe. Numerous joints of the same size casing are screwed together to form a long length of casing, called a *casing string*. The casing has an outer diameter of at least 2 in. (5 cm) less the wellbore diameter. The casing string is run into the well and cemented to the sides of the well (Fig. I–9) in a *cement job*.

Casing stabilizes the well and prevents the sides from caving into the well. It protects fresh water reservoirs from the oil, gas, and salt water brought up the well during production. Casing also prevents the production from being diluted by waters from other formations in the well.

Casing is seamless pipe made to API standards. It ranges in lengths from 16 ft (5 m) to greater than 42 ft (13 m) but is commonly about 30 ft (10 m) and is called a *joint* of casing. Diameters range from 4½ to 36 in. (11½ to

91½ cm) but are commonly 5½ to 13¾ in. (14 to 35 cm). Casing is graded by the API for 1) outer diameter and wall thickness, 2) weight per unit length, 3) type of coupling, 4) length, and 5) grade of steel.

The end of each casing joint has male threads that are protected by a plastic or metal cap called a *thread protector* until the casing is ready for use. A *collar* or *coupling*, a short section of cylindrical, steel pipe with female threads on the inside and a diameter slightly larger than the casing is used to connect joints of casing.

Before a cement job, the well is conditioned by running a drillstring with a used bit into the well. Mud is then circulated for a period of time to remove any remaining cuttings. In order to scrape the filter cake off the well sides to prepare it for the cement, *wall scratchers* with protruding wires (Fig. 20–1) are attached by collars or clamps to the casing string and run up and down or rotated in the well.

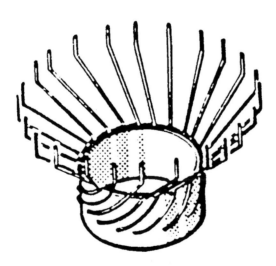

Fig. 20–1 Wall scratcher

A service company *casing crew* with one to five members usually cases the well. They have specialized equipment such as hydraulic casing tongs and a casing stabbing board. A *guide shoe*, a short section of rounded pipe with a hole in the end, is screwed into the end of the casing string to guide the casing string down the well. Spring-like devices called *centralizers* (Fig. 20–2) are attached to the outside of the casing as it is run into the well to position the string in the center of the well.

The casing is stored horizontally on the *pipe rack* on the ground next to the rig (Plate 20–1a). It is pulled a joint at a time up the pipe ramp, through the V-door in the derrick and onto the derrick floor (Plate 20–1b). Each casing joint is lifted by *casing elevators* on the traveling block and guided (*stabbed*) into the casing string already being suspended in the well. A thread compound is applied to the threads to make a tight seal. *Casing tongs* hung by cable above the drill floor are used to screw the casing joint onto the string. To assist in stabbing the joint of casing, a derrickman from the casing crew stands on the *stabbing board* located in the derrick. The casing string is run into the well (Plate 20–1c) and finally *landed* by transferring the weight of the string to the casing hangers in the casinghead on the top of the well. *Casing hangers* use either slips (wedges) or threads to suspend the casing in the well.

Fig. 20–2 Centralizer

Plate 20–1 a) casing racked in front of a well
(courtesy Parker Drilling)

The service company that provides the cementing equipment (Plate 20–2) prepares the wet cement called *slurry* for the cement job by mixing sacks of dry cement and water together at the wellsite in a *hydraulic jet mixer, recirculating mixer,* or *batch mixer.* The cement is Portland cement made to one of eight API classes with additives for various situations.

Common *cement additives* that come in sacks include *accelerators* that shorten cement setting time, *retarders* that lengthen the time, *lightweight additives* that decrease the slurry density, and *heavyweight additives* that increase the density. Other additives are used to change the cement's compressive strength, flow properties, and dehydration rate and to cause the cement to expand, to reduce the cost of the cement (*extenders*), and to prevent foaming (*antifoam*). *Bridging materials* can be used to plug lost circulation zones.

The *thickening time* or *pumpability time* of the cement is the time the cement is fluid and can be pumped before it sets. Special cements are available for very corrosive environments and for *setting* (hardening) at deep depths under high temperature.

After the casing has been run into the well to just above the bottom, an L-shaped fitting (*cementing head*) is attached to the top of the wellhead to receive the slurry through a line from the pumps. It contains two wiper plugs. A *wiper plug* (Fig. 20–3), made of cast aluminum and rubber, is designed to wipe the inside of the casing and separate two different liquids to prevent mixing.

Plate 20–1 b) casing being run up the pipe ramp onto the drill floor
(courtesy American Petroleum Institute)

Plate 20–1 c) casing being run into the well
(courtesy American Petroleum Institute)

Plate 20–2 *Cement job (courtesy Halliburton)*

The first wiper plug pumped down the casing is the *bottom plug*. It separates the drilling mud from the cement slurry. The cement slurry is pumped down the casing until the bottom plug is caught in a float collar located one or more casing joints above the guide shoe (Fig. 20–4). The *float collar* is a short pipe with the same diameter as the casing. It contains a constriction to stop the wiper plug and also acts as a one-way valve. This allows the casing string to be run in the well without the drilling mud flowing up the inside of the casing. Because of this, the casing string floats in the drilling mud to partially support its weight. It does, however, allow liquid cement to be pumped down the casing.

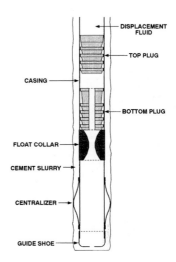

DISPLACEMENT FLUID

TOP PLUG

CASING

BOTTOM PLUG

FLOAT COLLAR

CEMENT SLURRY

CENTRALIZER

GUIDE SHOE

Fig. 20–3 *Wiper plug*

Fig. 20–4 *Bottom of well showing cement plug*

After the bottom plug has landed in the float collar, the pump pressure is increased until the cement slurry ruptures a diaphragm in the bottom plug and flows through it. The slurry flows out the guide shoe and up the outside of the casing string. After a predetermined volume of slurry has been pumped down the well, the *top plug* is pumped down the casing followed by a displacement fluid, which is usually drilling mud. When the top plug hits the bottom plug and all the slurry has been displaced between them, the pumps are shut down, and the cement is allowed to set (*waiting on cement-WOC*) for 8 to 12 hours.

The wiper plugs, guide shoe, and cement on the bottom are then drilled out. A temperature log can be run to locate the top of the setting cement behind the casing by the heat given off by setting cement.

Multistage cementing is used on long casing strings when the required pump pressures would be too high if the entire length of casing was cemented at once. Two or more sections of a single string are cemented separately.

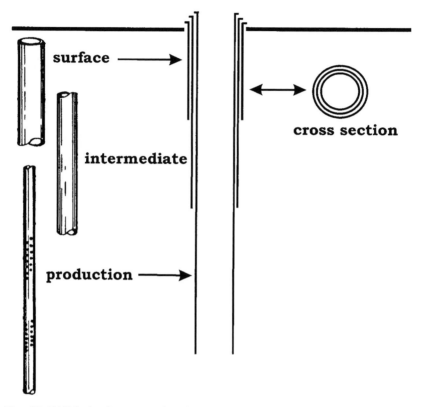

Fig. 20–5 *Well showing types of casing*

First the lower section is cemented. Next, holes in a coupling located on the casing string are opened for the cement to flow through, and the section above the coupling is cemented. This can be repeated several times at various levels up the casing string.

A well is drilled in stages called a casing program. During the *casing program*, the well is drilled to a certain depth and then cased, drilled deeper and cased again, drilled deeper and cased again. Each time the well is drilled deeper, a bit is used that is at least ½ in. (1 cm) smaller in diameter than the casing. The casing program defines the grades, lengths, and sizes of casing that are going to be used before the well is drilled. The casing is ordered and delivered to the drillsite as the well is being drilled.

A well commonly contains three or more concentric casing strings (Fig. 20–5). Shallow wells can have two or even just one casing string. Deeper wells have more casing strings. Each casing string runs back up to the surface. The largest diameter and shortest length string is on the outside. The smallest diameter and longest length string is on the inside. The outside string is cemented first and the inside string last.

Conductor pipe is the largest diameter casing string (30 to 42 in. or 76 to 107 cm offshore and 16 in. or $40\frac{1}{2}$ cm onshore) and is often several hundred feet long. Either a hole for the conductor pipe is drilled into hard rock, or the conductor pipe is pile-driven into soft ground. Conductor pipe serves as a route for the drilling mud coming from the well to the mud tanks, prevents the top of the well from caving in, and isolates any near surface, fresh water, and gas zones. The blowout preventers are attached to the top of the pipe if shallow gas is present.

The next casing string is *surface casing*, often $13\frac{3}{4}$ in. (35 cm) in diameter and several hundred to several thousand ft. in length. It prevents soft, near-surface sediments from caving into the well. It also protects fresh water reservoirs from further contamination by drilling mud.

Protection or *intermediate casing* can be set to isolate problem zones in the well such as abnormal high pressure, lost circulation, or a salt layer. It is typically $8\frac{5}{8}$ in. (22 cm) in diameter. In some wells there is no intermediate casing string.

The final string of casing is the *production* or *oil string* that runs down to the producing zone. It is typically $5\frac{1}{2}$ in. (14 cm) in diameter.

Instead of a casing string, a liner string can be set on the bottom of the well to save money. A *liner string* is similar to a casing string and is made of liner joints that are the same as casing joints. A liner string, however, never runs all the way up the well to the surface as a casing string does. It is suspended in the well by a *liner hanger* using slips in a casing string and can be cemented in. A liner string is often used in deep wells to save the cost of running a long casing string.

In New Mexico, special care must be taken during drilling into the Pennsylvanian age Morrow sands for gas. Drilling into the Morrow sands with heavy drilling mud causes extensive formation damage. Overlying the Morrow sands, however, is the Atoka Formation that has abnormal high pressure. Heavy drilling mud must be used when drilling the Atoka Formation to prevent blowouts. After the Atoka Formation is drilled, a string of intermediate or protection casing is cemented into place to seal off the Atoka Formation. Lighter drilling mud is then used to drill into the Morrow sands and complete the well.

The casing and cement is pressure-tested during a *mechanical integrity test (MIT)*. The well is shut in, and a liquid (water or drilling mud) is pumped into the well until the maximum casing pressure is reached. The pumps are then stopped, and the pressure is monitored. A pressure drop indicates a leak in the casing.

After the Yates field in west Texas (Fig. 7–17) was developed in the late 1920s, several new oil seeps were observed in the adjacent Pecos River. It became apparent that the substandard casing used to complete the wells was leaking. Until the casing was repaired, an estimated 70 million barrels (11 million m³) of oil was lost.

Bottom-Hole Completions

The bottom of the well is completed with either an open-hole or cased-hole completion. A *open-hole, top set*, or *barefoot completion* (Fig. 20–6) is made by drilling down to the top of the producing formation and the well is then cased. The well is then drilled through the producing formation, leaving the bottom of the well open. This completion is used primarily in developing a field with a known reservoir and reduces the cost of casing. Open-hole completions are commonly used in coal bed gas wells, where the bottom of the wells are enlarged by reaming.

An open-hole completion, however, cannot be used in soft formations that might cave into the well and doesn't isolate selective zones in the producing formation. Because the casing is *"set in the dark"* in an open-hole completion before the pay is drilled, the casing cannot be salvaged if the pay proves to be unproductive.

If the producing formation is composed of unconsolidated sands that can cave into the well (a *sand control problem*), a gravel pack completion can be used. A *gravel pack completion* starts similar to an open-hole completion

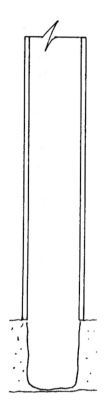

Fig. 20–6 *Open-hole completion*

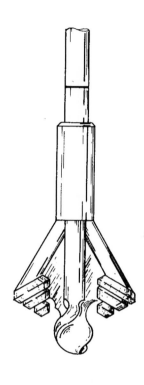

Fig. 20–7 *Underreamer*

with casing set to the top of the producing formation, and the producing formation is then drilled through. A tool called an *underreamer* is then run on a drillstring down the hole and rotated in the soft formation. The underreamer (Fig. 20–7) has arms on it that are expanded and used to ream out a cavity in the formation.

The cavity is then filled with very well-sorted, coarse sand, called a *gravel pack*. It is pumped down the well suspended in a carrier fluid that leaks off into the formation, leaving the gravel pack on the bottom of the well. The gravel pack is very porous and permeable. A section of slotted or screen liner is then run into the gravel pack (Fig. 20–8). A *screen liner* has holes in the liner wall and wire wrapping around the liner to prevent loose sediments

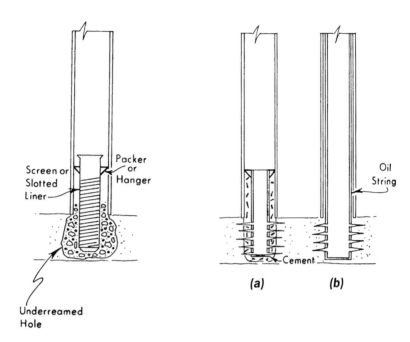

Fig. 20–8 Gravel pack completion

Fig. 20–9 a) perforated liner completion, b) perforated casing liner

from flowing through the holes into the liner. A *slotted liner* has several long openings (slots) cut into it to allow fluids, but not sediments to flow into the liner. A *prepacked slotted liner* is a liner that is filled with a gravel pack that is held together with a resin coating.

Another completion technique is an *uncemented slotted liner* that is set in an open hole. It is used if there is a sluffing shale problem. Horizontal drain wells are commonly completed either open-hole or with a slotted liner.

A *set-through completion* in which a liner string (Fig. 20–9a) or a casing string (Fig. 20–9b) is cemented into the producing reservoir is very commonly done today. Holes called *perforations* are then shot through the liner or casing and cement and into the producing formation.

A *perforating gun* is run into the well on a wireline or tubing string to make the perforations. The original perforating guns used steel bullets, but *jet perforators* that use shaped explosive charges are more commonly used today. When detonated, the cone-shaped explosives produce extremely fast jets of gasses that blow the perforations into the casing, cement, and producing formation. The explosives can be shaped to give either a maximum diameter or maximum length of penetration.

Perforating guns are either *expendable*, which disintegrates and leaves the debris in the well, or *retrievable*, which can be removed from the well. Perforations are described in shots per foot (*density*) and angular separation (*phasing*). Perforated completions are commonly used for multiple completions.

Tubing

Small-diameter pipe called *tubing* is run into the well to just above the bottom to conduct the water, gas, and oil (produced fluids) to the surface (Fig. 20–10). Tubing is special steel pipe that ranges from $1\frac{1}{4}$ to $4\frac{1}{2}$ in. (3 to $11\frac{1}{2}$ cm) in diameter and comes in lengths of about 30 ft. (10 m) long (Plate 20–3). The API grades tubing according to dimensions, strength, performance, and required threads.

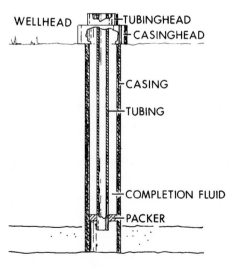

The thicker section with male joints on the ends of tubing joints is called the *upset.* Tubing joints are joined together with *collars* (short steel cylinders with female threads) to form a tubing string. A *tubing* or *completion packer* is used near the bottom of the tubing string (Fig. 20–10). The packer is made of hollow rubber that is compressed to seal the casing-tubing annulus. It keeps the tubing string central in the well and prevents the produced fluids from flowing up the outside of the tubing string.

Fig. 20–10 Completed well

Plate 20–3 *Tubing being run in a well*

Tubing protects the casing from corrosion by the produced fluids. Because the casing has been cemented in the well, it is very difficult to repair the casing. The tubing string, however, is suspended in the well and can be pulled from the well to repair or replace it during a workover. A *completion fluid*, commonly treated water or diesel oil, can be used to fill the annulus between the tubing and casing string to prevent corrosion.

If the well needs to be pumped, a downhole pump is put on the end of the tubing string. A *tubing anchor* that uses slips can be used to secure the tubing to the casing on the bottom of the well. A *seating nipple* is a special joint of tubing used near the bottom of the tubing string. It has a narrow inner diameter designed to stop any equipment that is dropped or run down the tubing string. A *tubingless completion*, used on very high volume wells, brings the produced fluids up the casing string without tubing.

On offshore wells, a *subsurface safety valve* is used in the tubing to stop flow during an emergency. The valve is held open by pressure. A drop in pressure automatically closes the valve. The *surface-controlled subsurface safety valve* is operated with hydraulic lines that run down the well.

Wellhead

The *wellhead* is the permanent, large, forged or cast steel fitting on the surface of the ground on top of the well (Fig. 20–10). It consists of casingheads and a tubing head.

The larger, lower *casingheads* seal off the annulus between the casing strings and contain the casing hangers for the top of each casing string. *Casing hangers* are used to suspend the casing strings in the well. They either screw onto the top of each casing string or are held by slips (wedges). There is a casinghead and casing hanger for each casing string.

The casinghead for the surface casing is the largest in diameter and is located on the bottom. The deepest and longest string casing is hung in the uppermost casinghead with the smallest diameter. Each casinghead is bolted to the one below and has a gas outlet to provide pressure relief. Each time the well is drilled deeper and cased, the BOP stack is removed and reinstalled (*nippled up*) on the new casinghead. A *unitized wellhead* uses only one casinghead for all the casing strings.

The smaller *tubinghead* located on the casingheads, suspends the tubing string down the well and seals the casing-tubing annulus. A *tubing hanger* uses slips or a bolted flange to hold the tubing string and fits as a wedge in the tubinghead.

The wellhead is bolted or welded to the conductor pipe or surface casing. *Wellhead equipment* is the equipment attached to the top of the tubing and casing. It supports the strings, seals the annulus between the strings and controls the production. Wellhead equipment includes the casingheads, tubinghead, Christmas tree, stuffing box and pressure gauges.

Chokes

A flowing well is seldom produced at an unlimited rate. It could result in a too-rapid depletion of reservoir pressure and a decrease in ultimate production. An excessive production rate creates a large pressure drop between the reservoir and wellbore, causing gas to bubble out of the oil and block the reservoir rock pores adjacent to the wellbore. Flow rates are limited by surface or subsurface *chokes*, which are valves that cause the fluid to flow through a small hole called an *orifice*. The smaller the orifice, the lower the flow rate. Chokes can be either *positive* with a fixed orifice size or *adjustable* to permit changing the flow rate.

Surface Equipment

Gas wells flow to the surface by themselves. There are some oil wells (4% in the United States), early in the development of an oil field, in which the oil has enough pressure to flow up the tubing string to the surface. For gas wells and flowing oil wells, a series of pipes, fittings, valves, and gauges are used on the wellhead to control the flow (Fig. 20–11). This plumbing is called a *Christmas* or *production tree*, and is bolted to the wellhead (Plate 20–4).

All Christmas trees have a *master valve* sticking out of the lower part to turn the well off during an emergency. The plumbing going off the side of the Christmas tree to the flowline is called the *wing*. If there is only one producing zone in the well, it is a *single wing tree*. Two producing zones require a *double wing tree* with two wings on opposite sides. On the wing is a *wing* or *flow valve* that is turned to regulate flow through that flowline. A *swab valve* on the upper part of the tree is used to open the well to allow wireline equipment to be lowered down the well during a workover. A *pressure gauge* at the top of the tree measures tubing pressure. Most Christmas trees are machined out of a solid block of metal.

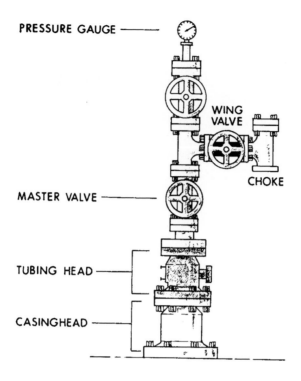

PRESSURE GAUGE ———

WING VALVE

CHOKE

MASTER VALVE ———

TUBING HEAD ———

CASINGHEAD ———

Fig. 20–11 *Christmas tree (modified from Baker, 1983)*

In many oil wells (96% in the United States), the oil does not have enough pressure to flow all the way up the tubing to the surface. The produced water and oil has to be lifted to the surface in one of several methods called *artificial lift* (Plate 20–5). Even with a flowing oil well, as more fluids are produced from the subsurface reservoir, the pressure on the remaining oil decreases until it no longer flows to the surface. When this happens, the Christmas tree has to be removed, and a surface pumper installed in a process called *putting the well on pump*.

Plate 20–4 *Christmas tree*

Plate 20–5 *Artificial lift with a sucker–rod pumping unit*

The most common artificial lift system is a sucker-rod pump. A *sucker-rod pump* or *rod-pumping system* uses a sucker rod pump on the bottom of the tubing string, a surface pumping unit and a sucker-rod string that runs down the well to connect them. The *sucker-rod pump* is built to API specifications and has a standing valve and traveling valve (Fig. 20–12).

The *traveling valve* moves up and down while the *standing valve* remains stationary. Both valves consist of a ball, a seat (a plate with a hole) and a cage to hold the ball over the seat. The steel ball allows the oil to flow up but not back down through the valve. Fluid flowing upward lifts the ball off the seat and opens the valve (Fig. 20–13a). Fluid cannot flow down because gravity holds the ball in the seat. Each upward stroke of the traveling valve lifts the oil and water up the tubing (Fig. 20–13b). There are commonly 10 to 20 strokes per minute.

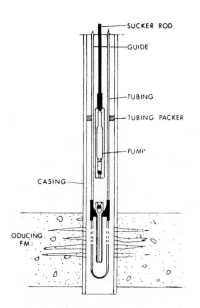

Fig. 20–12 *A sucker-rod pump on the bottom of a well (modified from Baker, 1983)*

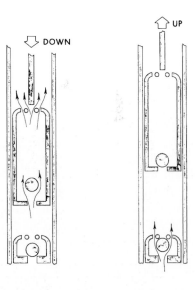

Fig. 20–13 *A sucker-rod pump showing oil flowing during the (a) downstroke and (b) up stroke*

Three types of sucker-rod pumps are insert or rod pump, tubing
pump, and casing pump. The *insert* or *rod pump* (the most common of the
sucker-rod-pumps) is run in the well as a complete unit on the sucker-rod
string through the tubing string. It is usually held in place by a bottom
anchor. The insert pump is the smallest of the pumps and has the lowest
capacity. The *tubing pump* is run as part of the tubing string. The plunger
and traveling valve are run on the sucker-rod string. A *casing pump* is a
relatively large version of an insert pump that pumps the produced fluids
up the casing. It is held in position by a packer and has a much larger
volume than an insert pump.

Fluid pound is a problem caused when the produced liquid is pumped
faster than it is flowing into the well. Gas enters the pump, and the pump can
be damaged. *Gas lock* is an extreme case of fluid pound. Gas accumulates in
the pump and prevents the pump from working.

The most common surface pumping unit is the *beam pumping unit*
(Fig. 20–14). It is mounted on a heavy, steel I-beam base or a concrete
base. The beam pumping unit has a steel beam (*walking beam*) that pivots
up and down on bearings on top of a *Samson post*. It is usually driven by
an electric motor but could also be driven by a motor that uses natural
gas produced from the well.

If an electrical motor is used, it can have a timer to periodically turn the
pumping system on and off. This conserves electricity and prevents the pump
from working when there isn't enough liquid in the well. The motor is connected by a V-belt drive to a *gear reducer* that decreases the rotational speed.
The gear reducer rotates a *crank* on either side. Two steel beams (*pitmans*) and
a cross bar (*equalizer*) connect the cranks on the gear reducer to the walking
beam and cause the end of the walking beam to rise and fall.

Attached to the opposite side of the walking beam is the *sucker-rod string*
that runs down the tubing string to the downhole pump. Sucker rods
(Fig. 20–15a) are solid steel alloy rods between $\frac{1}{2}$ and $1\frac{1}{4}$ in. (1 to 3 cm) in
diameter and are usually 25 ft (8 m) long. The API grades sucker rods accord-
ing to their alloy composition, recommended well depth, maximum allowable stress and environment. The rods have male threads on both ends. They
are connected together with a short, steel cylinder with female threads on the
inside called *sucker-rod couplings* (Fig. 20–15b). Flat areas (*wrench flats*) allow
a wrench to grip the coupling without harming it.

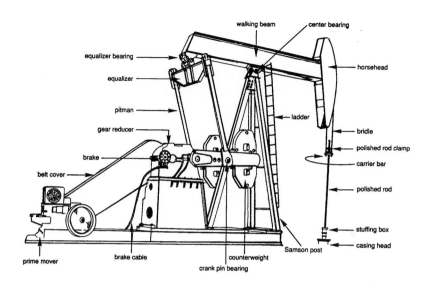

Fig. 20–14 Beam pumping unit (modified from Gerding, 1986)

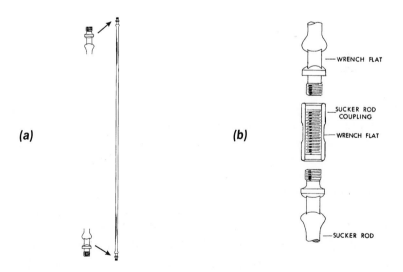

(a)

(b)

Fig. 20–15 (a) sucker rod (b) sucker-rod couple

The sucker-rod string is kept central in the tubing string by *sucker-rod guides* (Fig. 20–12) that are made of rubber, plastic, nylon, or metal. They move up and down with the sucker-rod string as produced fluids flow up the tubing through slots in the guides. Shorter lengths of sucker rods (*pony rods*) can be used to adjust the length of the sucker-rod string. The string can be either *untapered* (all one diameter) or *tapered* with a decrease in rod diameter down the well.

Counterweights of steel are used to balance the weight of the sucker-rod string on the walking beam of a beam pumping unit during the upstroke. Two rotating counterweights are located on both sides of the rotary crank on a *crank-balanced pumper* (Fig. 20–14). The counterweights on a *beam-balanced pumper*, used for shallow wells, are located on the walking beam opposite the wellhead (Fig. 20–16). They are adjustable by moving then along the walking beam.

A large, curved steel plate called a *horsehead* is used on the well side of the walking beam to keep the pull on the sucker rod string vertical (Fig. 20–14). Two wire ropes (*bridals*) and a steel bar (*carrier*) are used to connect the horsehead to the top of the sucker-rod string.

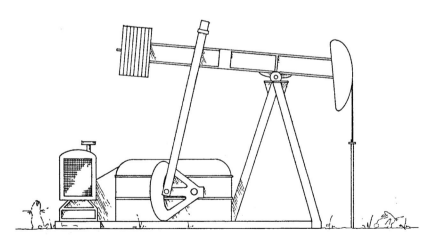

Fig. 20–16 Beam-balanced rod pumping unit

The *polished rod* is used at the top of the sucker-rod string. It is a smooth length of brass or steel rod with a diameter of $1\frac{1}{4}$ to $1\frac{1}{2}$ in. (3 to 4 cm) that ranges from 8 to 22 ft ($2\frac{1}{2}$ to $6\frac{1}{2}$ m) long. The polished rod moves up and down through the *stuffing box*, a steel container on the wellhead that contains flexible material or packing such as rubber that provided a seal around the polished rod.

A variation is the *air-balanced beam pumping unit* that uses a piston and rod in a cylinder filled with compressed air to balance the weight of the sucker-rod string (Fig. 20–17). It is more compact and lighter than the crank and beam balanced units. Another type, the *Mark II* uses a lever system to counterbalance the sucker-rod string weight (Fig. 20–18).

In *gas lift*, another type of artificial lift, a compressed, inert gas (*lift gas*) is injected into the annulus in the well between the casing and tubing (Fig. 20–19). The lift gas is usually natural gas that was produced from the well. *Gas lift valves* are pressure valves that open and close and are spaced along the tubing string. They allow the gas to flow into the tubing where it dissolves in the produced liquid and also forms bubbles. This lightens the liquid density, which, along with the expanding bubbles, forces the produced liquid up the tubing string to the surface where the gas can be recycled.

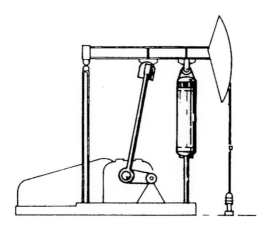

Fig. 20–17 Air-balanced beam pumping unit

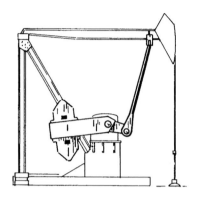

Fig. 20–18 *Mark II rod pumping unit*

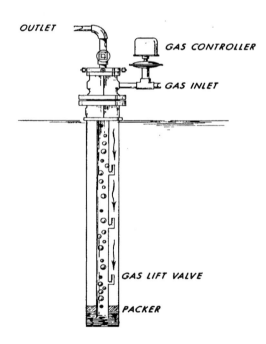

Fig. 20–19 *A gas lift well*

The advantages of gas lift are that there is very little surface equipment and few moving parts. Gas lift is a very inexpensive technique when many wells are serviced by one central compressor facility. However, it is effective only in relatively shallow wells. Offshore oil wells that need artificial lift are usually completed with gas lift. Gas lift is either *continuous* or they are *intermittent* (periodically on and off) for wells with low production.

An *electric submersible pump (ESP)* uses an electric motor that drives a centrifugal pump with a series of rotating blades on a shaft on the bottom of the tubing (Fig. 20–20). An armored electrical cable runs up the well, strapped to the tubing string. The electricity comes from a surface transformer. The electric motor has a variable speed that can be adjusted for lifting different volumes of liquids. Electric submersible pumps are used for

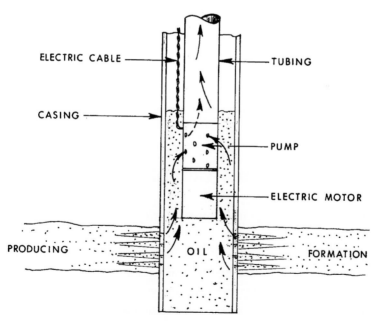

Fig. 20–20 A electric submersible pump on the bottom of a well

lifting large volumes of liquid up the well and for crooked and deviated wells. A gas separator is often used on the bottom of the pump to prevent gas from forming in the pump and decreasing the pump's efficiency.

A *hydraulic pump* is identical to a sucker-rod pump except it is driven by hydraulic pressure from a liquid pumped down the well. It uses two reciprocating pumps. One pump on the surface injects high-pressure *power oil* or *fluid* (usually crude oil from a storage tank) down a tubing string in the well. The power fluid drives a reciprocating hydraulic motor on the bottom of the tubing. It is coupled to a pump, similar to a sucker-rod pump, located below the fluid level in the well. The pump lifts both the spent power fluid and the produced fluid from the well up another tubing string. The power fluid causes the upstroke, and the release of pressure causes the downstroke. This type is called a *parallel-free pump*.

In another variation, the *casing-free pump*, the power fluid is pumped down a tubing string, and the produced liquid is pumped up the casing-tubing annulus. The stroke in a hydraulic pump is very similar to a sucker-rod pump stroke except it is shorter.

Hydraulic pumps can be either *fixed* (screwed onto the tubing string) or *free* (pumped up and down the well). They can be either *open*, with downhole mixing of power and produced fluids, or *closed*, with no mixing. Most hydraulic pumps are free and open.

Artificial lift in the United States consists of 82% beam pumper, 10% gas lift, 4% electric submersible pump, and 2% hydraulic pump.

Multiple Completions

Production from two or more zones in a well can be mixed (*commingled*) and brought up the same tubing string. Usually, however, the production is kept separate. If there are two producing zones, two packers and two tubing strings are used in a *dual completion* (Fig. 20–21). Sometimes only one packer and tubing string are used in a dual completion, and the production from one zone is brought up the casing-tubing annulus. If it is a flowing well,

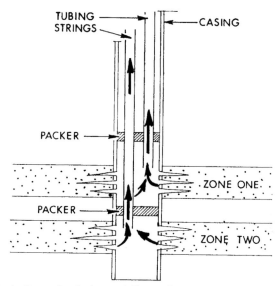

Fig. 20–21 *The bottom of a dual completion well*

Plate 20–6 *Double-wing Christmas tree*

there will be a double-wing tree (Plate 20–6). If beam pumping is used, there will be two beam pumpers and sucker-rod strings per well. Triple completions are the most completions that can be made in one well.

Intelligent Wells

An *intelligent* or *smart well* is a well that has downhole sensors that can measure well flow properties such as rate, pressure, and gas-oil ratio. An adjustable choke on the bottom of the well can be either automatically or manually adjusted, usually by hydraulics, to obtain an optimum production rate.

twenty-one

SURFACE TREATMENT AND STORAGE

Flowlines

A *flowline* is a pipe made of steel, plastic, or fiberglass that conducts the produced fluids from the wing on a Christmas tree or the *tee* on the wellhead of an artificial lift well (Fig. 21–1) to the separation, treatment and storage equipment. The flowline is located either on the surface or buried in the soil below the freeze line for protection from the weather.

Each well can have its own separation and storage facilities, or a *central processing unit (CPU)* can be used to service a group of wells. A gathering system of flowlines connects the wells to a central processing unit. *Headers* are larger pipes that collect the fluids from several smaller flowlines. A *radial gathering system* has the flowlines converging on a central processing unit (Fig. 21–2a). An *axial* or *trunk-line* gathering system has flowlines emptying into several headers that flow into a larger trunk line that takes the produced fluids to the central processing unit (Fig 21–2b).

Valves are gates on the flowlines that are designed to regulate the rate of flow through the line and open and close the line (Fig. 21–3). They can be either manual or automatic. Valves are named by their construction such as butterfly, needle, plug, ball, gate, bellows, or globe or by their use such as metering, check, safety, relief, regulating, pilot, or shut off.

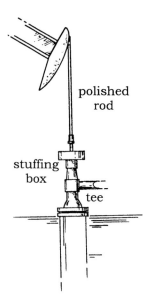

polished
rod

stuffing
box

tee

Fig. 21–1 *Wellhead with tee for flowline*

If both natural gas and water vapor flow through the flowline, the expanding gas cools with dropping pressure. This can chill the water to form a hydrate that blocks the flowline. *Hydrates* are similar in appearance to snow and contain methane molecules trapped in the ice crystals. They form between 30° and 70°F (-1° to 20°C), depending on the pressure. Hydrates can clog the flowline. Drying the gas before it enters the flowline can prevent the formation of hydrates. Heaters can be installed on the flowlines or chemicals such as glycol or methanol can be added to the produced fluids to prevent the formation of hydrates.

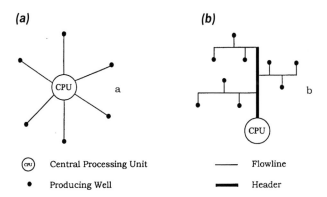

(a)

CPU

a

(b)

CPU

b

| (cpu) | Central Processing Unit | —— | Flowline |
| • | Producing Well | ▬▬ | Header |

Fig. 21–2 *Flowline patterns (a) radial (b) axial or trunk*

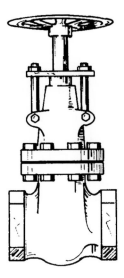

Fig. 21–3 Gate valve

Separators

Most oil wells produce salt water along with gas bubbling out of the oil. They are separated in a long, cylindrical steel tank called a *separator*. On each separator there is an inlet for fluids from the flowline and separate outlets at different elevations for each of the separated fluids. Every separator has a *diffuser section* that makes an initial separation of the gas and liquid from the inlet. The gas rises to the *gas-scrubbing section* at the top of the separator where most of the remaining liquid is removed from the gas before it goes out the *gas outlet*. The liquid falls to the bottom where the *liquid-residence section* removes most of the remaining gas from the liquid before it goes out the *liquid outlet*.

The separator can be either *vertical* (up to 12 ft or 3½ m high) or horizontal (up to 16 ft or 5 m in diameter). It is either a *two-phase separator* that separates gas from liquid (Fig. 21–4) or a *three-phase separator* that separates gas, oil, and water.

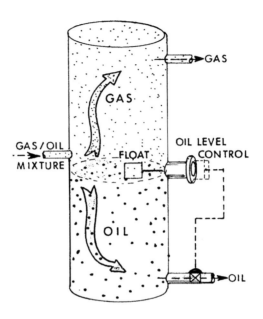

Fig. 21–4 *Vertical two-phase separator (modified from Baker, 1983)*

If gravity readily separates the produced oil and water, the water is called *free water*. In contrast, an *emulsion* has droplets of one liquid that are completely suspended in another liquid. A *water-in-oil* or *reverse emulsion* that has droplets of water suspended in oil is the most common emulsion produced from a well. Less common is an *oil-in-water emulsion* that has droplets of oil suspended in water. The *tightness* of an emulsion is the degree to which the droplets are held in suspension and resists separation. An emulsion can be *tight* and resist separation or *loose* and readily separate.

All separators use gravity to separate gas, oil, and water. In the diffuser section of a vertical separator, the fluids from the inlet are spun around the shell of the vessel to let centrifugal force help the initial separation. In a horizontal separator, the fluids from the inlet strike a flat, metal plate or angle irons to slow and divert the flow direction to let impingement help the initial separation.

In the gas-scrubbing section, the *mist extractors* are often used to coalesce and remove liquid droplets before the gas flows out the gas outlet. The wire-mesh pad type of mist extractor uses finely woven mats of stainless steel wire packed in a cylinder. *Vanes* that are parallel metal plates with collection pockets for the liquid can also be used.

The liquid residence section is often a relatively empty chamber that lets gravity make the separation. *Baffles*—flat plates over which the liquid flows as a thin film—can be utilized. A *weir* is a dam in the lower, liquid-residence section of the separator that impounds liquid behind it. It aids in separation by allowing only the lightest liquid, such as oil floating on water, to flow over the weir.

Separators have *liquid level controls,* which are floats on the oil/gas and water/oil surfaces that control the level of those fluids by opening and closing valves. Controls on a separator include liquid level, high- and low-pressure, high- and low-temperature, a safety relief valve, and a *safety head* or *rupture disc* that breaks at a set pressure. A *back-pressure valve* on the gas outlet maintains gas pressure in the separator.

A *free water knockout (FWKO)* is a three-phase, horizontal, or vertical separator used to separate gas, oil and free water by gravity (Fig. 21–5). Water is drawn out the bottom, oil from the middle and gas from the top.

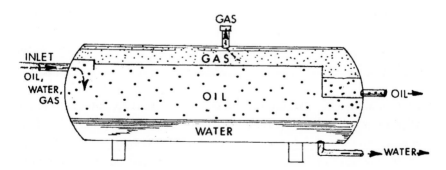

Fig. 21–5 *Horizontal free-water knockout (modified from Baker, 1983)*

A *double-barrel* or *double-tube-horizontal separator* has two horizontal vessels mounted vertically (Fig. 21–6). The produced fluid enters into the upper vessel where it flows over baffles to make an initial gas-liquid separation. The liquid then flows to the lower vessel to complete the oil-water separation. Oil-free gas flows out the upper barrel and gas-free oil flows out the bottom. A double barrel separator can process a higher volume of produced fluids than a single horizontal separator.

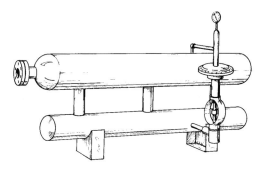

Fig. 21–6 *Double-barrel separator*

To separate or break an emulsion, the emulsion has to be heated. It is treated in a *heater treater*, a vertical or horizontal separator that has a *fire tube* (Fig. 21–7) where natural gas is burned. The fire tube can be either in contact with the emulsion (*direct-fired*) or in contact with a water bath that transfers the heat to the emulsion (*indirect-fired*). Gravity then separates the heated emulsion.

Electrode plates and electricity can also be used in an *electrostatic precipitator* to separate emulsions. If the emulsion is not very stable (loose), a large, settling tank called a *gun barrel* or *wash tank* is used for gravity separation. A *demulsifier* is a chemical that can be injected into a treating vessel to separate emulsions.

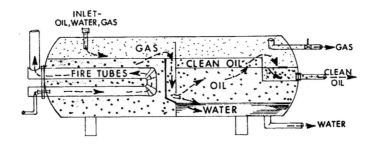

Fig. 21–7 *Heater treater (modified from Baker, 1983)*

Separators are rated by *operating pressures* that are between 20 and 1500 psi (1½ and 150½ kg/cm²). To maximize the retention of highly volatile components of oil, *stage separation* with several separators operating at decreasing pressures is used. The produced fluids flow first into a high-pressure separator and then through progressively lower pressure separators. The stock tanks are considered one stage in the separation. *Three-stage separation* (Fig. 21–8a) uses a high- and a low-pressure separator along with the stock tanks. *Four-stage separation* (Fig. 21–8b) uses high-, medium-, and low-pressure separators and the stock tanks. *Retention time*, the time that the fluids are in the separator, varies from one minute for light oils to five or six minutes for heavy oils in three phase separators.

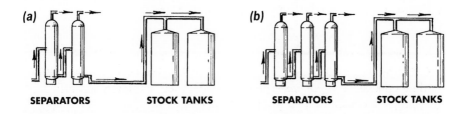

Fig. 21–8 *Stage separation (a) three-stage (b) four-stage*

Gas Treatment

Gas with minimal processing in the field so that it can be transported to a final processing plant is called *transportable gas*. Gas that has had final processing and is ready to be sold is called *sales-quality* or *pipeline-quality gas*. Natural gas is sold to a pipeline and must meet the specifications of the gas pipeline purchase contract. Pipeline-quality gas is dry enough so that liquids will not condense out in the pipeline and does not contain corrosive gasses. It has a minimum heat content (calorific value) that is usually between 900 and 1050 Btu/cf. The pressure must at least match pipeline pressure that is usually between 700 and 1000 psi (49 to 70 kg/cm^2). A maximum dew point for both water and hydrocarbons is specified. *Dew point* is the temperature at which a gas becomes a liquid as the temperature decreases.

The impurities in natural gas are removed by *gas conditioning* in the field before the gas can be sold to a pipeline (Plate 21–1). Removal of a liquid from a gas is called *stripping*.

Plate 21–1 Christmas tree and gas conditioning equipment

Gas conditioning equipment usually includes *dehydration* in which water is removed down to pipeline specifications that are often 7 lb of water vapor per MMcf. To remove the water, the gas can be bubbled up through bubble trays (Fig. 1–4) filled with a liquid desiccant such as glycol in a vessel called a *glycol absorber tower*. Triethylene glycol (TEG) is commonly used. When the glycol becomes saturated with water, the water is removed by heating in a vessel called a *reboiler*. The gas can also be dehydrated by passing it through solid desiccant beds of silica or alumina gel in a steel vessel called a *contactor*. Two contactors are used so that one can be kept on line while the other has water being removed from the desiccant by heating.

Corrosive gasses such as carbon dioxide (CO_2) and hydrogen sulfide (H_2S), called *acid gasses*, are removed by *sweetening*. The natural gas is passed through a *sweetening unit* that contains *iron sponge* (wood chips with impregnated iron) or chemical organic bases called *amines*. After the iron sponge or chemicals have become saturated, they are regenerated with heat to be reused. Pipeline specification often require the H_2S content to be reduced to 4 ppm by volume and CO_2 to 1 to 2% of the volume.

Solids in the gas are removed by filters which are metal cylinders packed with very fine glass fibers. They are designed to remove just solids or both solids and liquids on mist extracting baffles.

If the gas is wet gas, the valuable hydrocarbon liquids can be removed in a *natural gas processing plant* or *gas plant*. The gas plant uses either cooling or absorption to remove condensate, butane, propane, and ethane, called *natural gas liquids* (*NGLs*). In one cooling method, a heat exchanger vessel (*chiller*) uses propane as a refrigerant to cool the wet gas. Expansion also causes the gas to cool. A *low-temperature separator (LTX) unit* passes the gas through an expansion or choke valve. *Expansion turbines* or *turboexpanders* in a *cryogenic* or *expander plant* cause the gas to rotate turbine blades to cool the gas.

Absorption of natural gas liquids can also be done in an *absorption tower* as the gas is bubbled up through bubble trays (Fig. 1–4) containing a light hydrocarbon liquid that is similar to gasoline and kerosene. The natural gas is then removed by distillation. A *hydrocarbon recovery unit* uses beds of silica, activated charcoal or molecular sieves made of zeolites to remove the liquids.

The gas remaining after the natural gas liquids have been removed is called *residue* or *tail gas* and is primarily composed of methane that goes to the pipeline. A *straddle plant* on a high-pressure pipeline recovers condensate from the gas and returns the residual gas to the pipeline.

The salt water and/or condensate that is separated from natural gas is stored in the field next to the gas wells in steel tanks. They are similar but smaller in size than oil stock tanks.

Compressors are used to increase gas pressure and cause the gas to flow. A *positive displacement compressor* uses reciprocating pistons in cylinders. A *centrifugal compressor* uses spinning, turbine-like propellers on a shaft driven by a gas engine, gas turbine, or an electrical motor. The compressor is described by the *compression ratio (CR)*. This is the gas volume before compression divided by the gas volume after compression, such as 10:1. *Multiple stage compressors* are compressors that are joined together to compress gas in increasing increments. They are necessary for high pressures.

Storage and Measurement

Oil from the separator goes through a flowline to the *stock tanks* to be stored (Plate 21–2). Stock tanks are made either of bolted or welded sheets of carbon steel in sizes holding 90 to several thousand barrels of oil. They can be made to API specifications.

Bolted steel tanks are usually made of 12- or 10-gauge steel that is galvanized or painted. They come in many sizes and have the advantage that they can be assembled and repaired on the site. *Welded steel tanks* are prefabricated of 3/16 in. ($\frac{1}{2}$ cm) or thicker steel and transported to the site. The tank has either a flat or cone-shaped bottom to collect sediments.

Stock tanks can have a *vapor recovery system* to prevent loss of volatile hydrocarbons. The system is often a compressor vacuum line that takes the vapors from the stock tank to a suction separator where hydrocarbons liquids are separated from the gas. The liquids are then returned to the stock tank. The gas can be used in the field for an engine or sent to a pipeline.

Plate 21–2 Stock tanks

A minimum of two and usually three or four stock tanks are connected by pipe in a *tank battery* (Fig. 21–9) and are filled in sequence. The tank battery should hold a minimum of 4-days production. When the tanks are filled, either a service company will truck the oil to the refinery or the oil is transferred into a pipeline. About 1 ft (1/3 m) above the bottom of each tank is an *oil sales outlet* that allows the oil to be drained from the tank.

A steel stairway leads up to the *catwalk* that runs along the top of the stock tanks (Plate 21-2 and Fig. 21–9). This allows a person called a *gauger* to measure (*gauge*) the amount and quality of the oil in each stock tank. On the top of each tank is a *thief hatch* that can be opened. The gauger lowers a *gauge tape* marked each $\frac{1}{8}$ or $\frac{1}{10}$ in. with a brass weight on the end, down into the tank from a reel (Fig. 21–10). A *gauge* or *tank table* is used to relate the height of oil to the volume of oil in the stock tank. The gauger also samples the oil by using a *thief*, either a brass or glass cylinder about 15 in. (38 cm) long or with a 1-quart bottle with a stopper that is lowered in the tank. Samples can also be obtained from valves (*sample cocks*) that are located on the shells of some tanks.

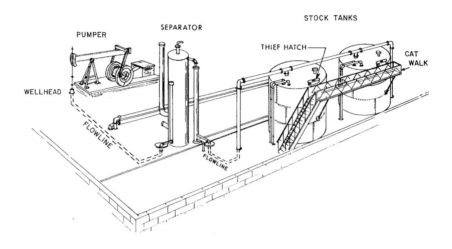

Fig. 21–9 *Pumper, separator and tank battery connected by flowlines*

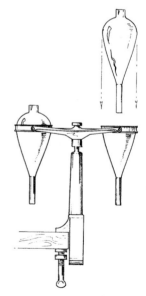

Fig. 21–10 *Gauge tape*   **Fig. 21–11** *Shake-out machine*

The temperature of the oil is measured and the sample is centrifuged in a glass tube with a *shake-out machine* (Fig. 21–11) during a *shake-out test*. The sediment and water are spun to the bottom of a glass container as the lighter oil rises. These impurities are called the *basic sediment and water* (*BS&W*) content of the oil. Generally, a pipeline or refinery will not accept oil with greater than 1% BS&W. Oil that is acceptable is called *clean* or *pipeline oil*.

Turbine meters can also be used to measure the volume of oil being transferred. They measure volume by the number of spins on the turbine shaft as the oil flows through the turbine and turns the blades on shaft. A *positive displacement* meter can also measure the volume of oil in separate, equal units.

When the oil is transferred from a tank battery to a tank truck or pipeline, a *run ticket* (a receipt) is made in triplicate by the gauger and witnessed by the pumper. A copy goes to the oil purchaser, and it is the legal instrument by which the operator is paid. It includes the well number, operator, °API of the oil, tank size, and gauge readings, BS&W content and temperature of the oil, seal number on the outlet, and time of transfer. *Division orders* listing each well owner's name, address, and percent interest are used by the operator for payments to each party.

Natural gas is also sampled, tested and measured. Common gas tests include compression and charcoal tests and a fractional analysis. A *compression test* uses a truck with a compressor and refrigerant to remove and measure the amount of condensate in the gas. A *charcoal test* also measures the condensate content of the gas by passing the gas over activated charcoal to adsorb the condensate. The condensate is then removed by distillation and measured. A *fractional analysis* of natural gas from a well is made by a laboratory. It is a chemical analysis of the gas sample by a gas chromatograph. The analysis includes a percentage of each hydrocarbon and water component, the GPM (gallons per 1000 cu ft.) of propane and higher gasses, and the heating value.

Gas volume is measured by a gas meter on the flowline. An *orifice gas meter* is commonly used. It measures the difference in gas pressures on gas flowing though an *orifice* (a round hole in a plate) on both sides of the orifice. The higher the flow rate, the greater the pressure drop across the orifice. The gas velocity is recorded on a circular *gas meter* chart in a *chart house*. The gas velocity has been calibrated to gas volume by a *meter prover.* Meter provers are used for both gas and liquid meters. They compare the volume of fluid flowing through the accurately calibrated meter prover with the same volume of fluid flowing through the meter being tested. The gas meter chart also shows the location of the meter, size of the orifice, date, and time. It is used to determine gas payments to the operator.

On many modern gas wells, equipment monitors the well's performance and the amount of gas being transferred (*electronic flow measurement – EFM*). It can be powered by a solar panel on the well.

A *field superintendent* is in charge of a producing field and gives orders to *production foremen* who work for him. Production foremen direct the crews that work in the field. A *pumper* is responsible for keeping the production equipment in operating order. *Roustabouts* help the pumper with the general labor.

A *lease automatic custody transfer (LACT) unit* uses automatic equipment to measure, sample, test, and transfer oil to a pipeline in the field and eliminate some of the labor. A probe in the LACT unit records the temperature and °API and BS&W content of the oil. If the BS&W content is too high, it can automatically send the oil back through the separation system. A positive displacement meter measures the volume of oil. The unit can also shut down the operation by closing a divert valve to the pipeline if there is too high a BS&W content. The LACT unit keeps records of all this.

The data collection and well control for both gas and oil wells has been modernized by the use of the Internet. Information such as electronic flow meter rates, compressor status, tank levels, run times, and casing and tubing pressures are measured automatically at the wells. The system is often powered by solar panels at the wellsite. The data can be automatically (or on command) sent digitally by telephone line or satellite to a server. This data is then made available with limited access to people who work on those wells. The system can issue an *alarm call-out* if equipment such as a compressor is down or not working at optimum. Commands can be sent back to the well to activate or adjust equipment such as speeding up or slowing down a compressor, opening or closing a choke or dumping a tank.

twenty-two

OFFSHORE DRILLING AND COMPLETION

During 1998 in the United States, the American Petroleum Institute reported that the average offshore oil well was drilled to 10,845 ft. (3306 m) at an average cost of $585/ft. whereas the average offshore gas well was drilled to 11,135 ft (3395 m) at an average cost of $632/ft. Offshore drilling operations and equipment are similar to those on land. The major difference is a top drive and the platform upon which the rig is mounted.

Preliminaries

In an offshore area in the United States, the states own the mineral rights out to three nautical miles from the shoreline (Fig. 22–1). Mineral rights for the states of Texas and Florida extend out to three leagues (nine nautical miles). The federal government owns the sea-bottom mineral rights from the state limit out to 200 nautical miles. This federal, offshore land is called the *outer continental shelf (OCS)*. Blocks on the OCS, often nine square nautical miles in size, are offered in periodic, closed-bid sales, usually going to the highest bonus offer. There is a one-sixth royalty and a 5-year primary term except for deep-water leases that have a 10-year term. In Canada, the offshore extent of mineral rights ownership varies considerably between different provinces.

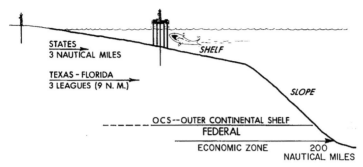

Fig. 22–1 *Mineral rights in offshore United States waters*

In 1959, the Gronigen Gas Field that has 58 trillion cu ft (1.8 trillion m³) of recoverable natural gas was discovered on land in the Netherlands. The trap is a faulted dome. The reservoir rock is Permian age sand dune sandstone (Rotliegendes Formation), the caprock is Permian salt (Zechstein Salt), and the source rock was the underlying Pennsylvanian age coals. The same rock layers underlie the North Sea, and it was immediately realized that there could be gas and oil fields in the North Sea. In 1964, mineral rights in the North Sea were divided amongst the countries that border the North Sea (Fig. 22–2).

Before the offshore rig is positioned, a *subsea site* or *soil investigation* of the bottom slope, composition, and load-bearing capacity is made to make sure it can support the rig. Buoys are then used to mark the site.

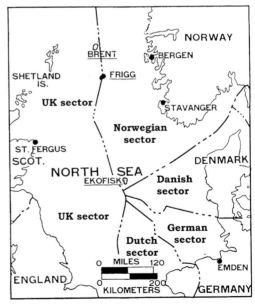

Fig. 22–2 *North Sea mineral rights*

Top Drive

The drilling rig on an offshore rig is similar to a drilling rig on land. A major difference is the top drive used on an offshore drilling rig (Plate 22–1). A *top drive* is a power swivel located below the traveling block that drives the drillstring. It is either a large electrical or hydraulic motor that generates more than 1000 horsepower. It is hung from the hook on the traveling block and turns a shaft into which the drillstring is screwed. The top drive moves up and down vertical rails to prevent it from swaying with any motion from waves. Drillpipe is added to the drillstring three joints at a time when using a top drive. The top drive saves time during making a connection. Slips are still used in the master bushing on a stationary rotary table to prevent the drillstring from falling down the well.

Plate 22–1 *Top Drive*

Offshore Crews

There are usually three crews on an offshore rig (Plate 22–2), two working on the rig and one off duty ashore. An onshore and an offshore crew are rotated once every two weeks. The crews work 12 hours from 11 a.m. to 11 p.m. and 11 p.m. to 11 a.m.

An offshore crew can have a driller, assistant driller, derrickman, roughnecks, motorman, diesel engine operator, pump operator, mud man, crane operator, and roustabouts. The *driller* handles the rotary controls, drawworks and pumps, while the *assistant driller* handles the pipe and racking system and the iron roughnecks. The assistant driller can also relieve the driller when necessary. The *crane operator* operates the crane used to lift equipment and supplies from supply ships and barges onto the offshore rig. He is also usually the *head roustabout* in charge of the roustabouts who handle the supplies and equipment. A tool pusher and company man will also be aboard. On some offshore rigs, a *pit watcher* can be responsible for the drilling mud and circulating equipment. There are often rig mechanics and electricians to maintain the rig.

On a large drillship, there can be quarters for 200 people. If the production platform is close to a port, the crew can be ferried out to the platform on fast crew boats for a crew change every two weeks. On many, however, the crew and other personnel are transported to and from the production platform by helicopters that land on the *helideck* (a flat platform).

Exploratory Drilling

In shallow, protected waters such as lagoons and canals up to about 25 ft (8 m) deep, the rig can be mounted on a *drilling barge*. Drilling barges have been very successful in canals on the Mississippi River delta marshes. A *posted barge* is a drilling barge designed to be sunk and rest on the bottom while drilling. The drilling deck is mounted on posts to keep it above the surface of the ocean.

Offshore, a *mobile offshore drilling unit (MODU)* is used. Three types are jackup, semisubmersible, and drillship.

Plate 22–2 *Fixed production platform (courtesy American Petroleum Institute)*

A *jackup rig* usually has two barge-like hulls and at least three vertical legs through the hulls (Fig. 22–3). The legs are either 1) *open-truss* with tubular steel members that are cross braced, or 2) *columnar* made of large-diameter steel tubes. The *cantilevered jackup rig* is most common with the derrick mounted on two large steel beams that protrude over the edge of the deck. Occasionally the derrick is mounted on the deck over a slot or key-way in the deck.

The jackup rig is usually towed into position although some are self-propelled. While moving, both hulls are together and float like a barge with the legs raised high. At the drillsite, the lower hull (*mat*) is flooded and positioned on the seafloor. On each of the legs is a *jack house* that uses a

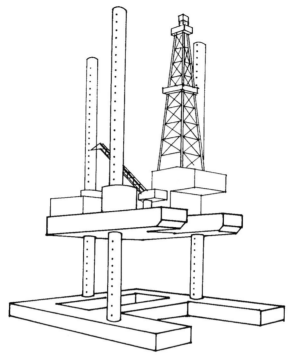

Fig. 22–3 *Jackup rig (modified from Exploration Logging Inc., 1979)*

rack-and-pinion arrangement powered by an electric or hydraulic motor to raise and lower the upper hull. The upper hull is jacked up on the legs until usually about 25 ft (8 m) above the sea surface. The drilling rig on the upper hull is then secure above the waves, and the mat acts as a stable foundation, even with a soft bottom. If the ocean bottom is relatively hard, smaller cylinders (*cams*) with a point on the bottom can be used on the bottom of each leg instead of a mat. After drilling, the hulls can be joined again and the rig towed or steamed to another site.

Legs on these rigs are constructed up to 550 ft (170 m) high. Jackups are generally used in water depths up to 300 ft (91 m) although some can drill in deeper waters.

For deep-water drilling, a *floater*, a semisubmersible, or drillship is used. A *semisubmersible* (or *semi* as it is commonly called) is a floating, rectangular-shaped drilling platform (Fig. 22–4). The most common type is the *column-stabilized semisubmersible.* Most of the rig flotation is in the pontoons, located 30 to 50 ft (9 to 15 m) below sea level when on station. Square or circular columns connect the drilling platform to the pontoons.

If it is relatively shallow water, the semisubmersible is anchored on station with a mooring pattern of anchors and chains radiating out from the rig. In deeper waters, it uses dynamic positioning. Because most of the flotation is below sea level in the pontoons, the rig is very stable even during high seas and winds.

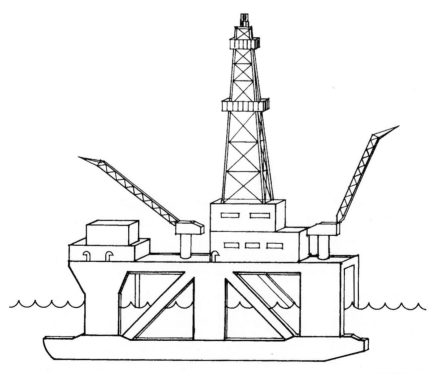

Fig. 22–4 *Semisubmersible rig (modified from Exploration Logging Inc., 1979)*

Once drilling is completed, the semisubmersible can be towed to another drillsite. To move the semisubmersible, the pontoons are emptied, and the rig floats high in the *transit mode* for easier towing. A *ballast-control specialist* supervises the raising and lowering of the semi and keeps it stable. For long-distance transport, the semi can be carried on the deck of a special ship during a *dry tow*. Some semis can drill in water depths up to 10,000 ft. (3000 m).

A *drillship* is a ship with a drilling rig mounted in the center (Fig. 22–5). The ship steams out to the drillsite and then drills through a hole in the hull, called the *moon pool*. The ship floats over the drillsite.

On the drillship, a satellite dish is used to track navigational satellites. A computer aboard the drillship constantly recalculates the drillship's location. If the drillship drifts off the drillsite, the computer engages the ship's propellers and puts the drillship back on location. Drillships have propellers on the side of the ship (*bow* and *stern thrusters*) and can move both back and forth and sideways. Keeping on station by computer is called *dynamic positioning.*

The driller on a semisubmersible or drillship usually sits in a chair at the *driller's station* in a temperature-controlled enclosure with large windows facing the drill floor. In front of the driller are computer monitors showing drilling parameters. Weight on bit, revolutions per minute, and other drilling factors can be adjusted from the driller's station using a joystick and touch screen. Mechanical devices, such as the pipe-handling unit that consists of elevator and torque wrench systems to manipulate the drillpipe, are controlled by the assistant driller who also uses a joystick and touch screen.

Drillships are very expensive. For efficiency, some modern drillships have the equipment and ability to drill two wells at the same time from one derrick. The derrick contains two traveling blocks and top drives, and the ship has two independent driller's and assistant driller's stations and two set-back areas to rack the pipe. Some drill ships have the capability of drilling in 35,000 ft (10,700 m) water depth.

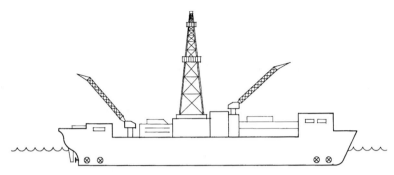

Fig. 22–5 Drillship (modified from Exploration Logging Inc., 1979)

Spudding an Offshore Exploratory Well

On a well drilled by a jackup rig, several hundred feet of large diameter (26 or 30 in.—66 or 76 cm), *conductor casing* is set into the sea bottom. On a very soft sea bottom, the conductor casing is jetted into the bottom by pumping seawater through the center. On a harder bottom, the conductor casing is pile-driven into the bottom. On a very hard bottom, a hole is drilled, and then the casing is run into the hole and cemented.

The conductor casing extends above sea level to just below the drilling deck. A smaller diameter hole is then drilled through the casing into the seafloor to several hundred feet below the bottom of the conductor casing. Surface casing is run into the hole and cemented. Next, a blowout preventer stack is bolted to the top of the surface casing. The rest of the well is then drilled and cased similar to a well on land.

On a well drilled by a floater (a semi or drillship), a *temporary guide base* or *drilling template* is installed on the sea bottom. The temporary guide base is a hexagonal-shaped steel framework with a hole in the center for the well. It is attached to bottom of a drillstring and lowered to the sea bottom. Four steel *guidelines* run from the sides of the temporary guide base up to the floater. They are used to lower and position other equipment in and on the well.

The drillstring is then raised back up to the floater leaving the temporary guide base on the sea bottom. A *guide frame* is then attached to the bottom of the drillstring. It has two or four arms through which the guidelines run. The drillstring and guide frame is then lowered down the guidelines to the temporary guide base. A large diameter hole (30 or 36 in.—76 or 91 cm.) is drilled through the center of the temporary guide base to about 100 ft (30 m) below the seafloor.

The drillstring and guide frame are then raised back to the floater. The guide frame is then attached to the lowest joint on the *foundation pile,* the first casing string run into the well. A *foundation pile housing* and *permanent guide structure* is attached to the top foundation pile joint. The foundation pile is then run into the hole and cemented. The permanent guide structure is attached to the temporary guide base on the sea bottom. The hole is then drilled deeper, and a string of conductor casing is run and cemented into the hole.

A subsea blowout prevent stack that can be activated from the floater is then lowered and locked onto the wellhead with a hydraulic wellhead connector. The drilling rig is then connected to the blowout preventers by a flexible, metal hollow tube (*marine riser*). The drillstring goes through the marine riser into the well. A *tensioner system* of wire rope and pulleys on the floater supports the upper part of the marine riser. The marine riser completes a closed system to circulate drilling mud down the drillstring and up the annulus between the drillstring and marine riser.

To compensate for the up-down motion (*heave*) of the floater on the surface, a *telescoping joint* that expands and contracts is used on the top of the marine riser. A *heave compensator* is also located between the traveling block and hook on the drilling rig. The compensator has pistons in cylinders that hold the hook and drillstring stationary as the floater heaves.

During an emergency such as severe weather, the blowout preventers can be closed and the marine riser disconnected from the BOP stack. The floater can then be moved off station to safety. After the emergency has passed, the well can be relocated and reentered.

Developmental Drilling and Production

After a commercial, offshore field has been discovered, it can be developed with a fixed production, tension leg, or compliant platform. A *fixed production platform* has legs and sits on the bottom. One type is called a *gravity-base platform* because it has a large mass of steel-reinforced concrete on the bottom of the legs, and gravity holds it in position (Fig. 22–6). The massive base has hollow cells that can be used for floatation when assembling and towing the platform into position. On location, the cells can be used for ballast or storage of crude and diesel oil. It is constructed in a sheltered, deep-water port along the shoreline and then towed into position. This type is used in areas of very rough seas.

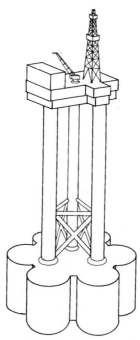

Fig. 22–6 *Gravity-base production platform (modified from Exploration Logging Inc., 1979)*

A more common type is the *steel-jacket platform* that has legs, the *steel jacket*, that sit on the bottom (Plate 22-2 and Fig. 22–7). It is constructed on land and either floated horizontally or carried on a barge out into position. It is then flooded and rotated vertically. Piles are driven into the sea bottom and bolted, welded, or cemented to the legs to hold it in position. A crane is used to lift the deck and modules such as power generation, crew quarters, and mud storage off deck barges and position them on the platform.

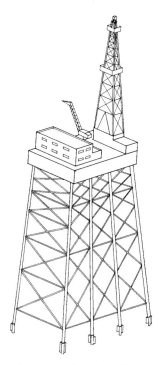

Fig. 22–7 *Steel-jacket production platform (modified from Exploration Logging Inc., 1979)*

Offshore platforms often have several *decks* (flat surfaces) on top of each other to serve various functions such as power and drilling (Plate 22–2). Wellheads are usually located on the lower *cellar deck*. Separators, treaters, and gas compressors are located on the platform. The treated oil or gas is then usually sent ashore through a submarine pipeline.

386

A crane is used to lift supplies and equipment aboard the platform. Usually one or two derricks are left on deep-water platforms after the wells have been drilled to use for workovers.

The deepest water depth for a production platform is 1350 ft. (411 m). In relatively shallow waters, there can be a separate *quarters platform* for the crew next to the *production platform* as a safety precaution. A *bridge* connects the platforms.

A *tension-leg platform (TLP)* floats above the offshore field. It is held in position by heavy weights on the seafloor (Fig. 22–8). The weights are connected to the tension-leg platform by hollow, steel tubes, 1 to 2 ft (0.3 to 0.6 m) in diameter, called *tendons.* The tendons pull the platform down in the water to prevent it from rising and falling with waves and tides.

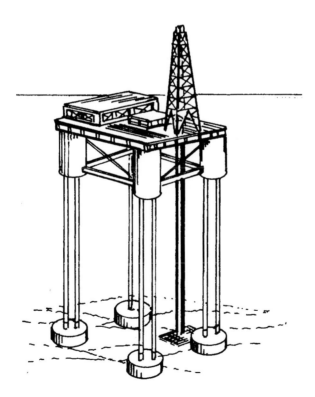

Fig. 22–8 Tension-leg platform

Tension leg platforms can be installed in water depths up to about 3500 ft (1065 m). A *tension-leg well platform*, in contrast, has only wellheads and no production treating facilities onboard. The produced fluids are sent by seabed pipeline to a production platform in shallow water for treatment.

A *compliant platform* (Fig. 22–9) is a relatively light production platform that is designed to sway with wind, waves, and currents. One type, a *guyed-tower*, is attached to a pivot on the ocean bottom. Another type, a *spar*, is a floating production platform in the shape of a closed, vertical cylinder like a buoy. The spar is designed to not rise and fall with the waves. Both the guyed-tower and spar are held in position with radiating guy wires and ocean bottom weights.

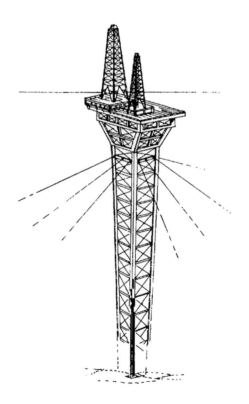

Fig. 22–9 Compliant tower

Wells, often 32 to 40, are drilled through a *well template* on the ocean bottom that is used to position and separate the wells. The template is a steel frame with slots for each well. It can be either 1) on the bottom of a leg and the wells are drilled through the leg or 2) on the sea bottom between the legs or tendons. It supports the equipment necessary to drill and produce the wells. Each slot on the template locates and standardizes the instillation of a well. There are usually one or two extra slots left undrilled on the template for any further field development.

The offshore field is developed by deviation drilling from one platform. If the offshore well flows, it is completed with a Christmas tree. An offshore oil well that needs artificial lift is usually completed with gas lift. Offshore wells are required by law to be equipped with *storm chokes*. The choke is installed on the bottom of the well and is closed either manually or automatically during an emergency.

Subsea Work

Installation and work on a well underwater can be done by three methods: a diver, a diver in a one-atmosphere diving suit and a remotely operated vehicle. A *saturation diver* breathing a helium and oxygen mix can work down to 1000 ft (300 m). The *one-atmosphere diving suit (ADS)* is a hard diving suit with one atmosphere air pressure in it and human-powered limbs. It can operate down to 2300 ft (700 m). A *remotely operated vehicle (ROV)* is an unmanned submersible that can effectively operate down to 15,000 ft (4600 m). It is connected to a mother ship on the surface by a cable (*umbilical*). A closed-circuit television camera on the ROV allows operators on the surface to manipulate the ROV with thrusters and do work with manipulator arms. ROVs used to work on offshore wells are very similar to those used to discover and explore sunken ships.

Subsea Completions and Wells

A *subsea completion* consists of a wellhead and production equipment such as a Christmas tree or gas lift on the bottom of the ocean. The *subsea well* is drilled from a floater instead of a production platform. The completion can be either *dry*, with an atmospheric chamber surrounding the equipment (Fig. 22–10a), or *wet*, which is exposed to sea water (Fig. 22–10b). Divers are used to service dry completion wells.

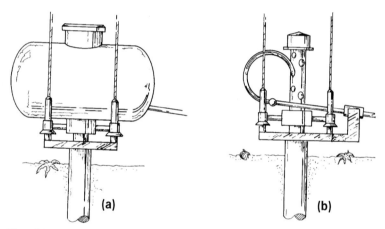

Fig. 22–10 *Subsea wells (a) dry (b) wet*

The production from a subsea well can be taken by flowline to a *subsea manifold* where it is commingled with production from other subsea wells. It is then taken by a flowline to a production platform in shallow water, a tension leg platform, a semisubmersible facility, a spar tower, or up a production riser to a floating production, storage, and offloading vessel (Fig. 22–11) for processing. The *floating production, storage, and offloading (FPSO)* vessel is a converted tanker, a semisubmersible or a specially built ship that contains separation and treating facilities. It can be kept on position by an anchoring system or by dynamic positioning. The treated oil is then transferred from the FPSO to a *shuttle tanker* to be brought ashore. Ultra deep-water wells are done with subsea completions.

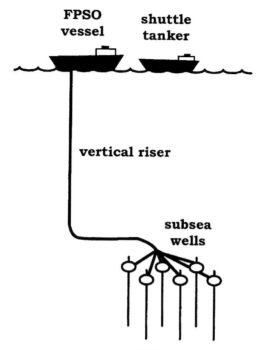

Fig. 22–11 *Subsea wells tied in to an FPSO vessel*

Remote portions of an offshore field and smaller fields that are not economic by themselves can be developed by *satellite wells*. The satellite well is drilled by a mobile offshore drilling unit and completed as a subsea well. The satellite well is then tied to an existing production platform by flowline.

Unstable Sea Bed

The seabed is very unstable off deltas where landslides called *submarine mudflows* are common in the loose sediments. During a large hurricane off the Mississippi River delta in 1969, several offshore oil platforms failed. At first it was thought that the large hurricane waves and high winds were directly responsible for the failure of the rigs. It was later shown, however, that the hurricane waves caused large submarine mudflows that flowed down the Mississippi River delta bottom, knocking the legs out from under the rig. Because of the instability of the sea bottom off the Mississippi River delta, a total of 23 rigs have failed and, in an average year, 110 pipelines fail.

twenty-three

WORKOVER

There are times in the life of a producing well in which the well must be *shut in* (production stopped) and remedial work done on the well to maintain, restore, or improve production by a *workover*. Workover includes both solving mechanical problems and cleaning out the well.

Equipment

In the past, when a well was drilled and completed, the derrick was often left standing above the well to raise and lower equipment into the well during a workover. Today, a service company does a workover with a *production* rig that is either a workover rig for more extensive work or a smaller service or pulling unit.

A *workover rig* looks similar to a drilling rig. It can drill using a workstring of drillpipe or tubing and circulate with either a water-base or oil-base mud or foam made of air or nitrogen. A *service, well-servicing,* or *pulling unit* uses hoisting equipment mounted on a truck body (Fig. 23–1) or on a trailer. A winch and mast system is mounted on the truck or trailer. The winch is driven off the truck engine. There is a crown block, traveling block, and hoisting line similar to a small drilling rig.

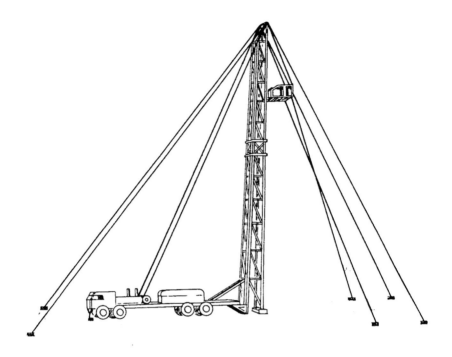

Fig. 23–1 Service unit

The hoisting line is either a *wireline* made of steel braided wire or a *slick line* of one solid, steel wire. Elevators on the traveling block are used to clamp onto joints of tubing and sucker rods. Hydraulic slips and hydraulic tongs (both sucker rod and tubing) are used to make up and break out connections. A swivel and rotating head with a small kelly can be used to rotate a workstring.

Two types of service unit masts are pole and structural. A *pole mast* is made of tubular steel that can both pivot and telescope up and down on a wire rope system as the unit is rigged up and torn down. It is either *single-pole* or *double-pole*, depending on the number of tubulars. A single-pole unit requires that any tubing or sucker rods that are pulled from the well be laid on the ground. A double-pole unit has either a single *racking platform* or

Plate 23–1 *Well service unit pulling tubing. Note the tubing platform (lower) and the rod basket (higher)*

Fig. 23–2 *Coiled tubing unit*

both a rod basket and hanger and a tubing platform. A person can stand on the racking platform and rack the tubing or sucker rods vertically in metal fingers on the racking platform. A *rod basket, hanger,* or *rack* is located further up on the mast and is used to rack three sucker rods at a time. The *tubing platform,* used to rack tubing, is located lower on the mast (Plate 23–1).

A *structural mast* is made of angular steel. It can also telescope up and also has a single racking platform or both rod basket and tubing platform.Masts are stabilized with radiating guy wire attached to anchors in the ground.

A service unit has a three- or four-man crew consisting of a service unit operator, derrickman, and one or two floormen. Unlike a drilling rig, work is done on a service unit only during daylight, except offshore.

A *carrier unit* is a highway vehicle designed for the workover engine-hoist-mast system. The hoist is driven by one or two diesel engines on the unit. Unlike a service unit, the power to drive the carrier unit on the highway comes from the hoist engines. Hydraulic pumps pivot and raise the mast. Some carrier units have the mast on the back (*back-in unit*), and others have the mast and cab on the same end (*front-in unit*) for better driver visibility. Hydraulic leveling jacks adjust the unit on uneven ground.

A *coiled tubing unit* (Fig. 23–2) is a service unit that uses tubing wound around a reel on the truck to raise and lower equipment in the well. Coiled tubing is a continuous length of flexible, steel tubing up to 19,000 ft (5800 m) long with a diameter that is often $1^{1}/_{4}$ in. (3 cm). It is unwound from a reel, gripped by friction blocks and fed into an *injector* as it goes into the well.

A coiled tubing unit reduces trip time compared to using straight tubing joints that have to be screwed on or off the workstring each 30 ft (10 m) on a common service unit. Because the tubing is injected into the well through a control head that maintains a pressure seal, coiled tubing can be used on high-pressure wells. Drilling can also be done with a coiled tubing unit using larger diameter tubing. Coiled tubing, however, cannot be rotated, and a turbine mud motor is used to drive the bit.

A *concentric tubing workover* uses smaller diameter and lighter equipment than normal that is run down the tubing string. A lightweight *macaroni rig* is designed to run $^1/_4$ and 1 in. (2 and $2^1/_2$ cm) diameter tubing into the well.

Snubbing units are designed for workovers in wells under high pressure. They use unidirectional slips to grip and force tubulars into or out of a well as one of the two preventers are alternately engaged or while using a solid rubber stripper head. The tubulars are forced into the well using blocks and wireline or hydraulic power.

Well Intervention on Offshore Wells

Well intervention is work on a producing well. It can include repairing, replacing, or installing equipment, workover, well stimulation, and production logging. Well intervention on offshore wells can be very expensive, and an interruption in production can be very costly if the well is shut in. Offshore well intervention is designed to be very efficient and, if possible, done without shutting down production.

For offshore well servicing, a service company is often contracted. On a small production platform, a wireline unit and small hoist can be used. A socket with a *sinker bar* (a weight) is attached to the end of the wireline to run it down the well. On larger platforms, satellite wells and subsea wells, a barge, jackup, or semisubmersible with workover equipment and living quarters for the workover crews is employed. The mast is located on two steel beams that project over the side of the vessel to position the mast over the well to be serviced.

Deep-water production platforms have one or two derricks that are left on the platform for workovers and the barge, jackup, or semi is not needed. For highly deviated and subsea wells, *through-the-flowline (TFL)* or *pump-down* equipment has been developed to workover wells. The equipment is circulated down the flowline into the production tubing, activated by pressure, and then recirculated out the tubing. For well simulation, a specially designed ship with frac equipment can be used.

Preparing the Well

The well is *killed* (flow is stopped) by filling the well with a *kill fluid* such as brine, drilling mud, oil, or a special liquid before most workovers. Often the kill fluid is pumped down the casing-tubing annulus and back up the tubing string. After the well has been killed, blowout preventers are usually installed, and the tubing string and other downhole equipment are removed. The blowout preventers may not be necessary on a low-pressure well.

Well Problems

Sand Cleanout

Loose sand from unconsolidated sandstone reservoirs can clog the bottom of the well, causing a *sand control problem*. Without removing the production tubing, either a coiled tubing or macaroni rig can be used to run a small diameter workstring of tubing down the production tubing. Salt water pumped down the workstring tubing picks up the sand and circulates back up the workstring tubing–production tubing annulus.

If the production tubing is removed, a workstring of larger diameter, tubing is run down the well. The salt water is pumped down the casing-tubing annulus, and the sand and water is circulated up the tubing. A *bailer* or *sand pump* on a wireline can also be run down the tubing string to remove the sand.

Loose sand grains in a reservoir can be stabilized by pumping an epoxy resin into the well to glue the sand grains together adjacent to the well. The Wild Mary Sudik Well in the Oklahoma City field is a famous example of a sand control problem

Well Cleanout

Because of the rapid drop in temperature and pressure between the reservoir and the bottom of the well, calcium carbonate, barium sulfate, calcium sulfate and magnesium sulfate can precipitate out of oilfield brine to form *scale* (a salt coating) in the tubing. Chemicals, called *scale inhibitors,* can be pumped down the well to dissolve and remove the scale.

Tubing can be clogged with waxes from a waxy crude. A *paraffin knife* (Fig. 23-3a) or *paraffin scraper* (Fig. 23-3b) run on a wireline through the tubing can be used to remove the wax. A *hot oil treatment* uses heated oil, usually from the separators, that is pumped down the well by a service company (*a hot oiler*) to dissolve the wax. The oil is then pumped back out of the well. Chemicals (*paraffin solvents*) can be pumped into the well and flowlines to remove wax.

Pulling Rods

The sucker-rod string on a beam-pumping unit can break due to corrosion or wear. The intact, upper portion of the sucker-rod string is pulled (*pulling rods*) and unscrewed using with a *power rod tong* or manually with a metal circle called a *back-off wheel* or *circle wrench*. A fishing tool, either a

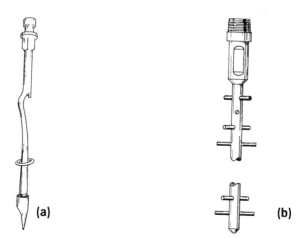

(a) (b)

Fig. 23–3 (a) paraffin knife (b) paraffin scratcher

sucker-rod overshoot or *mousetrap* is used to remove the lower, broken part of the sucker-rod string.

The rods also have to be pulled when repairing the tubing or downhole pump. When they are pulled, they are either laid on the ground when using a single-mast unit or are racked vertically, three at a time, in the rod hanger when using a double-pole or structural mast unit. If the well is under pressure, a *rod blowout preventer* is attached to the top of the well for safety when pulling rods.

Pulling and Repairing Tubing

When production in a well falls, it could be due to a leak in the tubing string caused by corrosion or stress or abrasion on the tubing string from the sucker-rod string. The tubing is pulled and inspected. As the tubing is being run back into the well, it can be pressure-tested with a portable hydraulic pressure rig. Collapsed tubing can be opened with a *tubing swage* (Fig. 23–4) run on a wireline several times through the tubing string.

Downhole Pump

Falling production in a well can also be due to a malfunctioning downhole pump. The sucker-rod string is pulled to retrieve a rod or insert pump. Both the sucker-rod and tubing string are pulled to retrieve a tubing pump. The pump seals, and parts are then inspected and repaired if necessary.

Fig. 23–4 *Tubing swage*

Casing Repair

Collapsed casing in the well can be opened with a *casing roller* (Fig. 23–5a) that uses a series of rollers on the sides. It can also be reamed out with a *tapered mill* (Fig. 23–5b), that is run on a workstring and rotated. If the collapsed casing cannot be opened, the well will have to be *sidetracked* (drilled out around the collapsed casing). Leaks in the casing can be located with pressure tests in the well. Casing holes are repaired with a metal *casing patch* glued in place with epoxy resin.

If the upper part of the casing string is damaged but not cemented into the well, it can be cut with a *chemical cutter* (Fig. 17–6) and retrieved. The chemical cutter is run into the well on a wireline. A chemical propellant in the tool is activated by an electrical signal, and high-temperature, corrosive fluid jets out of the cutter ports ressure to slice through the casing. The casing is then pulled from the well. A casing overshot or patch tool and new casing is then run in the well.

Secondary Cementing

Primary cementing is the cement job done on casing when it is originally run. *Secondary cementing* is done on a well during a workover. A *cement bond log*, a type of sonic log, can be run in a cased well to determine where and how well the cement has set behind the casing.

(a) (b)

Fig. 23–5 *(a) casing roller (b) tapered mill*

Gaps in the cement behind the casing are called *holidays* and can be filled by *squeeze cementing* (Fig. 23–6). The casing adjacent to the holiday is perforated, and the zone is isolated with packers. Cement is then pumped under pressure down the well, through the perforations and into the holiday. Cement squeezing can also be used to repair casing leaks.

A cement squeeze job is either a bradenhead squeeze or packer squeeze. A *bradenhead squeeze* is a relatively low-pressure cement squeeze job in which the cement is pumped down a tubing string or drillstring (workstring). The workstring is positioned just above the zone to be squeezed. The workstring-casinghead (*bradenhead*) annulus is closed. Pressure is applied through the

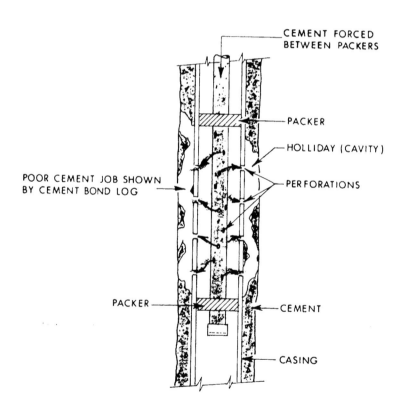

Fig. 23–6 Cement squeeze job

workstring to squeeze cement through the perforations. A *packer squeeze* is a relatively high-pressure cement squeeze job. A packer is used to seal the workstring-casing annulus above the zone to be squeezed. The cement is pumped down the workstring, and pressure is applied.

Swabbing

Swabbing is the removal of water or drilling mud from a well (*unloading*) so that the oil and gas can flow into a well. A *swab job* is done both after the well is completed to remove the last of the drilling mud or completion fluid and to restore production in a producing well.

A truck-mounted *swabbing unit* with a short mast is used to lower a swab tool down the tubing string on a wireline (Fig. 23–7). A *swab tool* is a hollow steel rod with rubber swab cups. When the swab tool is raised, the swab cups seal against the tubing to act as a piston and lift (*swab out*) the liquid out of the well. A *lubricator* (a length of casing or tubing) is temporarily attached above the valve on the tubing head or casinghead to provide a pressure seal. The swab tool can be run into a well under pressure through the lubricator so the well doesn't have to be killed during swabbing. An *oil saver* is used on top of the lubricator to retain any oil coming up on the wireline.

Sometimes a gas well will not flow because of water filling the well. Soap sticks can be dropped into the tubing (*soaping the well*) to form gas bubbles in the water to help lift the water out of the well.

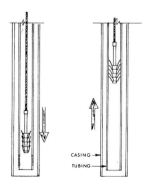

Fig. 23–7 *Swab job*

Replacing Gas-Lift Valves

Gas-lift valves can stick in an open or closed position. If the gas lift valve was installed in a *gas-lift mandrel,* a side pocket in the tubing, it can be retrieved and run back in on a wireline. If not, the tubing string must be pulled to retrieve the valve.

Replacing Packers

Packers are designed to seal the casing-tubing annulus. A tubing or completion packer is commonly used in well completion and sometimes has to be repaired or replaced. Packers are retrievable or nonretrievable. The *retrievable packer* is easily removed by pulling the tubing string. The *nonretrievable packer* is made of drillable material such as soft metal and has to be milled out.

Recompletion

A well is *recompleted* by abandoning the original producing zone and completing in another zone. The well can be either *drilled deeper* to complete in a deeper zone or *plugged back* to complete in a higher zone. The original producing zone must be sealed with cement in one of three methods. A cement squeeze job can be used to plug perforations in the depleted zone. Cement can be pumped down a workstring until it fills the well to the desired level (a *cement plug*). Also, a *bridge plug* can be used to mechanically seal that level of the well, and then a *dump bailer,* a long cylinder filled with cement can be run into the well to place the cement on top of the bridge plug.

twenty-four

RESERVOIR MECHANICS

Reservoir Drives

Pressure on the fluids in a reservoir rock causes the fluids to flow through the pores into the well. This energy that produces the oil and gas is called the *reservoir drive* or *reservoir energy* and comes from fluid expansion, rock expansion and/or gravity. The type of reservoir drive controls the production characteristics of that reservoir.

There are four different types of reservoir drives for oil reservoirs. Every oil reservoir has at least one and sometimes two of these reservoir drives. The relative importance of each reservoir drive can change with production. Gas reservoirs have only one of two types of reservoir drives.

Oil Reservoir Drives

A *dissolved-gas, solution-gas,* or *depletion drive* is driven by gas dissolved in the oil. In the subsurface, the oil is under high pressure and has a considerable amount of natural gas dissolved in it. When a well is drilled into the reservoir and production is initiated, pressure on the oil in the reservoir decreases, and gas can bubble out of the oil. Expanding gas bubbles in the pores of the reservoir force the oil through the rock into the well. The expanding volume of oil and rock as the pressure drops also helps the drive.

A dissolved-gas drive reservoir has a very rapid decline in both reservoir pressure and oil production rate as the oil is produced (Fig. 24–1). Because of the rapid reservoir pressure drop, any flowing wells have to be put on pumps early. Little or no water is produced during production from this type of reservoir. There is a very rapid gas/oil ratio increase near the end of production. A dissolved gas drive is very inefficient and will produce relatively little of the original oil in place from the reservoir. A *secondary gas cap* located on the subsurface oil reservoir can be formed by gas bubbling out of the oil.

A *free gas-cap expansion drive* reservoir is driven by gas pressure in the free gas cap above the oil. The expanding free gas cap pushes the oil into the wells. Any solution gas bubbling out of the oil adds additional energy. A free gas-cap expansion drive reservoir has a moderate decline in both reservoir fluid pressure and production rate as the oil is produced (Fig. 24–2). A sharp rise in the gas/oil ratio as the oil is produced from a well shows that

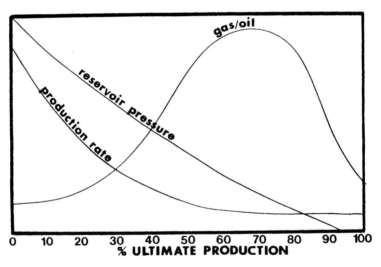

Fig. 24–1 *Characteristics of a dissolved gas drive oil field with time (modified from Murphy, 1952)*

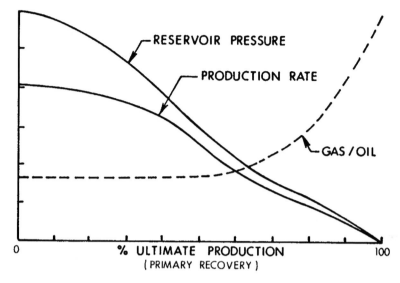

Fig. 24–2   *Characteristics of a free-gas cap expansion drive oil field with time (modified from Murphy, 1952)*

the expanding free gas cap has reached the well, and further oil production will be very limited from that well. This type of reservoir is best developed with wells producing only from the oil portion of the reservoir, leaving the gas in the free gas cap to supply the energy. Usually little or no water is produced. The recovery of oil in place from this type of reservoir is moderate.

Water drive reservoirs are driven by the expansion of water adjacent to or below the oil reservoir. The produced oil is replaced in the reservoir pores by water. The water can either come from below the oil reservoir in a *bottomwater drive* or from the side in an *edge water drive*. An active water drive maintains an almost constant reservoir pressure and oil production through the life of the wells (Fig. 24–3). The amount of water produced from a well sharply increases when the expanding water reaches the well and the well *goes to water*. The recovery of oil in place from a water-drive reservoir is relatively high.

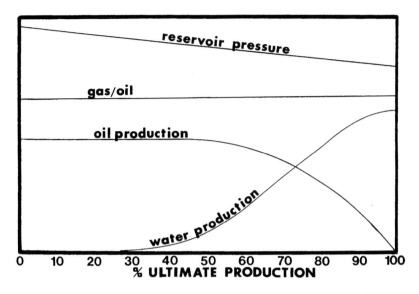

Fig. 24–3 *Characteristics of a water drive oil field with time (modified from Murphy, 1952)*

Gravity is also a drive mechanism. It is present in all reservoirs, as the weight of the oil column causes oil to flow down into the well. It is most effective in very permeable reservoirs with a thick oil column or with a steep dip. Gravity drive is common in old fields that have depleted their original reservoir drive. Down dip wells will have higher production rates than those updip. In a *gravity drainage pool,* the rate of oil production is usually low compared to other drives but oil recovery can be very high over a long period of time.

Many oil reservoirs have several reservoir drives and are called *combination* or *mixed-drive reservoirs.* The relative importance of the reservoir drive will change with time during production. In the later stages of oil production in a dissolved-gas drive reservoir, gravity drainage becomes significant. The most efficient reservoir drive system is a combination of free gas and water drives sweeping the oil from both above and below into the wells.

The East Texas oil field has an active water drive. The oil reservoir is in contact with a very extensive aquifer below it in the Woodbine Sandstone.

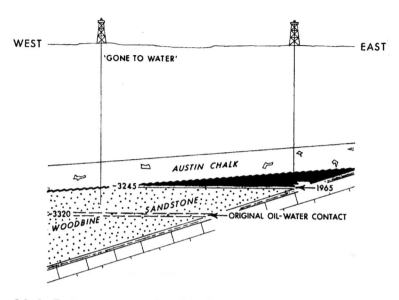

Fig. 24–4 *East-west cross section of the East Texas oil field showing the movement of the oil-water contact with production from 1930 to 1965 (modified from Landes, 1970)*

The expanding water forces the oil up into the wells. The original oil-water contact in the East Texas field, when the field was discovered in 1930, was at 3320 ft (1012 m) below sea level (Fig. 24–4). As oil was produced, the bottomwater drive caused the oil-water contact to rise to a level of 3245 ft (989 m) below sea level by 1965.

Wells on the west side of the field went to water first. Wells on the east side will have the longest production history and go to water last (Fig. 24–5). Water invading a reservoir is called *water encroachment*. By 1993, the field had produced 5135 million bbls (816 million m³) of oil. Only 4% of the field's original reserves, about 210 million bbls (33 million m³) of oil, was left to be produced in the next 10 to 15 years. The recovery will be more than 82% of the oil in place. This extremely high oil recovery is due to 1) the strong water drive and 2) the high porosity (about 30%) and permeability (1000's of md) of the Woodbine Sandstone reservoir, and 3) the low-viscosity oil.

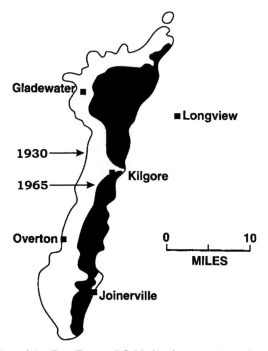

Fig. 24–5 *Map of the East Texas oil field showing west to east water encroachment from 1930 to 1965 (modified from Landes, 1970)*

The Turner Valley field, located to the southwest of Calgary, Alberta, is
formed by a large drag fold on a thrust fault along the disturbed belt (Fig.
24–6a). The reservoir rock is primarily Mississippian age limestone. The
field was discovered in 1913 by drilling with a cable tool drilling rig next to
a gas seep. Because the oil reservoir was 2000 ft (600 m) deeper and located to the west of the free gas cap, it was first believed that it was just a gas
field. The gas was wet and was produced in large quantities. The condensate was removed in a gas processing plant at Turner Valley and sold for gasoline. It was called skunk gas because it contained sulfur and stunk like a
skunk when burned in an automobile. There was little use of natural gas.
Almost all the dry gas from the gas processing plant was flared in a location
called Hell's Half Acre.

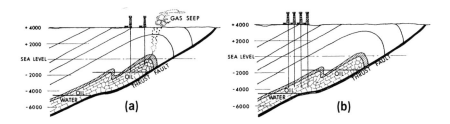

Fig. 24–6 *East-west cross section of Turner Valley field, Alberta (a) during gas production from free gas cap (1913-1930) (b) during oil production (after 1930) (modified from Gallup, 1982)*

In 1930, oil was discovered below the gas cap (Fig. 24–6b). Unfortunately, although Turner Valley holds about 1 billion bbls (160 million m³) of oil, the free gas cap, which was the oil reservoir drive, was depleted. Less than 12% of the oil will ultimately be produced. The reservoir is too deep to economically repressure it. The gas in a free gas cap should never be produced before or during oil production as it supplies the energy to produce the oil.

The reservoir drive of an oil field can be determined from both the nature of the reservoir and from production characteristics. Isolated reservoirs that are encased in shales such as shoestring sandstones and reefs or those cut by sealing faults often have dissolved gas drives. If the reservoir has a large free gas cap, it has a free gas cap drive; if not, it probably has a dissolved gas drive. Extensive sandstones and other reservoirs that connect to large aquifers often have water drives. Abnormally high pressure suggests that the reservoir is isolated and does not have a water drive.

Reservoir pressures and oil production will also indicate the type of reservoir drive. A rapid decrease in both reservoir pressure and oil production is characteristic of a dissolved gas drive. Shutting in wells will not cause the reservoir pressure to build up. An active water drive has almost constant reservoir pressure and oil production. If the reservoir pressure does decrease, shutting in the wells allows the reservoir pressure to increase to almost its original pressure.

Gas Reservoir Drives

Gas reservoirs have either an expansion-gas or water drive. An *expansion-gas* or *volumetric drive* is due to the pressure on the gas in the reservoir. The high-pressure gas in the pores of the reservoir expands out into the well. This drive recovers a relatively large amount of original gas in place in the reservoir.

A *water-drive* gas reservoir is similar to a water-drive oil reservoir and is due to expanding water adjacent to the reservoir. It is not as effective as an expansion-gas drive because the water flows around and traps pockets of gas in the reservoir. It has a moderate recovery of gas in place.

Maximum Efficient Rate

The *maximum efficient rate (MER)* is the maximum rate at which a well or field can be produced without wasting reservoir energy or leaving bypassed oil in the reservoir. It generally ranges from 3 to 8% of the recoverable oil reserves per year. This rate provides for an even rise in the oil-water contact (water encroachment) along the bottom of the reservoir and holds the gas-oil ratio to a minimum. The MER of a field can be accurately determined only after reservoir drives have been identified and productivity tests have been run.

twenty-five

PETROLEUM PRODUCTION

In 1999, there were 914,127 producing oil wells in the world. Of those, 554,385 (61%) were in the United States, 48,258 (5%) in Canada, 1685 in Iraq (<0.2%), 1560 in Saudi Arabia, 1120 in Iran, and 790 in Kuwait. Of the 861,834 producing wells in the United States, 36% are gas wells and 64% are oil wells.

A *petroleum engineer* is an engineer who is trained in drilling, testing, and completing a well and producing oil and gas. A *reservoir petroleum engineer* is in charge of maximizing the production from a field to obtain the best economic return.

Well and Reservoir Pressures

Tubing pressure is measured on the fluid in the tubing, whereas *casing pressure* is measured on the fluid in the tubing-casing annulus. The pressure gauge at the top of a Christmas tree measures tubing pressure. *Bottomhole pressure* is measured at the bottom of the well. The pressure is measured either as *flowing*, with the well producing, or *shut-in* or *static*, after the well has been shut-in and stabilized for a period of time such as 24 hours (Fig. 25–1). *Downdraw* is the difference between shut-in and flowing pressure in a well.

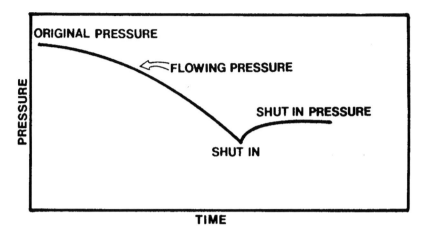

Fig. 25–1 *Flowing and shut-in pressure in a well*

The original pressure in a reservoir before any production has occurred is called *virgin, initial,* or *original pressure.* During production, reservoir pressure decreases. Reservoir pressure can be measured at any time during production by shut-in bottomhole pressure in a well. A *pressure bomb,* an instrument that measures bottomhole pressure, can be run into the well on a wireline. A common pressure gauge consists of a pressure sensor, recorder, and a clock-driven mechanism for the recorder. It is contained in a metal tube about 6 ft (2 m) long. The chart records pressure with time as the test is being conducted. Temperature can also be recorded on a similar instrument. Another type, an *electronic pressure recorder,* can be run on a conductor wire.

Well Testing

Tests are run on a well to determine the optimum production rate. The tests are run by the well operator, a specialized well tester, or a service company. They can use equipment available on the site or portable test equipment.

After the well has been completed, a potential test can be run. The *potential test* determines the maximum gas and oil that the well can produce in a 24-hour period. It uses the separator and tank battery on the site to hold the produced fluids. Potential tests can also run periodically during production and can be required by some government regulatory agencies.

A *productivity test* is run to determine the effect of different production rates on the reservoir. It is made with portable well test equipment (Fig. 25–2) that measures the fluid pressure at the bottom of the well when it is shut-in and then during several different stabilized rates of production. The measurements are used to calculate the absolute open flow rate and the maximum production rate that the well can produce without damaging the reservoir.

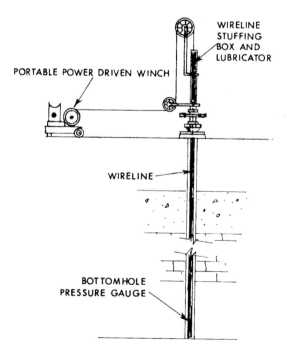

Fig. 25–2 Productivity test equipment

For wells that have a central processing unit, periodic *production tests* can be made to determine how much each well is producing. These tests are run manually or automatically. Oil well test data typically includes oil production, water production, gas rate, gas-oil ratio and flowing tubing pressure. Gas well test data typically includes gas rate, condensate production, water production, flowing tubing pressure and condensate-gas ratio.

Pressure transient testing on a well involves measuring pressures and flow rates. One type, a *drawdown test* measures the shut-in bottomhole pressure and then the pressure change as the well is put on production and the pressure drops to a stable, flowing pressure. A *buildup test* measures the flowing bottomhole pressure and then the pressure change as the well is shut-in and the pressure rises to a stable, shut-in pressure. A *multirate test,* such as a four-point test, measures the flowing bottomhole pressure at different, stabilized flow rates.

Deliverability is the ability of the reservoir at a given flowing bottomhole pressure to move fluids into the well. *Maximum potential flow* or *absolute open flow* (*AOF*) is the maximum flow rate into a well when the bottomhole pressure is zero. It is a theoretical flow rate that is calculated from a multivariate test. The *production index* (*PI*) of a well is the downhole pressure drawdown in psi divided by the production in bbls/d. Wells on land usually have a PI of greater than 0.1 psi/bbl/day whereas offshore wells have a PI greater than 0.5. *Inflow performance relationship* (*IPR*) is similar to PI because it plots drawdown against production but is more accurate in that it also accounts for reservoir drive, increasing gas-oil ratios, and relative permeability changes with production.

Gas wells are tested with routine production tests that measure the amount of gas, condensate and water produced. A *back-pressure test* measures the shut-in pressure and the pressures at different stabilized flow rates to determine the well deliverability.

Cased-hole Logs

After a reservoir in a well has been depleted, a decision must be made to either plug and abandon or recomplete the well. To recomplete, an oil or gas reservoir must be identified behind the casing. Only the natural gamma ray and neutron porosity logs can be run in a cased-hole. A *pulsed neutron log* is a type of neutron log that emits pulses of neutrons into the formation and measures returning gamma rays. It can distinguish gas and oil from water in the reservoir and is used to find gas and oil located behind the casing.

Production Logs

Production logs are run in producing wells to evaluate a problem. They are run either on a wireline through the tubing or on a tubing string. There are several types of production logs.

Tracer logs are used to detect fluid movement in a well. A radioactive tracer is injected into the well at a specific location, and its movement is tracked by recording gamma rays. A *continuous flowmeter* uses propellers on a vertical shaft to measure fluid flow up a well to make a continuous record of flow versus depths in the well. A *packer flowmeter* uses a packer to seal the well at that depth to ensure that all the fluids flow up through the flowmeter in the packer and are measured.

A *noise log* uses a microphone to detect and amplify any sounds in a well. The log can locate where fluids are flowing into the well, and the frequency of the sound can be used to distinguish between liquid and gas. A *temperature log* measures the temperature of fluid filling a well. Before the temperature log is run, the well is shut in for a period of time to allow the temperatures to come to equilibrium. Because expanding gas cools when entering a well, it can be located by a temperature log.

A *manometer* measures pressure in the well at a specific depth, and a *gradiometer* measures a continuous profile of the pressure gradient. A *watercutmeter* measures the amount of water in the fluid filling the well. A *collar log* has a casing-collar locator that uses either a magnetic detector or scratcher to locate the casing collars in a well. It is used to accurately find locations in the well. A collar log is used with a natural gamma ray log to locate where to perforate the casing.

Decline Curves

A *decline curve* is a plot of oil or gas production rate with time made for a single well or an entire field (Fig. 25–3). Production rate will decline with time as the reservoir pressure decreases.

The *initial production* (*IP*) of a well is the first 24 hours of production and is usually the highest. As the production rate declines, the well eventually becomes a *stripper well* that is barely profitable. Stripper wells are defined in the United States as producing less than 10 bbls (1.6 m³) of oil or 60 mcf (2000 m³) of gas per day and receive special tax advantages. In 1999, there were 422,730 stripper oil wells in the United States (76% of the total wells). These stripper wells represent 46% of the total producing gas and oil wells in the world.

The *economic limit* of a well is when production costs equal net production revenue. It depends on how deep the well is, how much water it produces, where the well is located and several other factors. When the economic limit of a well is reached, it is either plugged and abandoned, or a waterflood or enhanced oil recovery is initiated. Most wells are designed with a 15- to 20-year life.

The shape of the oil decline curve depends on the reservoir drive (Fig. 25–4). Solution-gas reservoirs have a very sharp decline, whereas water-drive reservoirs have almost constant production for the life of the wells. The shape of a free gas cap drive curve is between the curves for solution gas and water drive reservoirs.

The decline curves for wells producing from a fractured reservoir in a tight sandstone or dense limestone such as the Spraberry field of Texas are very distinctive (Fig. 25–5). The well can have a high initial production as oil drains through the very permeable fractures. As the fractures drain, the production rapidly drops. Within a short period, the production settles to a long and steady rate as the oil drains slowly from the relatively impermeable rock into the fractures.

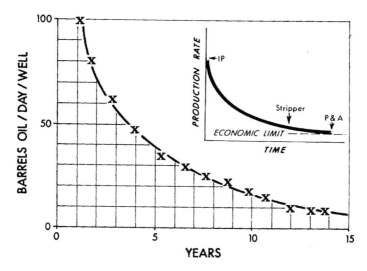

Fig. 25-3 Decline curve

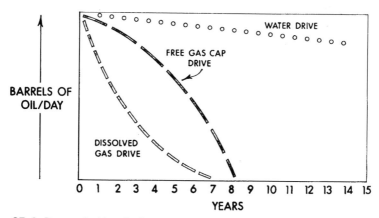

Fig. 25-4 Reservoir drive decline curves

Bypassing and Coning

Drilling and completing a well is an economic investment. The best return on that investment is to produce the gas and oil as fast as possible to recover costs and make a profit as soon as possible.

Many reservoirs, however, are not homogenous, and there are pockets of oil or gas in less permeable areas. In a water-drive reservoir, the water flows in to replace the oil or gas as it is being produced. If the oil and gas is produced too fast, the water can flow around pockets of oil and gas in less permeable areas in a process called *bypassing* (Fig. 25–6). Bypassing seals the oil (*bypassed oil*) and gas (*bypassed gas*) in that area and prevents it from being produced from existing wells. To prevent significant bypassing and have maximum ultimate production, the oil and gas should be produced at a slower rate to allow less permeable zones time to drain.

Coning is caused by oil being produced too fast. The oil-water contact is sucked up in a bottom water-drive reservoir (Fig. 25–7), or the gas-oil contact is sucked down in a free gas cap drive reservoir. This can cause permanent damage to the well. Horizontal drain wells can be used to prevent coning.

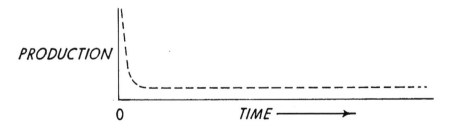

Fig. 25-5 *Decline curve for a fractured reservoir*

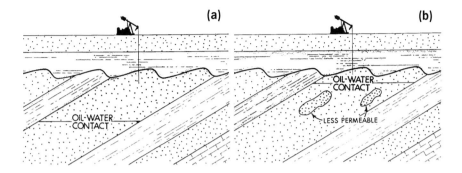

Fig. 25–6 *Bypassing (a) before production (b) after production*

Cycling

As reservoir pressure drops during gas production from a retrograde gas reservoir, condensate separates out of the gas in the reservoir. The liquid coats the pore surfaces and is very difficult, if not impossible, to recover. To

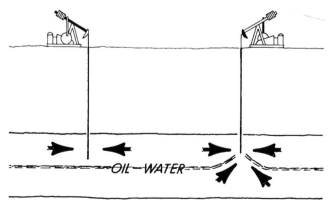

Fig. 25-7 Coning

prevent condensate from separating in the reservoir, *cycling* is used. Produced gas is stripped of natural gas liquids on the surface. The dry gas is then reinjected through injection wells into the reservoir to maintain the reservoir pressure.

Well Stimulation

Several well stimulation methods can be used to increase the well production rate. These include acidizing, explosive fracturing, and hydraulic fracturing.

Acidizing

The well can be *acidized* or given an *acid job* by pumping acid down into the well to dissolve limestone, dolomite, or any calcite cement between sediment grains. HCl (*regular acid*), HCl mixed with HF (*mud acid*) and HF (*hydrofluoric acid*) are acids that are commonly used. HCl is effective on limestones and dolomites and HF is used for sandstones. For formations with high temperatures, acetic and formic acids are used. To prevent the acid from corroding the steel casing and tubing in the well, an additive called an *inhibitor* is used. A *sequestering agent* is an additive used to prevent the formation of gels or precipitates of iron that would clog the pores of the reservoir during an acid job.

Two types of acid treatment are matrix and fracture acidizing. During *matrix acidizing*, the acid is pumped down the well and enlarges the natural pores of the reservoir. During *fracture acidizing*, the acid is pumped down the well under higher pressure to fracture and dissolve the reservoir rock. After an acid job, the spent acid, dissolved rock and sediments are pumped back out the well during the *backflush*. An acid job used to remedy skin damage on a wellbore is called a *wash job*.

Explosive Fracturing

From the 1860s until the late 1940s, explosives were commonly used in wells to increase production. *Well shooting* or *explosive fracturing* was done with liquid nitroglycerin in a tin cylinder called a *torpedo*. It was lowered down the well and detonated. The explosion created a large cavity that was then cleaned out and the well was completed as an open hole. The person in charge of the nitro was called the *shooter*. The technique was both effective and dangerous.

Hydraulic Fracturing

Hydraulic fracturing was developed in 1948 and has effectively replaced explosive fracturing. During a *frac job* or *hydraulic fracturing* (Fig. 25–8), a service company injects large volumes of frac fluids under high pressure into the well to fracture the reservoir rock (Plate 25–1). Frac jobs are done either in an open-hole or a cased well with perforations.

A common *frac fluid* is a gel formed by water and *polymers*, long, organic molecules that form a thick liquid when mixed with water. Oil-based frac fluid and foam-based frac fluids using bubbles of nitrogen, or carbon dioxide can also be used to minimize formation damage. The frac fluid is transported out to the frac job in large trailers.

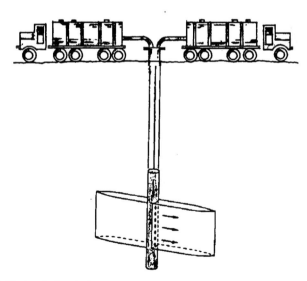

Fig. 25-8 *Hydraulic fracturing*

Plate 25–1 *Hydraulic fracturing an oil well*

A frac job is done in three steps. First, a pad of frac fluid is injected into the well by several, large, pumping units mounted on trucks to initiate fracturing the reservoir. Next, a slurry of frac fluid and propping agents are pumped down the well to extend the fractures and fill them with propping agents. *Propping agents* or *proppants* are small spheres that hold open the fractures after pumping has stopped. The propping agents are usually well-sorted quartz sand grains, ceramic spheres, or aluminum oxide pellets. The well is then *back flushed* in the third stage to remove the frac fluid.

Crosslinked frac fluids that have a high viscosity to carry the propping agents when pumped down the well can be used. A *breaker fluid* is injected into the well to make the crosslinked frac fluid more fluid during backflush.

Medium and hard formations are best for fracturing, as loose formations do not permit the propping agents to hold open the fractures. A frac job using propping agents is often called a *frac/pack*.

All the equipment used during the frac job is driven onto the site. The frac fluid is mixed and stored in *frac tanks*. The frac fluid is mixed with proppants in a blender. Pump trucks are connected to a manifold to pressurize the pad and the slurry and pump them down the well. A *wellhead isolation tool* can be connected to the top of the well to protect the wellhead from the high pressures and abrasive propping agents. The frac job is monitored and regulated from the *frac van*.

Frac jobs are described by the amount of frac fluid and proppants used. A typical frac job would use 43,000 gallons of frac fluid and 68,000 pounds of sand. A *massive frac job* is a very large frac job using more than 1,000,000 gallons (4 million liters) of frac fluid and 3000,000 lbs (1¹/₂ million kilograms) of sand (Plate 25–2).

Hydraulic fracturing is a very common well stimulation technique that increases both the rate of production and ultimate production. It increases the production rate from 1¹/₂ to 30 times the initial rate with the highest increases in tight reservoirs. Ultimate production is increased from 5 to 15%. About 50% of the gas wells and 30% of the oil wells drilled in the United States are fraced. It is used in all tight gas sand reservoirs and as a common remedy for skin damage in a wellbore.

Plate 25–2 *Aerial photograph of a massive frac job. The well is in the center with lines of pumping units and frac fluid trailers on either side. (courtesy Halliburton)*

A well can be fraced several times during its life. In some instances, however, hydraulic fracturing can harm a well by *fracing into water*. The hydraulically induced fractures extend vertically into a water reservoir that floods the well with water.

Oilfield Brine and Solution Gas Disposal

Natural gas produced with oil is often a disposal problem. It comes from the separators at very low (atmospheric) pressure, and there is often no market for it. In the past, it was usually burned (*flared*) in the oil fields. This is against the law today in many countries. It still occurs in some situations when any other gas disposal method is not practical or during well testing.

The natural gas can be used to increase the ultimate oil production from the field by reinjecting it into the subsurface reservoir in a *pressure maintenance system* (Fig. 25–9). Produced wet gas is first gathered and is usually stripped of valuable natural gas liquids before injection. It is then compressed and pumped into an *injection well*. In a saturated oil field, the gas is injected into the free gas cap. In an undersaturated oil field, the gas is injected into the oil reservoir.

Gas from the separators can also be given to the landowner to heat his home and operate irrigation pumps. This *farmer's gas* can be part of the lease agreement before any wells are drilled. Gas from the separator could also be used to operate equipment in the field such as the engine used to drive a beam pumping unit.

Oilfield brine from the separators can be pumped down another injection well into the subsurface reservoir below the oil-water contact as part of the pressure maintenance system. If there is no injection well system available for the well or field, the oilfield brine or water removed from natural gas is stored in a metal or fiberglass tank. A *disposal well* is used to pump the brine into a subsurface reservoir rock.

The disposal well has to be permitted by a government agency and must meet specific criteria. The oilfield brine cannot be injected into a subsurface,

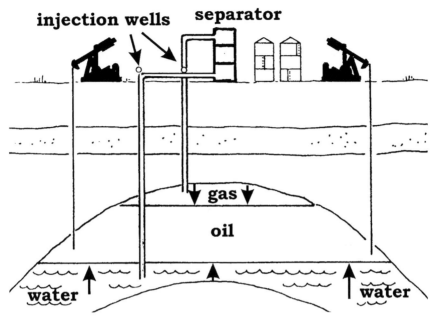

Fig. 25-9 *Pressure maintenance system*

fresh water reservoir. The reservoir must already contain naturally saline waters that cannot be used for drinking or irrigation. The reservoir also must also be able to sustain the increased pressure of the injected water without leaking into other fresh water reservoirs.

If there is no disposal well, the brine is stored in an open, fiberglass, or metal tank to evaporate and reduce the volume. When the tank is filled, a service company (a *water hauler*) is used to transport the brine to a salt water disposal well.

Surface Subsidence

During production, reservoir pressure decreases, and water usually flows in from the sides and bottom to replace the produced fluids. If water does not replace the produced fluids, the subsurface reservoir rock can compact and the surface of the ground subsides (Fig. 25–10).

This has happened in the Wilmington oil field in Long Beach, California that has been producing since the 1930s. Beginning in the 1940s, surface subsidence in the shape of a bowl was noted. The center of the bowl has now subsided a total of 29 ft. (9 m) leaving much of the city below sea level. A massive water injection program has stopped the subsidence, and the city is now protected from seawater flooding by a dike.

The bottom of the North Sea above the Ekofisk oil field has subsided several tens of feet because of compaction of the Ekofisk Chalk reservoir. The subsidence was first noticed in 1984 after the casing in several wells had collapsed, and the level of the boat dock on the platform became submerged. The elevation of the Ekofish production platform had dropped to a dangerous level. In 1987, the legs of the platform had to be cut, the deck jacked up, and extensions spliced into the legs to raise the deck.

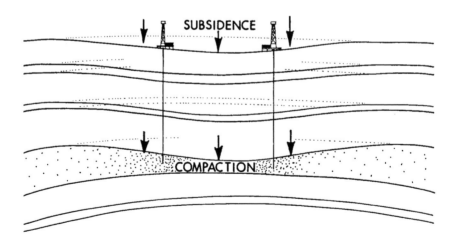

Fig. 25-10 Surface subsidence due to production

Corrosion

Corrosion is the chemical degradation of metal. It can be a problem during both drilling and production. Corrosion occurs when metal is exposed to air, moisture or seawater or by chemicals such as oxygen, carbon dioxide (*sweet corrosion*), or hydrogen sulfide (*sour corrosion*) in the produced fluids. *Total acid number* is a measure of the acidity and corrosiveness of a crude oil. It is a number expressed in mg KOH/g. Higher numbers are more corrosive.

Exposed metal surfaces on equipment are painted for protection. *Inhibitors* are chemicals that are injected to coat steel in the well and the production facilities with a thin film. The inhibitor can be injected either automatically or manually in periodic batches into the casing-tubing annulus of the well. A concrete coating can be used to protect the insides of flowlines. The tubing in injection and disposal wells is often lined with plastic. Large metal structures such as pipelines and offshore production structures can be shielded by *cathodic protection*. It involves charging the structure with an electrical charge to prevent corrosion.

Production Maps

A *well status map* is used to analyze production from a field and identify problem wells. The map shows the location of all wells in a field. Producing wells have the well number, barrels of oil and water production per day, and the gas/oil ratio next to them. Injection wells have the well number, barrels of water injected per day, pressure, and cumulative injection in thousands of barrels. A *cumulative production map* lists the total amount of water, gas, and oil that each well has produced up to a specific date. *Bubble maps* are used to show how much wells have produced (Fig. 25–11). A circle is drawn around each well with the radius of the circle (bubble) proportional to the well's cumulative production of gas, oil, or water.

Stranded Gas

Stranded gas is natural gas that has no market. Large reservoirs of stranded gas occur in western Siberia, northwestern Canada, Alaska, and the Middle East. Natural gas can be transformed into a liquid to decrease its volume and transport it to a market either as liquefied natural gas or synthetic crude oil.

When methane is compressed and cooled to –269°F (-167°C), it becomes a liquid called *liquefied natural gas* (*LNG*). LNG occupies 1/645th the volume of natural gas. Special tankers can then be used to transport the LNG across the sea to markets.

Gas-to-Liquid involves mixing natural gas and air in a reactor to form *synthesis gas* (CO and H). The synthesis gas is then put in another reactor to form synthetic crude oil.

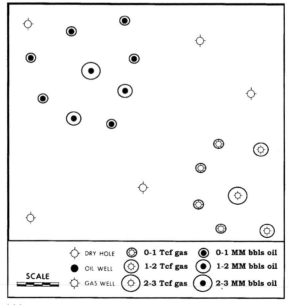

Fig. 25-11 *Bubble map*

twenty-six

RESERVES

There are 1,016,041 million bbls of proven oil reserves in the world. Of these, 675,636 million bbls (66%) are in the Middle East, 21,034 million bbls (2%) in the United States and 4931 million bbls (0.5%) in Canada. There are 5179 trillion cubic feet of proven gas reserves in the world. Of these, 1749 trillion (34%) are in the Middle East, 164 trillion (3%) are in the United States and 64 trillion (1%) are in Canada.

Recovery Factor

The amount of oil or gas in the subsurface reservoir is called *oil in place (OIP)* or *gas in place (GIP)*. *Recovery factor* is the percentage of OIP or GIP that the reservoir will produce. The recovery factor for an oil reservoir depends on 1) the viscosity of the oil, 2) the permeability of the reservoir, and 3) the reservoir drive. Typical values for oil and gas reservoir drives are shown in Table 26–1.

Shrinkage Factor and Formation Volume Factor

The amount of natural gas dissolved in crude oil in the subsurface reservoir is called the *dissolved, reservoir,* or *solution gas-oil ratio* and is expressed

431

Table 26–1
Recovery Factors

oil reservoir drives	recovery factors
solution gas	5 to 30%
free gas cap	20 to 40%
water	35 to 75%
gravity	50 to 70%
gas reservoir drives	
expansion gas	75 to 85%
water	60%

(modified from Sills, 1992)

in scf/bbl. It depends upon the temperature and pressure of the reservoir and the chemistry of the oil. In general, deeper reservoirs have higher dissolved gas-oil ratios. If the oil has dissolved all the gas it can hold under those conditions, it is *saturated.* An oil reservoir with a free gas cap is saturated. If there is no free gas cap, the oil reservoir is *undersaturated* and can hold more gas.

When the oil is produced, the high reservoir temperature and pressure decreases to surface conditions, and gas bubbles out of the oil. The amount of gas and oil produced on the surface is called the *producing gas-oil ratio.* It is similar to the dissolved gas-oil ratio but can be higher if some gas is being produced from the free gas cap.

As gas bubbles out of the oil on the surface, the volume of the oil decreases. The amount to which one barrel of oil decreases in volume on the surface is called the *shrinkage factor* (Fig. 26–1). It is expressed as a decimal that ranges from 1.0 to 0.6 and depends on the amount of gas that bubbles out of the oil. A stabilized barrel of oil under surface conditions (60°F temperature and 14.7 psi pressure or 15°C and 101.325 kPa) is called a *stock tank barrel of oil.* The number of barrels of oil under reservoir conditions that need to be produced to shrink to a stock tank barrel of oil is called the *formation volume factor (FVF)* (Fig. 26–1). It usually varies from 1.0 to 1.7 and is the inverse of the shrinkage factor. A formation volume factor of 1.4 is characteristic of high-shrinkage crude oil, and 1.2 is low-shrinkage crude oil. The formation volume factor can be estimated from the producing gas-oil ratio. The higher the producing gas-oil ratio: the larger the formation volume factor.

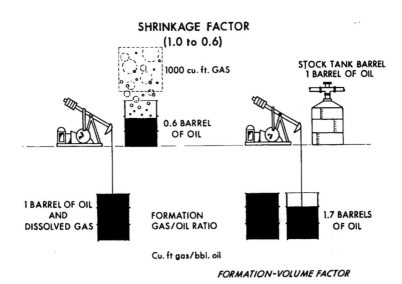

Fig. 26–1 Shrinkage factor and formation volume factor

Gas in a subsurface reservoir is under high pressure and temperature. When the gas is produced, the pressure and temperature decrease to surface conditions, and the gas expands in volume. The volume of natural gas in the subsurface reservoir that expands to one cu ft on the surface is called the *gas formation volume factor (B$_g$)*. It depends on reservoir temperature and pressure and gas composition. Natural gas is measured in *standard cubic feet (scf)*, the cubic feet of natural gas under surface conditions defined by law.

Reserve Calculations

Reserves are the amount of oil and gas that can be produced from a well or field in the future under current economic conditions using current technology. Reserves are always reported in stock tank barrels of oil and standard cubic feet of gas.

Oil Reserves

Oil reserves can be computed both volumetrically and by decline curves. The *volumetric* or *engineering formula* for oil reserves for a single well or an entire oil field is:

$$\text{stock tank bbls of oil} = \frac{V \times 7758 \times \emptyset \times S_o \times R}{FVF}$$

V is the volume of the oil pay zone drained by a well or wells expressed in units of acre-feet. One *acre-foot* is the volume generated by a surface one acre in area and one foot deep (Fig. 26–2). An acre-foot of volume can hold 7758 barrels of oil. The porosity *(∅)* of the reservoir is expressed as a decimal. It is usually determined from well logs or cores. S_o is oil saturation expressed as a decimal. It is determined from electrical resistivity well logs or cores. *R* is the recovery factor that is estimated from the reservoir drive, reservoir permeability, and oil viscosity, and it is expressed as a decimal. The *FVF* can be estimated from the producing gas-oil ratio or determined from laboratory analysis of produced fluids.

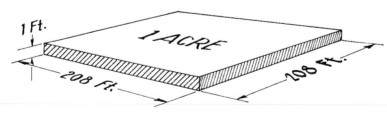

Fig. 26-2 Acre-foot

The *decline curve method* uses production data to fit a decline curve and estimate future oil production. It is assumed that the production will decline on a reasonably smooth curve with allowances for well(s) shut in or temporary production restrictions. The curve can be expressed by mathematics or plotted on graph paper with an arithmetic production scale to estimate future production (Fig. 26–3a). If the well produces with a dissolved gas or free gas cap drive, plotting the curve on a logarithmic production scale can yield a straight line for the decline curve (Fig. 26–3b).

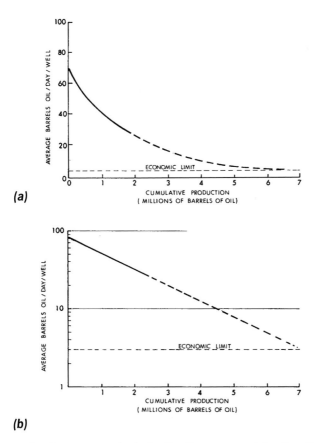

Fig. 26-3 *Decline curve method of estimating reserves (a) arithmetic scale (b) logarithmic scale*

Gas Reserves

Gas reserves are computed volumetrically and by P/Z plots. For a gas reservoir without any oil, the *engineering* or *volumetric formula* is:

$$\text{standard cubic feet of gas} = \frac{V \times 43{,}560 \times \emptyset \times S_g \times R}{B_g}$$

V is the volume of the gas reservoir in acre-feet (one acre foot can hold 43,560 cubic feet) $\emptyset$ is reservoir porosity, S_g is water saturation, R is the recovery factor that is determined from the reservoir drive and permeability, and B_g is the gas formation volume factor that is determined from tables of reservoir temperature and pressure and gas composition.

Gas from an associated gas reservoir with oil is more difficult to calculate because the gas comes from both the free gas cap and solution gas that bubbles out of the oil as the reservoir pressure drops.

For a single gas well, reserves are estimated from a *P/Z plot* (Fig. 26–4). P is the reservoir pressure measured in the well. It will decrease as gas is produced from the well. Z is the *compressibility factor* that compensates for natural gas not behaving as an ideal gas under high pressure and temperature conditions of the subsurface reservoir. It varies between 1.2 and 0.7 and is determined from tables of temperature, pressure, and gas composition. P/Z plotted against cumulative production is a straight line. Where it intersects the abandonment pressure is the ultimate gas production from that well.

Abandonment pressure is the lowest gas reservoir pressure at which the well is plugged and abandoned. It is usually the lowest pressure that the gas pipeline will accept. This is between 700 to 1000 psi (49 to 70 kg/cm^2). When economics permits, the life of a gas well can be extended with a compressor to increase the produced gas pressure to pipeline pressure.

Materials Balance Method

The *materials balance method* for a gas or oil field uses an equation that relates the volume of oil, water, and gas that has been produced from a reser-

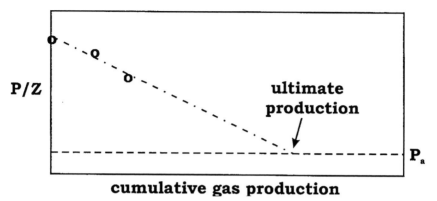

Fig. 26-4 *P/Z method of estimating gas reserves*

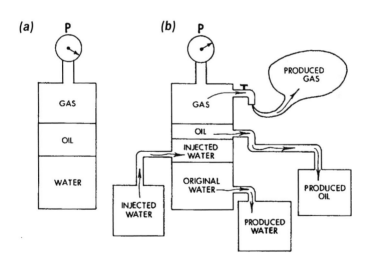

ORIGINAL VOLUMES		PRODUCED VOLUMES		CURRENT VOLUMES		MIGRATED VOLUMES
OIL, GAS, & WATER	=	OIL, GAS, & WATER	+	OIL, GAS, & WATER	+	OIL, GAS & WATER

Fig. 26-5 *Model of materials balance equation (a) reservoir before production (b) reservoir during production*

voir and the change in reservoir pressure to calculate the remaining oil and gas. It assumes that as fluids from the reservoir are produced, there will be a corresponding change in the reservoir pressure that depends on the remaining volume of oil and gas (Fig. 26–5).

Types of Reserves

There are several types of reserves. *Proven* reserves are reserves that can be calculated with reasonable certainty because the field has been defined by appraisal wells that have been tested. Proven reserves can be either developed or undeveloped. *Developed* reserves can be produced from existing wells. *Undeveloped* reserves will have to be produced from wells that have not yet been drilled or from zones that are *behind the pipe* (behind casing that is not perforated in that zone). They can be developed later by perforating the casing. *Unproven reserves* are calculated similar to proven reserves but because of technical or economic uncertainties, their production is not as certain as proven reserves. They can be either probable or possible reserves. Depending on the probability of the reserves existing, *proven reserves* are >90%, *probable reserves* are >50%, and *possible reserves* are >10%.

twenty-seven

IMPROVED OIL RECOVERY

Primary production is the oil produced by the original reservoir drive energy. It depends on the type of reservoir drive, oil viscosity, and reservoir permeability but averages 30 to 35% of the oil in place and can be as low as 5%. This leaves a considerable amount of oil in the reservoir after the pressure has been depleted. Because of this, *improved oil recovery* (engineering techniques that include waterflood and enhanced oil recovery) is often used to recover more oil. *Ultimate oil recovery* is the total production from a well or field by primary production, waterflood and enhanced oil recovery, if justified by economic conditions.

A typical gas reservoir will produce 80% of the gas by primary production. Because so little gas is left in the depleted reservoir, gas fields are plugged and abandoned after primary production.

Waterflood

A *waterflood* involves injecting water through injection wells (Fig. 27–1) into the depleted oil reservoir It can be initiated either before or after the reservoir drive has been fully depleted. The water sweeps some of the remaining oil through the reservoir to producing wells. A waterflood can recover 5 to 50% of the remaining oil in place.

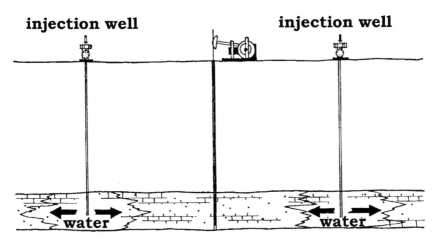

Fig. 27-1 *Waterflood*

Plate 27-1 *Water injection well, Burbank oil field waterflood, Oklahoma*

The injected water is often oilfield brine from the separators but can also be water from other sources that has been treated. The injected water must be compatible with the producing formations and not cause reactions that decrease the permeability of the formation being flooded. Suspended solids that can plug the pores are removed from the injection water by filtration. Organic matter and bacteria that produce slimes are neutralized by biocides. Oxygen is removed from the water to prevent corrosion.

The water is either pumped under pressure down the well or is fed by gravity from storage tanks on a higher elevation such as a hill. The injection wells can be either drilled or converted from producing wells (Plate 27–1).

Waterfloods are described by the aerial pattern of the wells and are either spot or line drives. The common *five-spot pattern* has four water-injecting wells located at the corners of a square with a producing well at the center (Fig. 27–2a). The pattern is repeated in the field so that four injection wells surround each producing well and four producing wells surround each injection well. A *seven-spot pattern* has six injector wells surrounding a producer (Fig. 27–2b); whereas an *inverted seven-spot pattern* has six producer wells surrounding an injector (Fig. 27–2c). A *line drive* has alternating lines of producers and injectors and can be either *direct* (Fig. 27–2d) or *staggered* (Fig. 27–2e). An *edge waterflood* uses injection wells along the margin of the field. The injected water drives oil up and toward the producing wells in the center.

The waterflood usually becomes uneconomical and is abandoned when the water cut reaches 90 to 99%. Waterfloods are most effective in solution-gas drive reservoirs where there is relatively little primary production. Some waterfloods may take up to two years of injection before any increase in production occurs. This *fill-up time* is caused by the gas bubbles in the pores of the reservoir being compressed and redissolved in the remaining oil.

In many oilfields, however, the reservoir is not homogenous, and the waterflood is not efficient. Fluids such as water will always flow along the route of least resistance. A reservoir rock might have a zone of high permeability, such as a well-sorted bed of sandstone or a porous or fractured zone in limestone. As the water sweeps through the reservoir, the injected water flows fastest through the most permeable zone (a *thief zone*) and reaches the producing well to cause a *breakthrough*. Once a breakthrough occurs, the rest of the water will tend to flow through that permeable zone bypassing oil in the less permeable portions of the reservoir. The sooner the water breaks through, the less efficient the waterflood.

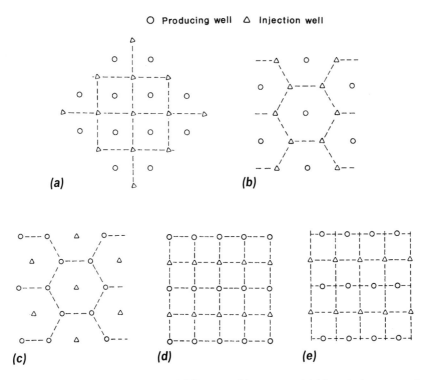

Fig. 27-2 *Waterflood patterns (a) five-spot (b) seven-spot (c) inverted seven-spot (d) direct line drive (e) staggered line drive*

Gravity also affects waterflooding. Because the water is heavier than oil, water tends to flow furthest along the bottom of the reservoir because of *gravity segregation*. This leaves oil untouched in the top of the reservoir.

In one variation of a waterflood, heated water is injected to make the oil more fluid. The water can also be treated with polymers (long, chain-like, high-weight molecules) that increase the viscosity of the water. *Alkaline* or *caustic flooding* uses an alkaline chemical such as sodium hydroxide mixed with the injected water. The chemicals react with the oil in the reservoir to improve the amount of recovery.

Enhanced Oil Recovery

During *enhanced oil recovery (EOR)*, substances that are not naturally found in the reservoir are injected into the reservoir. Enhanced oil recovery includes thermal, chemical, and gas miscible processes. It can be initiated after either primary production or waterflooding.

Miscible Gas Drive

A *gas miscible process* involves injecting a gas into the reservoir that dissolves in the oil. *Inert gas injection* uses either carbon dioxide (CO_2), nitrogen, or liquefied petroleum gas (LPG). The injected gas should not corrode metal equipment in the well, should not mix with natural gas in the reservoir to form an explosive combination, and should be relatively inexpensive.

During a *carbon dioxide flood*, carbon dioxide gas is usually brought to the project by pipeline from carbon dioxide wells or trucked in as a liquid. Large, natural reservoirs of carbon dioxide gas occur in many areas. It is also available as a byproduct of power, chemical and fertilizer plants, and coal gasification. When carbon dioxide is injected into the reservoir, it is *miscible* with the oil (dissolving in the oil), making the oil more fluid. The carbon dioxide gas then pushes the fluid oil through the reservoir toward producing wells. It can often recover about 35% of the remaining oil. The largest carbon dioxide flood project in the United States was initiated in 1972 on the Kelly-Snyder oil field in Texas.

Because of the very low viscosity of the carbon dioxide, it tends to finger and break through to producing wells leaving unswept areas in the reservoir. To prevent this, alternating volumes of water and gas can be injected into the reservoir in a *water-alternating-gas (WAG)* process.

Liquefied petroleum gas is also miscible with oil and is used in a *LPG drive*. The source of the LPG (propane, or propane-butane mixture) is usually wet gas. Under some reservoir conditions, nitrogen is used to flood the reservoir.

Chemical Flood

A *chemical flood* is a process in which different fluids are injected into the depleted reservoir in separate batches *(slugs)*. The fluids, each serving a different purpose, move as separate *fronts* from the injection wells, through the reservoir rock toward the producing wells (Fig. 27–3).

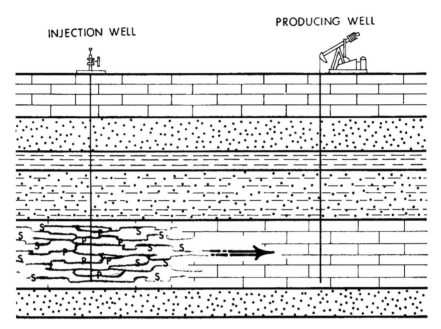

Fig. 27-3 Chemical flood   S is surfactant   P is polymer

In a *micellar-polymer* flood, a slug of reservoir water is first injected to condition the reservoir as it moves ahead of other slugs of injected chemicals. Next, a slug of surfactant solution is injected into the reservoir. The *surfactant* acts as a detergent, reducing the surface tension of the oil and washing the oil out of the reservoir pore spaces. The oil forms small droplets suspended in the water called a *microemulsion*. The next slug is water thickened by polymers. Pressure on the polymer water from the injection well drives the surfactant and oil microemulsion front ahead through the reservoir rock toward producing wells (Fig. 27-3).

A chemical flood can be used only for sandstone reservoirs because carbonates absorb the surfactants. It can recover about 40% of the remaining oil but is an expensive process.

Thermal Recovery

Thermal recovery techniques utilize heat to make heavy oil (< 20°API) more fluid for recovery. *Cyclic steam injection* or the *huff and puff* method uses single wells to inject steam into the heavy oil reservoir for a period of time such as two weeks during the *injection period* (Fig. 27–4a). During the following *soak period,* the well is shut in for several days to allow the steam to heat the heavy oil and make it more fluid. The same well is then used to produce the heated heavy oil with a sucker-rod pump during the *production period* for a similar period of time to the injection period (Fig. 27–4b). Steam injection and pumping are alternated for up to 20 cycles until it becomes ineffective.

A *steamflood* or *steam drive* uses both injection and production wells (Plate 27–2). The superheated steam in pumped down injection wells into a heavy oil reservoir. The steam heats the heavy oil to greatly reduce its viscosity. As the steam gives up its heat, it condenses into hot water that drives the oil toward producing wells. The pattern of injection and producing wells in a steam flood is similar to that of a waterflood but are very closely spaced. The recovery will vary between 25 to 65% of the oil in place.

(a)   **(b)**

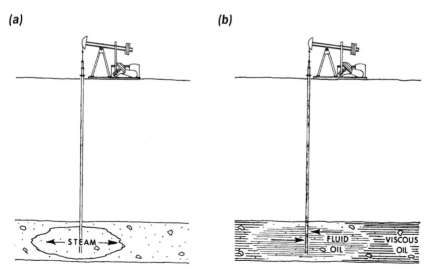

Fig. 27-4 *Cyclic steam injection (a) injection–huff (b) production–puff*

A steamflood is being used in the Kern River field, Bakersfield, California (Plate 27–2). The field was discovered in 1899 by digging a pit on the banks of the Kern River next to an oil seep. The oil is 12 to 16 °API. The reservoir, 500 to 1300 ft. deep (150 to 400 m) deep consists of unconsolidated sands with 1 to 5 D permeability and 28 to 30% porosity. Primary recovery was 15%, but with the steamflood, begun in the mid 1950s, the recovery will be 55%. Steamflooding is also being used on several Bolivar Coastal fields in Venezuela and in Alberta.

Plate 27-2 *Kern River field steamflood, California*

A *fireflood* or *in situ combustion* involves setting the subsurface oil on fire. If the well is shallow, the fire can be started with either a phosphorous bomb or a gas burner lowered down the well. Pumping air into the reservoir to start the fire by spontaneous combustion works in deeper reservoirs. Once the oil is burning, large volumes of air must by injected into the reservoir to sustain the fire. Air pumping is a large expense in a fireflood and increases with depth of the producing formation as more and bigger compressors are required.

The fire generates heat, causing the oil to become more fluid. The large volume of hot gasses generated by the fire drives the heated oil toward producing wells (Fig. 27–5). A fireflood will fail if there is not enough oil in place to sustain the fire.

The most common fireflood is *forward combustion* in which the fire and injected air originate at the injection well. The oil flows toward the producing wells. In *dry combustion,* only air is injected. In *wet combustion* or *combination of forward combustion and waterflooding (COFCAW),* water and air are injected either together or alternately. The generated steam from water helps drive the oil.

The recovery from a fireflood can be 30 to 40% of the oil in place. Corrosion of equipment is a problem because of the high temperatures and corrosive gasses that are generated. Time-lapse seismic methods can be used to trace the movement of the subsurface fire front.

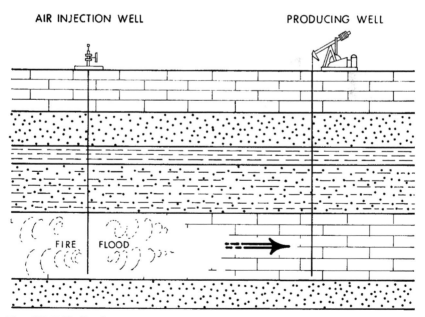

Fig. 27-5 *Fireflood*

Efficiency

The effectiveness of a waterflood or enhanced oil recovery project is described by sweep and displacement efficiencies. *Sweep efficiency* is a ratio of the pore volumes that are contacted by the injected fluid to the total reservoir pore volume. Both horizontal and vertical sweep efficiencies are computed. Sweep efficiency is strongly influenced by the *mobility ratio*, the ratio of the driving fluid viscosity to the oil viscosity being displaced. Mobility ratios close to one are most efficient. *Displacement efficiency* is ratio of the volume of oil that is swept by the process to the volume of oil in place before the process.

Unitization

When more than one company is operating in a field, the field can be *unitized* to coordinate a field-wide effort to increase ultimate production. A *unit operator* is appointed to direct a pressure maintenance, waterflood, or enhanced oil recovery project in that field. The costs and production are shared proportional to each member's acreage or reserve position in the field. Unitization can be either voluntary or forced by government decree.

The Prudhoe Bay oil field, Alaska, is unitized for pressure maintenance and waterflood. Produced gas and water, along with treated sea water, are being injected into the reservoir. The original reserve estimate for Prudhoe Bay was 9.6 billion bbls (1.5 billion m^3) of oil but because of the pressure maintenance and waterflood, along with reservoir management, the reserve estimate is now 13 billion bbls (2.1 billion m^3) of oil.

Plug and Abandon

Both dry holes and producing wells, either onshore or offshore, that have been depleted must be *plugged and abandoned (P & A)*. The procedure is required by law to prevent salt water from the well from polluting fresh ground waters and involves cementing the borehole.

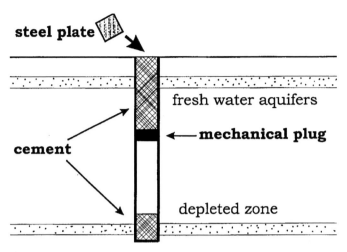

Fig. 27-6 *A plugged and abandoned well*

First, if possible, casing is cut and pulled for salvage. All depleted, producing formations are sealed by placing cement plugs at those levels (Fig. 27–6). Near surface fresh water reservoirs are also protected by cement. A mechanical plug is used to bridge the wellbore at a specific depth to control the cement level. The upper portion of the well containing the freshwater reservoirs is then cemented. The job can be as simple as cementing the upper 100 ft (30 m) of the well. A more complex job also might involve cementing above and below all high-pressure and all permeable zones in the well.

The casing is cut 6 ft (2 m) below the surface, and a steel plate is welded to the top of the casing. The hole is then filled with dirt, and a marker is installed.

Offshore wells are plugged and abandoned just like land wells. All the subsea equipment is retrieved. Any part of the well that sticks above the mudline, such as casing, is cut off to not leave a navigational hazard. On an abandoned offshore production platform, usually only the deck equipment such as the modules are salvaged. The jacket can be either tipped over to lie on the ocean bottom or cut off below sea level to prevent an obstruction to navigation. The abandoned structure on the ocean bottom makes an excellent fish reef.

glossary

A. 1) ampere or 2) area.

A/. acidized with.

ab. above.

abandonment pressure. the lowest gas pressure before a gas well must be abandoned.

abd. 1) abandoned or 2) abundant.

abnormal high pressure. pressure in a subsurface reservoir that is higher than expected from hydrostatic pressure at that depth.

absolute open flow. the maximum rate at zero bottomhole pressure that a well can produce. (AOF) (maximum potential flow)

absorption tower. a vertical, steel vessel where natural gas bubbles up through a light hydrocarbon liquid that removes the natural gas liquids.

abt. about.

ac. acid or acidizing.

accelerator. an additive that increases the rate of a process such as cement setting.

accumulator. skid-mounted, steel cylinders that contain hydraulic fluid under pressure. They are located next to a drilling rig and are used to operated the rams on the blowout preventers.

ac-ft. acre-foot.

acid gas. a gas such as hydrogen sulfide or carbon dioxide that forms an acid with water. It can cause corrosion of metal equipment.

acidizing or **acid job.** a well stimulation technique used primarily on limestone reservoirs. Acid is poured or pumped down the well to dissolve the limestone and increase fluid flow.

acoustic impedance. sound velocity times density of a rock.

acoustic velocity log. a wireline well log that measures sound velocity. The porosities of the rocks can be calculated from the log. (sonic log) (AVL)

acre. an area of 43,560 square feet.

acreage. leased land.

acre-foot. the volume formed by a surface area of one acre that is one foot deep. This volume can hold 7758 barrels of oil. Acre-feet is used to describe reservoirs and to calculate reserves. (ac-ft)

additive. a substance that is added to cause an effect.

aeromagnetics. the use of a magnetometer to measure the strength of the earth's magnetic field from an airplane.

AFE. authority for expenditure.

air balanced beam pumping unit. an oil well beam pumping unit with a piston in a compressed air cylinder that offsets the weight of the sucker-rod string.

air drilling. rotary drilling with air pumped down the drillstring instead of circulating drilling mud. (pneumatic drilling)

air gun. a seismic source used in the ocean. It is a metal cylinder that is towed behind a boat and is continuously being filled with high-pressure air. A high-pressure air bubble is periodically released into the water from the air gun for the energy impulse.

algorithm. the precise procedure for a numerical or algebraic procedure. Algorithms are used for computer processing of seismic data.

alkaline flood. an improved oil recovery method using alkaline chemicals in the injection water to improve oil recovery. (caustic flood)

allowable. the amount of gas or oil that a regulatory agency permits a well, lease, or field to produce during a period of time such as a month.

amine unit. natural gas processing equipment that uses organic bases (amines) to absorb and remove hydrogen sulfide and carbon dioxide.

amplitude anomaly. a brightening (bright spot) or dimming (dim spot) of a seismic reflector over a local area.

amplitude versus offset analysis. a seismic method in which the amplitude of a reflector is compared at different offsets (source to detector distances). (AVO analysis)

ang. angular.

angular unconformity. an ancient erosional surface with the sedimentary rock layers below the unconformity tilted at an angle to those above.

anhydrite. a salt mineral composed of $CaSO_4$. (Anhy or anhy)

annular preventer. a cylinder at the top of a blowout preventer stack containing rubber with steel ribs. Pistons compress the rubber to close around any size or shape pipe in the well.

annulus. the space between two concentric cylinders such as between the tubing and casing strings.

anoxic basin. a basin in which the bottom waters lack oxygen, and organic matter can be preserved.

anticline. a large, long, upward fold of sedimentary rocks. It can trap petroleum.

antifoam. an additive used to reduce foam.

AOF. absolute open flow.

API. American Petroleum Institute.

appraisal well. a well drilled out from the side of a discovery well to determine the area of a new field. (step out or delineation well)

aquifer. 1) a water-bearing rock or 2) a permeable rock.

aquitard. a rock through which fluids cannot pass. (impermeable)

arch. a long uplift in rocks.

arenaceous. sandy. (aren)

argillaceous. shaly. (arg)

arkose or **arkosic sandstone.** sandstone derived from the weathering of granite. (granite wash) (Ark or ark)

array. several geophones connected to a single channel. Arrays are described by their geometry such as in-line, perpendicular, cross, and diamond. Several arrays make a spread.

artesian well. a well from which water flows to the surface under its own pressure.

artificial lift. a system to raise oil up the tubing string in a well that will not flow by itself to the surface. Examples. rod pumping, electrical submersible pump, gas lift, and hydraulic pump.

asphalt. a brown to black, solid hydrocarbon formed by molecules of high molecular weight. (Asph or asph)

asphalt-base crude oil. a refiner's term for crude oil that contains little paraffin wax and has a residue of asphalt. It will yield a relatively high percentage of high-grade gasoline and asphalt when refined.

assoc. associated.

associated gas. natural gas that is in contact with crude oil in the reservoir.

atm. atmospheric.

atoll. a circular or elliptical reef with a central lagoon.

aulacogen. a long, narrow rift in a continent, often filled with thick sediments.

authigenic. a mineral that was formed by a chemical reaction in the sub-surface. (authg)

authority or **authorization for expenditure.** a cost estimate of something such as drilling a specific well, both as a dry hole and a completed well. (AFE)

Av or **av.** average.

AVL. acoustic velocity log

AVO. amplitude versus offset.

AW. acid water.

axis. the center of a fold.

asimuth. horizontal direction measured in degrees clockwise from north.

B or **b.** barrel.

B. formation volume factor.

B/. base of.

B/D. barrels per day.

BA. barrels of acid.

back off operation. a method used to remove stuck pipe from a well. A string shot located above the stuck point is exploded on the drillstring as torque is applied to unscrew the pipe. A fishing tool is then used to retrieve the pipe left in the well.

backflush. to pump an injected fluid back out of a well.

back-pressure test. a gas well test that measures pressures at different flow rates to determine well deliverability.

bacterial degradation. the removal of lighter, shorter molecules from crude oil by bacteria.

bailer. a device lowered down a well by a sand line on a cable tool rig to remove well cuttings and water from the hole.

bald-headed anticline. an anticline with production along the flanks but no production on the crest due to erosional removal of the reservoir rocks from the crest. (scalped anticline)

bald-headed structures. a petroleum trap in which erosion has removed the reservoir rocks from the top of the trap.

bar finger sand. a long, narrow sand body deposited by the distributary mouth bar of a prograding delta.

barefoot completion. a well with casing run and cemented down to the top of the reservoir rock, which is left uncased. (open-hole or top set completion)

barite. a mineral composed of $BaSO_4$. It is used to weight drilling mud.

barrel. the English system measure of crude oil volume. A barrel contains 42 U.S. gallons and is equivalent to 0.159 cubic meters. 7.5 barrels of average weight oil weighs about one metric ton. (B, b or bbl)

barrels of oil equivalent. the amount of natural gas that has the same heat content as an average barrel of oil. It is about 6000 cf of gas. (oil equivalent gas or energy equivalent barrels) (BOE)

Bas or bas. basalt.

basal conglomerate. a soil zone located on an unconformity.

basalt. the most common volcanic rock. It is very fine grained, dark in color and can contain gas bubbles. (Bas or bas)

base map. a map that shows the location of a) wells that have been drilled or b) seismic lines and shot points.

basement rock. unproductive rocks underlying sedimentary rocks. It is usually an igneous or metamorphic rock. (Bm or bsmt)

basic sediment and water. the solid and water impurities in crude oil. (BS&W)

basin. a large area with a thick accumulation of sedimentary rocks.

batch. a single treatment in contrast to continuous.

batholith. a large, irregular subsurface intrusion of igneous rock.

bbl. barrel.

bbl/D. barrels per day.

bbl/min. barrels per minute.

BC. barrels of condensate.

Bcf. billion cubic feet.

Bcf/D. billion cubic feet per day.

BCPMM. barrels of condensate per million cubic feet of gas.

Bd or **bd.** bedded.

Bdg. bedding.

Bdst. boundstone.

beam pumping unit. an oil well rod pumping unit with a walking beam that pivots on a Samson post.

bedding. layers in sedimentary rocks. (Bdg)

behind the pipe. located in the rock behind casing in a well.

benchmark crude oil. a crude oil used by a country as a standard for comparing the properties of other oils and for setting prices. West Texas Intermediate is the benchmark for the United States and North Sea Brent for Great Britain.

bent sub. a short section of pipe with an angle machined into it. It is used on a drillstring to kick off a deviated well.

bentonite. a clay mineral used to make common drilling mud. (Bent or bent)

BF. barrels of fluid.

BFO. barrels of frac oil.

BFPD. barrels of fluid per day.

BFW. barrels of formation water.

BHA. bottomhole assembly.

BHC. bottomhole choke.

BHCS. borehole-compensated.

BHP. bottomhole pressure.

BHT. bottomhole temperature.

bi-center bit. a drilling bit with both a pilot bit on the bottom and a reamer on one side. It is designed to drill and ream a larger diameter hole than the inner diameter of the casing through which it passes.

billing interest. *see* working interest.

bin. the area (square or rectangular) in which all seismic reflection midpoints that fall within it are used to make a common midpoint gather for a 3-D seismic survey.

bioclastic. composed of shell fragments. (biocl)

biogenic gas. methane gas produced by bacterial action on organic matter at shallow depths. (microbial gas)

biostratigraphy. the use of microfossils to study and identify sedimentary rocks.

biotite. black mica (Biot)

bioturbation. the disturbance and mixing of sediments by burrowing animals and plant roots.

birdfoot delta. a delta with several lobes protruding out into a basin.

Bit or **bit.** bitumen.

bit. the rock cutting tool used in drilling. A tricone bit is commonly used in rotary drilling.

bit breaker. a plate placed in the rotary table to grip the bit. It enables the rotary table to screw or unscrew the bit from the drillstring.

bitumen. solid hydrocarbons such as tar in sedimentary rocks. (Bit or bit)

BL. barrels of load.

black oil. an oil that contains a relatively high percentage of long, heavy, non-volatile hydrocarbon molecules.

Bld. boulder.

blind rams. two, large metal blocks with flat surfaces that are closed across the top of well to shut it in during drilling. They are used in a blowout preventer stack.

blk. black.

BLO. barrels of load oil.

block. a pulley on a drilling rig.

blowout. an uncontrolled flow of fluid from a well.

blowout preventer stack. a series of rams and spools mounted vertically on top of a well below the drill floor. They are designed to close the well when drilling. (BOP stack)

BLPW. barrels of liquid per day.

BLW. barrels of load water.

Bm. basement.

Bnd or **bnd**. banded.

BOE. barrels of oil equivalent.

BOL. barrels of oil load.

boll weevil. an inexperienced oilfield worker.

bonus. money paid to a mineral rights owner for signing a lease.

boot sub. a fishing tool. Drilling mud is circulated down the inside of the tool to flow out the bottom and pick up small pieces of junk on the bottom of the well. The mud then circulates up along the outside of the tool where the junk falls into a basket.

BOP. blowout preventer.

BOPD. barrels of oil per day.

bottomhole assembly. the drill collars, subs and bit on the bottom of the drill-string. (BHA)

bottomhole pressure. fluid pressure on the bottom of a well. It can be either static or flowing.

bottomwater. water located in the reservoir below the oil.

boundstone. a type of limestone formed by organisms still in their original positions such as reef rock. (Bdst)

box. a female-threaded connection that mates with a pin.

BPD. barrels per day.

brackish water. a mixture of fresh water and brine. (brksh)

braided stream. a stream with numerous, intertwining channels separated by gravel bars.

brak. brackish.

break. to separate an emulsion such as oil-in-water.

break out. to unscrew pipe.

breakthrough. to have injected water flow through a reservoir and reach a producing well.

breccia. a conglomerate with angular particles. (Brec or brec)

bridging material. *see* lost circulation material.

bright spot. an intense seismic reflection. It can be off the top of a gas reservoir.

brine. water that has more salt than sea water that has 35 parts per thousand.

brit. brittle.

British thermal unit. the English system unit used to measure the heat content of natural gas. It is equal to about one kilojoule. (Btu)

Brk or **brk.** break.

brksh. brackish.

brn. brown.

brt. bright.

BS. basic sediment.

BS&W. basic sediment and water.

bsmt. basement.

Btu. British thermal unit.

bu. buff.

bug picker. *see* micropaleontologist.

build angle. to increase the deviation of a deviated well.

bull wheel. a spool of drilling line on a cable tool rig.

Bur or **bur.** burrow.

butane. a hydrocarbon composed of C_4H_{10}. It is a gas under surface conditions and is found in natural gas. (C_4)

button tricone bit. a tricone bit in which holes have been drilled into the steel cones and buttons of hard tungsten-steel carbide have been inserted. (*see* insert tricone bit)

buttress sand. sand deposited on top of an unconformity.

BWL. barrels of water load.

BWPD. barrels of water per day.

bypassing. water flow around relatively impermeable rocks containing oil and gas in the reservoir.

c. core.

C. 1) concentration or 2) coal.

C & C. circulation and conditioning.

C/H. cased-hole.

C_1. methane.

C_2. ethane.

C_3. propane.

C_4. butane.

C_5. pentane.

C_6. hexane.

C_7. heptane.

cable tool rig. an older type of drilling rig that pounds a hole in the ground by raising and lowering a bit on a cable.

calcareous. containing calcium carbonate. (calc)

calcite. a common mineral composed of $CaCO_3$. Most sea shells and the rock limestone are composed of calcite. (Ca, Calc or calc)

caliper log. a wireline log that measures the diameter of the wellbore. (CL, CAP, CAL, cal or CALP)

calorific value. the heat content per unit volume of natural gas. It can be measured in Btus per cubic feet.

Cambrian. a period of geological time from 570 to 500 million years ago. It is part of the Paleozoic Era.

cantilevered mast. a mast that is assembled horizontally and then pivoted vertical into place by the drawworks and traveling block on the rig.

caprock. 1) an impermeable rock layer that forms the seal on top of an oil or gas reservoir (seal) or 2) the insoluble rock on the top of a salt plug.

carbonaceous. containing carbon. (carb)

Carboniferous. a time period from 365 to 290 million years ago. It is divided into Mississippian and Pennsylvanian periods in the United States and Canada and is part of the Paleozoic Era.

carrier bed. a permeable rock layer along which fluids can flow.

carrier unit. a self-propelled workover rig.

cased-hole log. a wireline well log that can be run in a well that has been cased. The natural gamma ray log is an example.

casing. relatively thin walled, large diameter (commonly between $5\frac{1}{2}$ to $13\frac{3}{8}$ inches), steel pipe. Joints of casing are screwed together to form a casing string, run into a well and cemented to the sides of the well. (csg)

casing point. 1) the depth to which a casing string has been set in a well (CP or csg pt) or 2) the time after drilling and testing a well that a decision has to be made to either complete (case) the well or plug and abandon.

casing pressure. pressure on the fluid in the casing-tubing annulus. It can be either flowing or static. (CP or csg prss)

casing program. the lengths, diameters and other specifications of different casing strings that are to be used in a well.

casing pump. a large sucker-rod pump held in position on the bottom of a well by a packer.

casing roller. a long, tapered tool with rollers on the sides that is run on a drill-string to roll out collapsed casing in a well.

casing string. a long length of casing made by screwing many joints of casing together.

casing-free pump. a hydraulic pump that uses only one tubing string. The power oil goes down the tubing string and the produced fluids come up the tubing-casing annulus.

casinghead. a forged or cast steel fitting on the lower part of the wellhead that seals the annulus between two casing strings. The casing hanger that suspends a casing string is located in the casinghead. Each casing string has a casinghead. (CH)

casinghead gas. natural gas that bubbles out of oil on the surface at the well. (CHG)

casinghead gasoline. *see* condensate.

catheads. a hub on a shaft (catshaft) on the drawworks of a drilling rig that is used to pull a line (catline) to lift or pull equipment.

cathodic protection. a method that applies an electrical charge to a metal structure such as a pipeline or offshore platform to prevent corrosion.

catwalk. a flat, steel walkway that is elevated and connects stock tanks or installations.

caustic flood. *see* alkaline flood.

Cav or **cav.** cavern.

cave. the collapse of well walls into the hole. (sluff)

cavings. small rock particles that have fallen off the well walls and down the well.

CB. core barrel.

CDP. common-depth-point.

cellar. a rectangular pit dug below the floor of a large drilling rig to hold the blowout preventers. It is usually lined with boards or cement.

cem. cement.

cement. 1) minerals that naturally grow between clastic grains and solidify a sedimentary rock or 2) Portland cement used to bind the casing strings to the well walls. (Cmt, cmt or cem)

cement bond log. a type of sonic log that determines where and how well cement has set behind casing. (CBL)

cement job. to cement casing into a well.

cementing head. an L-shaped fitting that is attached to a wellhead for a cement job. It conducts the wet cement from the cement pumps down the well.

Cenozoic. an era of time from 65 million years ago to today.

centipoises. a unit of viscosity (cp).

central processing unit. 1) a common separator and tank battery for several oil wells or 2) common gas conditioning equipment for several gas wells. (CPU)

centralizer. an attachment to the outside of a casing string that uses steel bands to keep the string central in the well.

cf. cubic foot.

cg. coring.

Cgl or **cgl.** conglomerate.

ch. choke.

CH. casinghead.

chalk. an extremely fine-grained limestone composed of microfossils such as coccoliths. (Chk or chk)

channel. 1) a single seismic recording unit. The geophones in an array that are recorded together on one channel or 2) a cavity in the cement behind casing in a well.

charcoal test. a test used to measure the amount of condensate in natural gas. Activated charcoal is used to absorb the condensate from a volume of natural gas.

chat. a driller's term for conglomerate.

check shot. a method used to determine the seismic velocities of rock layers in a well. The seismic source is located on the surface next to the well. A geophone is raised in the well to measure seismic velocities at various depths.

chemical cutter. a tool used to cut pipe in a well with jets of hot, caustic chemicals under high pressure.

chemical flood. an enhanced oil recovery method in which batches of chemicals are injected into a depleted oil reservoir. A micellar-polymer flood is an example.

chert. a sedimentary rock composed of amorphous quartz. (flint) (Cht or cht)

CHG. casinghead gas.

chiller. a heat exchanger vessel that uses cooling to remove natural gas liquids from natural gas.

chk. choke.

Chk or chk. chalk.

choke. a constriction in a line that restricts flow. It is described by its diameter in $\frac{1}{64}$th of an inch. (ch or chk)

choke manifold. a series of pipes and valves located next to a drilling rig. It is designed to guide fluids from the well to the mud tanks, reserve pit and other areas and to direct kill mud into the well.

CHP. casinghead pressure.

Christmas tree. the fittings, values and gauges that are bolted to the wellhead of a flowing well to control the flow from the well. (production tree) (Xtree)

Cht or cht. chert.

Circ or circ. circulate.

circulate. to pump drilling mud down the drillstring. (circ)

circulation. the movement of drilling mud down through the drillstring and back up through the drillstring-casing annulus.

Cl or **cl.** clay.

clastic. a sedimentary grain that has been transported and deposited as a whole particle such as a sand grain. (clas)

clastic ratio map. a map that uses contours to show the ratio of conglomerates, sandstones and shales to limestones, dolomites and salts in a formation.

clay. a fine-grained particle less than $\frac{1}{256}$ mm in diameter. (Cl or cl)

clay mineral. a fine-grained mineral formed by a layered molecular structure of aluminum, silicon and oxygen atoms. Bentonite is a clay mineral used to make drilling mud.

clean oil. crude oil that is below a maximum basic sediment and water content and meets pipeline specifications. (pipeline oil)

clean sands. well sorted sands.

clints. fractures in coal.

cln. clean.

closure. the vertical distance between the top of a reservoir rock down to the spill point in a trap.

Clvg. cleavage.

CMP. common-mid-point.

Cmt or **cmt.** cement, cemented or cementing.

cmtd. cemented.

cmtg. cementing.

CNL. compensated neutron log.

cntrt. contorted.

CO. 1) clean out or 2) circulate out.

coal. a sedimentary rock composed primarily of carbonaceous material formed by plant remains transformed by heat and time. (c)

coal bed or **coal seam gas.** methane gas generated during coal formation. It is adsorbed to the natural fracture surfaces in coal.

coastal plain. a plain with underlying thick sediments deposited along an ocean margin.

coccolith. A calcium carbonate plate from a small, single-cell animal (coccolithopore) that lives floating in the ocean and can only be identified by a scanning electron microscope. A pure coccolith deposit is called chalk.

COFCAW. combination of forward combustion and waterflooding.

COH. coming out of hole.

coiled tubing unit. a well service unit that has a reel of coiled tubing for running equipment down a well during workover or for drilling. Before the coiled tubing goes in the well it goes through a pipe straightener.

coiled tubing. high-strength, flexible, steel tubing that is often $1\frac{1}{4}$ inches in diameter. It comes in a long, continuous length wrapped around a reel.

collar. a short, steel cylinder with female threads. It is used to join pipe joints with male threads. (coupling or tool joint)

collar log. a production log that records the depth of each casing collar in a well.

com. common.

combination of forward combustion and waterflooding. a fireflood in which air and water are alternately injected into the reservoir. The steam generated from the water helps drive the oil toward producing wells. (wet combustion) (COFCAW)

commingle. to mix production from a) two or more zone in a well (subsurface commingling) or from b) two or more wells (surface commingling).

common-depth-point stacking or **common-mid-point stacking.** a seismic acquisition and processing method in which numerous, different reflections off the same subsurface point within a bin are combined to reduce noise and reinforce the reflector. (CDP or CMP stacking)

compaction. the volume decrease of sediments by pressure due to burial.

compaction anticline. an anticline formed by compaction of softer sediments over and along a harder reef or bedrock hill.

company man or **representative.** an employee of the operator who works with the tool pusher to make sure a well is being drilled to specifications.

compensated log. a wireline well log that has been adjusted to irregularities in wellbore size and roughness.

compensated neutron log. *see* neutron log. (CNL)

completion card. a form published by a commercial firm containing information on the drilling and testing history and the geological and producing characteristics of a specific well.

completion fluid. an inert fluid, usually treated water or diesel oil, used in the casing-tubing annulus of a well to prevent casing corrosion.

completion packer. a packer run on the bottom of the tubing to seal the space between the tubing and casing. (tubing packer)

completion rig or **unit.** a derrick and hoisting unit used after the drilling rig has been released to run the final string of casing.

compliant platform or **tower.** an offshore production facility that is anchored on the bottom but the upper end is free to move within a restricted radius. Spars and tension leg platforms are examples.

compounder. a system of pulleys, belts, shafts, chains and gears that transmit power from the prime movers to the drilling rig.

compressibility factor. *see* Z factor.

compression ratio. a ratio of the volume of uncompressed gas divided by the volume of the same gas compressed by a compressor such as 10.1.

compression test. a test used to determine the condensate content of natural gas. A gas sample is compressed and then allowed to expand to cool it. Condensate separates from the cooling gas.

compressor. a device that increases the pressure of gas and can cause the gas to flow. Compressors use pistons in cylinders, rotating vanes on a shaft and other methods.

computer generated log. a log made by a computer from two or more logging measurements. It can be generated at the wellsite (a quick-look log) or in a computing center.

concentric tubing workover. a workover that uses smaller than normal equipment that can be run down a tubing string.

concession agreement. *see* tax royalty participation contract.

condensate. A hydrocarbon mixture composed primarily of molecules with 5, 6 and 7 carbon atoms. It is a liquid under surface conditions but is a gas mixed with natural gas under subsurface reservoir conditions. Condensate is very light in density and is transparent to yellowish in color. It is almost pure gasoline in composition. (natural gasoline, drip gasoline, casinghead gasoline and white oil).

conditioning. 1) preparing and altering drilling mud or 2) circulating drilling mud in a well for a period of time to prepare the well for logging or another process.

conductor casing or **pipe.** a length of large diameter, steel pipe that is pile driven or drilled into the ocean floor before drilling. The blowout preventer stack is attached to it. (structural casing)

conductor hole. a large diameter, shallow hole drilled to hold the conductor pipe.

conductor pipe. large diameter pipe that is cemented into the conductor hole. It is used to stabilize the soil as a well is being drilled and to attach the blowout preventers.

confirmation well. a well drilled after the discovery well to prove the extent of a new petroleum deposit.

conglomerate. a poorly-sorted sedimentary rock with rounded, pebble- to clay-sized grains. (Cgl or cgl)

coning. the drawing of a) the oil-water contact up or b) the gas-oil contact down into an oil reservoir. The contact has the shape of a cone around the well and is caused by too rapid oil production.

connate water. saline, subsurface water.

cons. considerable.

consolidated sediments. sediments bound together into a relatively hard sedimentary rock. (indurated sediments) (consol)

contact. the boundary between two rock layers. (Ctc)

contactor. a metal vessel that causes a gas passing through it to come in contact with a chemical on a bed. A contactor can use a solid desiccant to remove water from natural gas.

Contam or **contam.** 1) contamination or 2) contaminated.

continental drift. a relatively old theory that the continents were all joined in one supercontinent that broke up during the Mesozoic with the fragments drifting across the earth.

continental rise. an enormous wedge of sediments at the base of a continent slope in water depths of 5000 to 13,000 feet.

continental shelf. a shallow platform that surrounds the continents. It extending from the beach out to an ocean depth of about 450 feet where the shelf break (an abrupt change of slope) is located. Most offshore drilling and production occurs on the continental shelf.

continental slope. the slope (3° to 4°) leading down from the shelf break on the edge of continental shelf to the deep ocean bottom.

continuous flowmeter. an instrument that measures fluid flow versus depth in a well.

contour. a line of equal value on a map.

contract depth. the depth to which a well is to be drilled as stated in a drilling contract.

controlled exploratory well. a well drilled to find new gas or oil reserves. It can be drilled to 1) test a trap that has never produced (new-field wildcat), 2) test a reservoir that has never produced in a field and is shallower or deeper than the producing reservoir(s) or 3) extend the known limits of a producing reservoir in a field. (wildcat well)

coquina. a sedimentary rock composed of broken shells. (Coq)

Cor. coral.

core. 1) a cylinder of rock drilled from a well. It can be either a) a full diameter ($3\frac{1}{2}$ to 5 in.) or b) a sidewall core (1 in.) (c or cr) or 2) to drill a core.

core barrel. a tubular used above a rotary coring bit. It has both an outer core barrel to rotate and cut the core and an inner barrel to remain stationary and receive the core. They are separated by ball bearings. (CB)

correlation. the matching of rock layers.

corrosion. the chemical degradation of metal. Sweet corrosion is due to CO_2 and sour corrosion is due to H_2S.

corrosion inhibitor. a chemical that is applied in batch or continuous form to prevent corrosion in a well. It usually coats the metal.

cost oil. oil that is available to a multinational company to reimburse previous expenditures.

coupling. a short, steel cylinder with female threads. It is used to connect pipe joints with male threads. (collar or tool joint)

cp. centipoises.

CP. 1) casing point or 2) casing pressure.

CPU. central processing unit.

cr. core.

cracking. a refining process that breaks long-chained hydrocarbons into more valuable short-chained hydrocarbons.

craton. land covered by sedimentary rocks surrounding a shield.

crest. the top of a structure such as an anticline.

Cretaceous. a period of geological time from 144 to 65 million years ago. It is part of the Mesozoic Era.

crevasse splay. sediments deposited to the side of a delta through a break in the levee.

crooked hole. a well with a large deviation along the wellbore that was not made on purpose.

crooked hole country. an area with dipping hard rock layers that cause wells to be drilled out at an angle.

cross beds. sedimentary rock layers deposited by a current at an angle to horizontal in dunes or ripples.

crossbedded. a sedimentary rock that displays cross beds. (x-bd, x-bdd, X-bdd or XBD)

crossover sub. a short section of pipe used in a downhole assembly to connect pipes of different sizes or thread types.

crown block. a fixed, steel frame with steel wheels on a horizontal shaft. It is located at the top of a derrick or mast.

crown land. land owned by the federal or provincial government in Canada.

crude oil. a liquid composed of over one hundred different types of hydrocarbon molecules. The molecules range from 5 to more than 60 carbon atoms in length. Crude oil colors range from black to greenish to yellowish to transparent.

crude stream. crude oil from a single field or a mixture from fields that is offered for sale by an exporting country.

cryogenic plant. an installation that uses a natural gas driven turbine to cool the gas and remove natural gas liquids. (expander plant)

CSA. casing set at.

csg. casing.

csg press. casing pressure.

csg pt. casing point.

Ctc. contact.

Ctgs. cuttings.

cubic foot. the English system unit of natural gas volume measurement. It is the volume of a cube, one foot on a side. (cf)

cubic meter. the metric system unit of natural gas and crude oil volume measurement. It is the volume of a cube, one meter on a side. One cubic meter is equal to 6.29 barrels of oil or 35.3 cubic feet of gas. (m^3)

cuesta. a ridge formed by a resistant, dipping rock layer.

cut. to dilute something.

cuttings. rock flakes made by the drill bit. (Ctgs)

Cvg. caving.

cyclic steam injection. an enhanced oil recovery method used for heavy oil. A well is used to first pump steam into the subsurface reservoir to heat the oil and make it more fluid. The same well is then used to pump the heated, heavy oil. (huff and puff)

cycling. the injection of produced gas back into a retrograde condensate reservoir to slow the drop of reservoir pressure and the separation of condensate in the reservoir.

cyclothem. alternating marine and nonmarine sedimentary rocks.

d. diameter.

D. depth.

D.O. division orders.

daily drilling report. a report made by the tool pusher or company man on the last 24 hours of activity on a drilling rig such as footage drilled and supplies used. (morning report) (DDR)

darcy. the unit of permeability. One darcy permeability in a porous medium allows one cubic centimeter of fluid of one centipoise viscosity to flow in one second through a pressure differential of one atmosphere through a cross section of one square centimeter and a length of one centimeter.

dat. datum.

datum. a level surface to which contours are referred to such as sea level. (dat)

daywork contract. a drilling contract based on a cost per day during drilling to contract depth.

DB. diamond bit.

dd. dead.

DDR. daily drilling report.

dead oil. crude oil in a rock that will not flow.

decline curve. a plot of oil production rate versus time.

deconvolution. a computer process that makes subsurface reflections sharper and reduces noise.

°API or °API gravity. A measure of the density or weight of a crude oil. Average weight oil is 25–35, heavy oil is below 25 and light oil is 35–45. (gr API)

dehydrator. a vessel that uses either a solid or liquid desiccant to remove water from natural gas.

delay rental lease. a type of lease in which a delay rental payment must be made each year during the primary term to the lessor if drilling has not commenced.

delineation well. a well drilled to the side of a discovery well to determine the extent of the new field. (step out or appraisal well)

deliverability. the ability of a reservoir to move fluids into a well at a given, flowing bottomhole pressure.

delta. sediments deposited by a river emptying into an ocean.

delta switching. a process in which a river abandons an old delta for a shorter route to the ocean and builds a new delta.

demulsifier. a chemical used to break an emulsion.

dense limestone. limestone with little or no permeability.

density. 1) weight per unit volume (ρ) or 2) perforations per foot.

density log. *see* formation density log. (DL)

depletion drive. *see* dissolved gas drive.

derrick. the steel tower with four legs that sits on the drill floor of a drilling rig. A derrick must be raised in sections in contrast to a mast.

derrick floor. *see* drill floor. (DF)

derrickman. a member of the drilling crew who is second in charge. The derrickman stands on the monkeyboard when making a trip. (monkeyman)

desander and desilter. metal cones used on a drilling rig to centrifuge drilling mud from a well to remove the well cuttings.

detrital. a sediment grain that has been transported and deposited as a whole particle such as a sand grain. (detr)

developmental geologist. a geologist who specialized in the exploitation of petroleum fields.

developmental well. a well drilled in the known extent of a field.

deviation. the angle of a wellbore from vertical. (drift angle)

deviation drilling. drilling a well out at an angle on purpose. (directional drilling)

Devonian. a period of geological time between 405 and 365 million years ago. It is part of the Paleozoic Era.

dew point. the temperature at which a liquid starts to separate out of a gas as it is being cooled.

DF. drill floor.

diagenesis. the processes that form sedimentary rock from loose sediments. (Diag or diag)

diamond bit. a steel bit with no moving parts. Hundreds of small, industrial diamonds have been attached to the bottom and sides of the bit in geometric patterns. (DB)

diatom. a single-cell plant that floats in water. It has a silicon dioxide shell.

diatomaceous earth. a sedimentary rock composed primarily of siliceous diatom shells.

differential wall pipe sticking. the adherence of a drillstring to the sides of a well due to suction.

dike. a layer of igneous intrusion that cuts preexisting layers.

DIL. dual induction log.

dim spot. a portion of a seismic reflector with a less intense reflection amplitude. A porous or gas-saturated reef overlain by shale can cause a dim spot.

dip. the angle and direction in which a plane such as a sedimentary rock layer or fault goes down in the ground. It is measured at right angles to the strike.

dipmeter or **dip log.** a wireline well log that measures the orientation of each rock layer in a well. (DM or DIP)

dip-slip fault. a fault with predominately vertical displacement. It can be either a normal or reverse dip-slip fault.

direct hydrocarbon indicator. a bright spot, flat spot or other evidence of petroleum on a seismic record.

directional drilling. drilling a well out at an angle on purpose. (deviation drilling)

directional survey. a well survey that measures the angle and orientation of the wellbore using a magnetic compass or gyroscope. (DS)

dirty sands. poorly sorted sands.

disconformity. an ancient, erosional channel. The sedimentary rock layers above and below the disconformity are parallel.

discovery well. a well that locates a new petroleum deposit, either a new field or a new reservoir.

displacement efficiency. the ratio of volume of oil sweep divided by the volume of oil in place in a reservoir during waterflood or enhanced oil recovery.

disposal well. a well used to inject an unwanted fluid, usually oilfield brine, into the subsurface

dissolved gas drive. a reservoir drive in which the drop in reservoir pressure during production causes dissolved gas to bubble out of the oil and force the oil through the rock. It has a very low oil recovery efficiency. (solution gas or depletion drive)

dissolved gas-oil ratio. the standard cubic feet of natural gas dissolved in one barrel of oil in the reservoir. (formation or solution gas-oil ratio)

distributary. a river channel outlet on a delta.

distributary mouth bar. a sand bar deposited in front of a distributary on a delta.

disturbed belt. a zone of thrust faults that moved during the formation of a mountain range. (overthrust belt)

division orders. a form that establishes the distribution of production revenues and the assessment of costs royalty and working interest owners. (D.O.)

dns. dense.

doghouse. the room or vehicle that houses the seismic recording equipment.

dogleg. a relatively sharp turn in a well.

dolomite. a mineral composed of $CaMg(CO_3)_2$. It is formed by the natural alteration of calcite. A rock composed of dolomite is called dolostone and can be a reservoir rock. (Dol or dol)

dolostone. a sedimentary rock composed primarily of dolomite mineral grains. It forms from the natural, chemical alteration of limestone and is often a reservoir rock.

dom. dominate.

dome. a circular or elliptical uplift in sedimentary rocks. It can form a petroleum trap.

double. two joints of drillpipe.

double barrel separator. a separator with two, horizontal steel cylinders mounted vertically. The upper cylinder receives the produced fluids and makes an initial gas-liquid separation. The lower cylinder completes the oil-water separation. (double tube separator)

double section. the same section of rock encountered twice by drilling through a reverse fault. (repeated section)

double tube separator. *see* double barrel separator.

down dip. in a direction located down the angle of a plane such as a sedimentary rock layer.

down-to-the-basin fault. a dip-slip fault that moves down on the basin side (see growth fault).

downdraw. the difference between static and flowing pressure in a well.

downhole mud or **turbine motor.** a motor that is driven by drilling mud pumped down the drillstring. It drives the bit located below it.

downthrown. the side of a dip-slip fault that moved down.

dpg. deepening.

drag fold. a fold formed along a fault plane. It is caused by friction of one side of the fault against the other when it moved.

drawworks. a drum in a steel frame used on the floor of a drilling rig to raise and lower equipment in a well. It is driven by the prime movers. Hoisting line is wound around the reel.

dress. 1) to mill a fish to prepare the surface for a fishing tool or 2) to sharpen a drag bit.

drift angle. the angle of a wellbore from vertical. (deviation)

drill break. a change in drill penetration rate; usually the result of drilling into a different rock such as from shale into limestone.

drill collar. a heavy, large-diameter pipe run on the bottom of a drillstring. It comes in 31 ft. sections with both a male-threaded and female-threaded end.

drill floor. the elevated flat, steel surface on which the derrick or mast sits and most of the drilling activity occurs. It is supported by the substructure. (derrick floor) (DF)

driller. the person in charge of the drilling rig crew on that tour. The driller operates the machinery.

driller's depth. *see* total depth.

driller's method. a technique used to control a well that has a kick. After the blowout preventers have been thrown, the kick-diluted mud is replaced with original mud under pressure during the first circulation. The original mud under pressure is then replaced with kill mud during the second circulation.

drilling and spacing unit. the area, such as 40 acres, upon which one producing well can be located. It is declared by a regulatory agency. (DSU)

drilling barge. a barge with a drilling rig mounted on it. It is used for exploratory drilling in shallow, protected waters.

drilling break. a sudden change in the rate of penetration.

drilling console. a metal panel on the drill floor. It contains the weight indicator, mud pump pressure, rotary table torque, pump strokes, rate of penetration and other indicators.

drilling contractor. a company that owns and operates drilling rigs.

drilling fluid engineer. a service or oil company employee who monitors the properties of drilling mud being used on a well. The engineer is responsible for making changes in the mud properties (conditioning) when necessary. (mud man)

drilling line. wire rope made of strands of steel, braided cable wound around a fiber or steel core. It is commonly between 1 and $1\frac{5}{8}$ inches in diameter. The line is used to raise and lower equipment on a drilling rig. (hoisting line)

drilling mud. a viscous mixture of clay (usually bentonite) and additives with either a) water (usually fresh), b) oil, c) an emulsion of water with droplets of oil or d) a synthetic organic fluid. Mud is circulated on a rotary drilling rig to cool the bit, remove rock chips, and control subsurface fluids.

drilling spool. a steel spool used between the rams on a blowout preventer stack to attach the kill and choke lines.

drilling time log. a log showing the rate of drill bit penetration with depth, usually in minutes per foot or meter.

drillpipe. steel pipe that comes in 18 to 45 ft sections but is commonly 30 ft long. It is threaded on both ends. Each section is called a joint.

drillship. a ship with drilling rig aboard that drills through a hole (moon pool) in its hull. It is kept on position by a computer.

drillsite. the location for a drilling rig.

drillstem test. a test made by running a drillstem with packers in a well. The packers isolate the zone to be tested. A valve is opened in the drillstem and fluids from the zone can flow into the drillstem. It is a temporary completion of the well. (DST)

drillstring. the rotating kelly, drillpipe, drill collars, subs and bit in the well.

drip gasoline. *see* condensate.

drl. drill.

drld. drilled.

drlg. drilling.

drop angle. to decrease the deviation of a well.

dry gas. pure methane (CH_4) gas. It is a gas under both subsurface reservoir and surface conditions.

dry hole. a well that was drilled and did not encounter commercial amounts of petroleum. (duster)

DS. directional survey.

DSI. drilling suspended indefinitely.

DST. drillstem test.

DSU. drilling and spacing unit.

Δt. The sound velocity through a rock measured by a sonic log. The units are in microseconds per foot. (interval transit time)

dual completion. a system to keep the production from two zone in a well separate. Usually two tubing strings and two packers are used. There will be two surface pumping units or a double wing production tree on the well.

dual induction log. an wireline log that gives a medium and a deep induction resistivity measurement. (DIL)

dune. a hill of sand shaped by blowing wind or flowing water.

duplex pump. a mud pump with two double-acting pistons in cylinders. The mud is pumped on both the forward and backward strokes of the pistons.

duster. a well that did not encounter commercial amounts of petroleum. (dry hole)

DWT. deadweight ton.

dynamic positioning. the use of a computer that continuously plots a drillship's location to keep the drillship on station.

ea. earthy.

earth pressure. the pressure on rocks at a specific depth. It is caused by the weight of the overlying rocks. (lithostatic pressure)

economic limit. the time in the history of a well in which the revenue from production equals production costs.

edge water. water located in the reservoir to the side of the oil.

EEB. energy equivalent barrel.

effective permeability. the permeability of a fluid when it shares the pore space with another fluid.

effective porosity. the percent porosity including only interconnecting pores.

elastomer. rubberlike material.

electric submersible pump. an electrical motor that drives a centrifugal pump with rotating blades on a shaft. It is located on the bottom of a tubing string and is used for oil well artificial lift. (ESP)

electrical log. a wireline resistivity log. It is often run with a spontaneous potential or natural gamma ray log. (EL or E log)

electrostatic precipitator. a separator that uses charged electrode plates to separate an emulsion.

emul. emulsion.

emulsion. droplets of one liquid suspended in a different liquid such as water-in-oil. (emul)

emulsion mud. drilling mud made with water containing suspended droplets of oil. The oil improves the lubricating qualities of the mud.

energy equivalent barrel. *see* barrels of oil equivalent. (EEB)

enhanced oil recovery. the injection of fluids that are not found naturally in a producing reservoir down injection wells into the depleted reservoir to recover more oil. (EOR)

Eocene. an epoch of geological time from 55 to 34 million years ago. It is part of the Tertiary Period.

eolian. formed by blowing wind.

EOR. enhanced oil recovery.

epoch. a time subdivision of periods such as Miocene.

era. a major time division of earth history such as Paleozoic.

ESP. electric submersible pump.

ethane. a hydrocarbon composed of C_2H_6. It is a gas under surface conditions and is found in natural gas. (C_2)

Evap or **evap.** evaporate.

ex. excellent.

expandable casing. casing that can be run into a well and then have its diameter expanded by hydraulically pumping an expander plug through it.

expander plant. an installation that uses natural gas to drive a turbine to cool the gas and remove natural gas liquids. (cryogenic plant)

expansion-gas drive. a gas field reservoir drive in which the expanding gas produces the energy to force the gas through the rocks. (volumetric drive)

exploration geologist. a geologist who specialized in the search for petroleum.

exploratory well. *see* controlled exploratory well.

explosive fracturing. to explode nitroglycerin in a torpedo at reservoir depth in a well to fracture the reservoir and stimulate production. (well shooting)

extended-reach well. a deviated well with a relatively large horizontal distance between the surface location of the well and the bottom of the well.

extender. an additive to a fluid such as bentonite in cement slurry that reduces its cost.

extension well. a well that significantly increases the productive area of a field. It is called an outpost well before it is successful.

extr. extremely.

f. fine.

F. factor.

F/. flowed or flowing.

F/GOR. formation gas-oil ratio.

facies. a distinctive part of a rock layer such as a sandstone facies.

fairway. the area along which a petroleum play occurs. (trend)

farmer's gas. the produced gas that goes to the lessor for his use as part of the lease agreement.

farmin. 1) a lease obtained from another company for drilling in return for a consideration such as a royalty or 2) to receive a farmin. (FI)

farmout. 1) a lease given to another company for drilling in return for a consideration such as a royalty or 2) to give a farmout. (FO).

FARO. flowed at rate of.

fault. a break in the rocks along which there has been movement of one side relative to the other side. Faults are either dip-slip or strike slip. (Flt or flt)

FBHP. flowing bottomhole pressure.

FCP. flowing casing pressure.

FDC. formation density compensated log.

fee land. private land that has both a surface and mineral rights owner.

feedstock. a chemical refined from hydrocarbons and used to produce petrochemicals.

fence diagram. a three-dimensional representation of wells and the geological cross sections between them. Cross sections between the wells are called panels that close to form the fence.

FI. farmin.

field. the surface area directly above one or more producing reservoirs on the same trap such as an anticline.

field print. the original copy of a wireline well log that is made in the logging truck.

field superintendent. an engineer who is in charge of production from a field. The field superintendent gives orders to the production foremen.

filter cake. the solid ring made of clay particles that were plastered against the sides of the well by drilling mud. (mud cake)

filtrate. *see* mud filtrate.

final print. the last copy of a wireline well log that has been cleaned up, computer processed and printed in the office.

fireflood. an enhanced oil recovery process in which the subsurface oil is set afire. The heat makes the oil more fluid and the gasses generated by the fire drives the oil to producing wells as air is pumped down injection wells. (in situ combustion)

fish. a) a tool or broken pipe that has fallen to the bottom of a well (junk), b) fishing.

fishing. a process in which fishing tools are used to retrieve an object (fish) on the bottom of the well. (fish or fsg)

fishing string. a length of tubulars that are run in a well with a fishing tool on the end.

fishing tools. tools leased from a service company to fish for a fish in a well.

five-spot pattern. a common pattern of injector and producing wells used for a waterflood. Four injecting wells are located at the corners of a square and the producing well is in the center.

fixed cutter bit. a steel bit with no moving parts. Man-made diamonds in blanks on the bottom and sides of the bit shear the rocks. (polycrystalline diamond compact bit)

fixed production platform. a relatively permanent offshore platform with treaters for produced fluids. It has legs that sit on the ocean bottom. Two types are a) steel jacket and b) gravity based.

fl. fluid.

fl/. flowed or flowing.

flare. 1) burning gas or 2) to burn gas.

flat spot. a flat, seismic reflector in sedimentary rock layers that are not horizontal. It can be off a gas-liquid contact.

flint. a sedimentary rock composed of amorphous quartz. (chert)

float collar. a short length of tubular that is run just above the bottom of a casing string. A one-way valve allows the casing string to float in the drilling mud as it is being run in the well. A constriction in the float collar stops the wiper plug as it is being pumped down the string during a cement job.

floater. a floating, drilling platform such as a semisubmersible or drillship.

flowing pressure. pressure on a fluid as the fluid is flowing. (FP)

flowline. a steel, plastic or fiberglass pipe that conducts a) produced fluids from the wellhead to the separators, b) oil from the separators to the stock tanks or c) gas to the treaters.

Flt or flt. fault.

fluid pound. a problem caused by gas in a downhole, sucker-rod pump.

fluid pressure. the pressure on fluids in the pores of rock at a specific depth. Normal fluid pressure is due to the weight of the overlying waters. (reservoir or formation pressure)

Fluor or fluor. fluorescent

Fm. formation.

fnly. finely.

FO. farmout.

foam drilling. air drilling with a detergent to form a foam and lift water from the well.

fold. 1) the number of reflections (traces) off the same subsurface point that are combined in common depth point stacking to form a single reflection (trace) or 2) a bend in sedimentary rock layers such as an anticline or syncline.

footwall. the side of the fault that protrudes under the other side.

footage contract. a drilling contract based on a cost per foot to drill to contract depth.

foraminifera. small, one-cell animals with primarily calcium carbonate shells that float in the ocean or live on the bottom of the ocean. They are common microfossils. (forams) (Foram or foram)

formation. a mappable rock layer. It has a sharp top and bottom. The formation is given a two-part name such as Bartlesville Sandstone. Formations can be subdivided into members and combined to form a group. (Fm)

formation damage. a decrease in the permeability of a reservoir rock adjacent to a wellbore. It can be caused by mud filtrate that is forced into the pores during drilling. (skin damage)

formation density log. a radioactive wireline well log used to determine the density and porosity of rocks. (FDL) (gamma-gamma log)

formation gas-oil ratio. *see* dissolved gas-oil ratio. (F/GOR)

Formation MicroScanner™. a wireline well log that images the wellbore using resistivity.

formation pressure. *see* fluid pressure.

formation volume factor. the number of barrels of reservoir oil that shrinks to one stock tank barrel of oil on the surface after the pressure has decreased and the gas has bubbled out. (FVF or B) (*see* reservoir and stock tank barrels of oil)

fossil. the preserved remains of an ancient plant or animal. Fossils can be either macrofossils or microfossils, depending on their size. (Foss or foss)

fossil assemblage. a group of fossils that identifies a particular geologic time or a rock zone.

4-C seismic. a seismic survey that records not only the usual compressional waves (P waves) but also shear waves (S waves). It is used to better determine rock types and locate fractures.

4-D seismic. the seismic differences between several 3-D seismic surveys run at different times over the same reservoir during production from that oil field. Changes in seismic responses from the reservoir such as amplitude can show the flow of fluids through the reservoir. (time lapse seismic)

FP. flowing pressure.

fpso. floating production, storage and offloading.

fpso vessel. a ship that is stationed above or near an offshore oil field. Produced fluids from subsea completion wells are brought by flowlines to the vessel where they are separated and treated.

fr. fair.

frac job. a well stimulation method in which high pressure liquid is pumped down a well to fracture the reservoir rock adjacent to the wellbore. Propping agents suspended in the liquid are used to keep the fractures open. (hydraulic fracturing, frac-pac, frac/pack, frac pack, and sand frac)

Frac or **frac.** fracture.

frac/pack. *see* frac job.

fractionating. the separation of crude oil by heating and boiling off different components at different temperatures.

free gas cap expansion drive. a reservoir drive in which the expanding gas in the free gas cap drives the oil through the rocks. It has a moderate oil recovery efficiency.

free gas cap. the uppermost portion of a saturated oil reservoir. The pores of the reservoir rock are occupied by natural gas.

free water. water that readily separates from oil by gravity.

freewater knockout. a horizontal or vertical separator that uses gravity to separate gas, oil and water. (FWKO)

freq. frequent.

fresh water. drinking quality water. It contains less than one part per thousand salt.

frs. fresh.

fsg. fishing.

ft/min. feet per minute.

ft/sec. feet per second

ft-lbf. foot-pound.

FTP. flowing tubing pressure.

fulcrum assembly. a downhole assembly that uses the sag caused by the weight of a drill collar between two stabilizers to lift the bit and increase the angle (make angle) on a deviated well.

full diameter core. a cylinder of rock that was drilled from a well. It is between $3\frac{1}{2}$ and 5 inches in diameter.

FVF. formation volume factor.

FWKO. freewater knockout.

g. gram.

G. gas.

G & OCM. gas and oil-cut mud.

G.W. granite wash.

G/L. gathering line.

gage. *see* gauge.

gal/min. gallons per minute.

gamma ray log. *see* natural gamma ray log (GR or GRL)

gamma-gamma log. *see* formation density log.

gas. *see* natural gas. (G)

gas cap. *see* free gas cap.

gas cap drive. *see* free gas cap expansion drive.

gas conditioning. the removal of impurities such as water, acid gasses and solids from natural gas to meet pipeline contract specifications.

gas cut. diluted with gas.

gas effect. a divergence of porosities calculated from the neutron porosity and formation density logs on the same rock. It is caused by gas in the pore spaces.

gas in place. the amount of gas in the pores of a reservoir. (GIP)

gas injection. injection of natural gas into an oil reservoir or a free gas cap to maintain reservoir pressure and produce more oil.

gas lift. an artificial lift method for oil wells. An inert gas called lift gas that is usually natural gas is injected into the casing-tubing annulus, through gas lift valves and into the tubing to form bubbles that raise the produced liquids.

gas lift valve. a pressure-activated valve in the tubing string of a gas lift well. It allows lift gas to flow into the tubing.

gas lock. the failure of a downhole, sucker-rod pump because of gas filling the pump.

gas plant. an installation that removes natural gas liquids from natural gas by cooling or absorption. (natural gas processing plant)

gas sand. a driller's term for sandstone that contains gas.

gas-oil contact. the boundary between oil and gas in a reservoir. (GOC)

gas-oil ratio. the amount of natural gas per barrel of oil. The amount of natural gas in standard cubic feet dissolved in a barrel of oil in the subsurface reservoir is the dissolved, reservoir, or solution gas-oil ratio. The amount of natural gas in standard cubic feet produced with each barrel of oil is the producing gas-oil ratio. (GOR)

gathering system. a system of flowlines that conducts produced fluids from wells to a central processing unit.

gauge. 1) to measure, 2) the diameter of a bit, wellbore or tubular or 3) a measuring instrument. (gage)

gauge hole. a wellbore that meets a minimum diameter.

gauge table. a table that relates the height of oil in a stock tank to the volume of the oil. (tank table)

gauge tape. a metal tape on a reel with a brass weight on the end. It is marked in $\frac{1}{8}$ in. increments and is used to measure the height of oil in a stock tank.

gauger. a person responsible for measuring the amount and quality of oil in stock tanks.

GC. gas cut.

GCM. gas-cut mud.

gd. good.

gel. *see* drilling mud.

gel strength. the ability of a fluid to suspend solids.

gen. generally.

geochemist. a geologist who uses chemistry to study rocks and search for petroleum.

geographic information system. computer manipulation of geographical map data. Geological, physical, cultural and political maps can be super-imposed. (GIS)

geologic map. a map showing where rock layers, usually formations, crop out on the surface of the earth.

geologist. a scientist who identifies and studies rocks. A petroleum geologist searches for and exploits oil and gas deposits.

geolograph. a drilling recorder used on the floor of a drilling rig.

geophone. a vibration detector used on land to detect subsurface echoes during a seismic survey. (jug)

geophysicist. a scientist who uses physics and mathematics to study the earth. A geophysicist uses surface methods such as seismic, magnetic and gravity to image the subsurface and explore for petroleum.

geopressured. a reservoir rock with abnormal high pressure.

geosteering. the use of a measurements-while-drilling system, a logging-while-drilling system and a steerable downhole assembly to drill a horizontal drain well that stays in the oil pay zone.

geothermal gradient. the rate of temperature increase with depth in the earth.

GI. gas injection.

GIH. going in hole.

GIS. geographic information system.

GL. ground level.

global positioning system. a worldwide navigation method that uses satellites to obtain precise latitude and longitude and less precise altitude. It is especially useful in locating seismic lines. (GPS)

glycol absorber tower. a vertical, metal vessel that causes natural gas to bubble up through a liquid desiccant (glycol) to remove water.

gn. green.

gneiss. a metamorphic rock that has light and dark bands of coarse grained minerals. It is a basement rock.

GOC. gas-oil contact.

gone to water. a producing well that started to produce large amounts of water.

GOR. gas-oil ratio.

gouge zone. the fractured mass of rocks along a fault plane.

GPM. gallons (natural gas liquids) per thousand standard cubic feet (natural gas).

GPS. global positioning system.

gr API. °API gravity.

Gr or **gr.** grain.

GR. gamma ray.

GR Log. natural gamma ray log.

graben. the valley formed by the down-dropped block between two normal faults.

grad. grading.

graded bed. a clastic sedimentary rock layer that is coarse on the bottom and grades upward to fine on the top. Turbidity currents can deposit a graded bed.

grainstone. a type of limestone in which large sand-sized grains are in contact with each other and fine-grained material is absent. (Grst)

gram. The unit of weight in the metric system. There are 454 grams in a pound. (g)

Gran or **gran.** granule.

granite. the most common igneous rock that crystallizes in the subsurface. Granite is coarse grained with a light, speckled texture. It is a common basement rock. (Grt)

granite wash. a sandstone composed of sand grains from weathered granite. (G.W.)

grav. gravity.

gravel pack completion. a well completion used for unconsolidated reservoirs. A large cavity is reamed out in the reservoir. It is then filled with very well sorted, loose sand (gravel pack). A slotted or screen liner is run in the gravel pack. (GVLPK)

gravimeter. *see* gravity meter.

gravity. *see* °API gravity.

gravity drainage pool. an oil field in which the reservoir drive is gravity pulling the oil down into the wells. It can be very effective over a long period of time.

gravity meter. an instrument that measures the acceleration of gravity. It is able to detect variations in the density of the earth's crust. The units of measurement are milligals. (gravimeter)

gravity-base production platform. a fixed production platform that has a large mass of steel and concrete on the bottom to hold it in position.

graywacke. a poorly sorted, dark-colored sandstone.

ground truthing. to compare surface observations with remote sensing measurements such as satellite photographs.

ground water. water that occurs in the subsurface pores of sedimentary rocks.

group. 1) several geophones that are connected together to record as a single channel or 2) several adjacent formations that are similar in rock type. It is given a formal name such as Chase Group.

group shoot. a seismic survey paid for and shared by several different exploration companies.

growth fault. a fault that occurs where sediments have been rapidly deposited in a basin. It is parallel to the shoreline and has a curved fault plane that is steepest near the surface and flattens with depth toward the basin. (down to the basin fault)

Grst. grainstone.

Grt. granite.

Grv. gravel.

gry. gray.

GS. gas show.

guide fossil. a distinctive fossil that represents a particular geologic time.

guide shoe. a short, metal cylinder with a rounded nose having a hole in the end. It is run on the end of a casing string to guide the string into the well.

gun. *see* perforating gun.

gun barrel separator. a settling tank that uses gravity to separate a loose emulsion. (wash tank)

GVLPK. gravel packed.

GWI. gross working interest.

gypsum. a common salt mineral composed of $CaSO_4 \cdot 2H_2O$. (Gyp or gyp)

h. thickness.

halite. a common salt mineral composed of NaCl. (Hal or hal)

hang. to arrange well logs according to a common level reference surface such as sea level going as a straight line through the logs.

hanger. a circular device that suspends a casing, tubing or liner string in a well. It is attached to the top of the tubular by threads or slips.

hanging wall. the side of the fault that protrudes over the opposite side.

HBP. held by production.

hd. hard.

header. 1) a large pipe into which several, smaller flowlines feed, 2) the well information at the top of a well log or 3) the seismic information at the side of a seismic record.

heater-treater. a separator that uses heat from a fire tube to separate emulsions.

heave. the horizontal displacement on a fault.

heavy oil. viscous, high-density oil with an °API less than 25.

heavyweight additive. an additive such as galena that is used to increase the density of a fluid.

heavyweight drillpipe. drillpipe that is intermediate in strength and weight between drill collars and drillpipe. It has the same outer diameter as drillpipe and comes in 30 $\frac{1}{2}$ ft. sections. It is run on the drillstring between the drill collars and drillpipe.

helirig. a drilling rig that can be broken down in modules and transported by a helicopter.

hertz. one cycle per second. Seismic source frequencies are described in hertz. (Hz)

hetr. heterogenous.

hoisting line. wire rope made of strands of steel, braided cable wound around a fiber or steel core. It is commonly between 1 and $1\frac{5}{8}$ inches in diameter. The line is used to raise and lower equipment on a drilling rig. (drilling line)

hole opener. a sub that uses roller cones to enlarge a well.

holiday. an area behind casing with no cement.

Holocene. an epoch of geological time from 10,000 years ago until the present. It is part of the Quaternary Period. (Recent)

hom. homogenous.

homocline. inclined sedimentary rock layers with a constant dip.

hook. a curved steel fastener located below the traveling block on a drilling rig. It is used to suspend the swivel and drillstring in the well.

hor. horizontal.

horizon. a surface.

horizontal drain well. a highly deviated well drilled along the pay zone and parallel to the bedding of a reservoir.

horizontal section. the relatively flat portion of a horizontal drain well.

horizontal slice. a flat, seismic section made at a specific depth in time from 3-D seismic data. It shows where each seismic reflector intersects the slice. (time slice)

horsehead. a steel plate used on the end of a walking beam to keep the pull on the sucker-rod string vertical.

horst. the ridge between two normal faults.

host company. an oil company owned by a federal government. It operates only in that country.

hot oiler. a service company that removes wax from tubing in wells. A heated tank truck is used to heat crude oil. The heated crude oil is pumped down the well to dissolve the wax and is them pumped out.

HP. 1) hydrostatic pressure or 2) horsepower.

hp-hr. horsepower-hour.

huff and puff. an enhanced oil recovery method used for heavy oil. A well is used to first pump steam into the subsurface reservoir to heat the oil and make it more fluid. The same well is then used to pump the heated, heavy oil. (cyclic steam injection)

hvy. heavy.

Hydc. hydrocarbons.

hydrate. a snow-like substance that can form from water in a flowline as the temperature of natural gas falls. It is composed of ice with methane in the ice crystals.

hydraulic fracturing. a well stimulation method in which liquid under high pressure is pumped down a well to fracture the reservoir rock adjacent to the wellbore. Propping agents are used to keep the fractures open. (frac job, frac-pac, frac/pack and sand frac)

hydraulic pump. an artificial lift system. A pump on the surface injects power oil into the well. The power oil drives a pump that is coupled to a sucker-rod pump on the bottom of the tubing string.

hydrocarbon recovery unit. a vessel that uses silica, activated charcoal or molecular sieves to remove natural gas liquids from natural gas.

hydrocarbons. molecules formed primarily by carbon and hydrogen atoms. Crude oil and natural gas are composed of hydrocarbon molecules. (Hydc)

hydrogen sulfide. H_2S; a poisonous and corrosive gas that can be found mixed with natural gas. (sour gas)

hydrogen sulfide embrittlement. the weakening of steel by contact with H_2S.

hydrophone. a vibration detector used at sea to detect subsurface echoes during seismic exploration.

hydrostatic pressure. fluid pressure in subsurface rocks due to the weight of the overlying fluids. (normal pressure) (HP)

Hz. hertz.

I.P. in part.

Ice Ages. an epoch of time from about 1.8 million years to 10,000 years ago during which glaciers occupied much of the land area. It is part of the Quaternary Period. (Pleistocene)

ID. inner diameter.

igneous rock. rock formed by cooling and solidifying a hot, molten liquid. Granite and basalt are examples. (Ig or ig)

immature oil. heavy oil generated at shallow depths in the oil window.

impermeable. rock that does not allow fluids to flow through it. (aquitard)

impression block. a tool used during fishing to determine the shape of a fish. It is a weight with wax or lead on the bottom.

improved oil recovery. the methods of waterflood and enhanced oil recovery that are used to produce more oil from a depleted reservoir.

in situ combustion. an enhanced oil recovery process in which the subsurface oil is set afire. The heat makes the oil more fluid and the gasses generated by the fire drives the oil to producing wells as air is pumped down injection wells. (fireflood).

in./sec. inches per second.

incr. increasing.

ind. indurated.

induction log. a wireline well log that measures the resistivity of the rocks and their fluids with an induced current. (I, IL or IEL)

indurated sediments. *see* consolidated sediments.

inert. a gas that doesn't burn such as steam, carbon dioxide or nitrogen.

inert gas injection. an enhanced oil recovery method in which an inert gas such as carbon dioxide or nitrogen is injected into a depleted reservoir to produce more oil.

infill drilling. drilling between producing wells in a developed field to produce petroleum at a faster rate.

inhibitor. an additive to a fluid to retard a reaction.

initial pressure. the original reservoir pressure before any production. (virgin or original pressure)

initial production. the first 24 hours of production from a well. (IP)

inj. 1) injection or 2) injected.

injection well. a well used to pump fluids down into a producing reservoir for pressure maintenance, waterflood or enhanced oil recovery. (IW)

insert pump. a common type of oil well downhole pump driven by a sucker-rod string. It is run as a complete unit on the sucker-rod string through the tubing string. (rod pump)

insert tricone bit. a tricone bit in which holes have been drilled into the steel cones and buttons of hard tungsten-steel carbide have been inserted. (button tricone bit)

insol. insoluble.

intangible drilling costs. expenditures for drilling and completing a well that cannot be salvaged or recovered. They receive a very favorable tax consideration. (IDC'S)

intbed. interbedded.

intelligent well completion. a well with downhole sensors for temperature, pressure and flow velocity. A surface-controlled, downhole adjustable choke is used to regulate flow based on downhole conditions.

interfinger. a boundary between two rock types in which both form distinctive wedges protruding into each other.

intermediate casing. *see* protection casing.

interval transit time. the sound velocity through a rock measured by a sonic log. The units are in microseconds per foot. (Δt)

intrusive body. an igneous rock mass that was injected in a molten state into preexisting rock. (Intr or intr)

Intvl. interval.

invade zone. the area in a reservoir rock adjacent to the wellbore that has been flushed and filled or diluted with mud filtrate.

IP. initial production.

IR. injection rate.

irreducible water. pore water that will not move. (residual water)

isochron map. a map that uses contours to show the thickness in time (milliseconds) between two seismic horizons. (isotime or time interval map)

isolith map. a map that uses contours to show the thickness of one rock type such as sandstone in a formation.

isopach map. a map that uses contours to show the thickness of a subsurface rock layer.

isotime map. *see* isochron map.

ISP. initial shut-in pressure.

IW. injection well.

J. productivity index.

jacket. the legs on an offshore production platform.

jackup rig. an exploratory offshore drilling system with a two hulls and at least three tall legs through the hulls. It is towed into position similar to a barge. The lower hull rests on the bottom of the ocean and the upper hull is jacked up the legs. The drilling rig is mounted on the upper hull.

jar. a tubular run on a drillstring or fishing string that is designed to impart a sharp, upward or downward blow to the string on command.

jet bit. a tricone drilling bit with one large and two small nozzles. The jetting action of the drilling mud from the large nozzle is used to start drilling the well out at an angle.

jnk. junk.

JOA. joint operating agreement.

joint. 1) a natural fracture in rock along which there has been no movement or 2) a section of a tubular such as drillpipe. (Jt or jt)

joint operating agreement. an agreement between several companies to explore, drill and develop a common area called the working interest area. The agreement defines how the costs and revenues are to be shared among the parties and who is the operator. (JOA)

jts. joints.

jug. a vibration detector used on land to detect subsurface echoes during a seismic survey. (geophone)

jug hustler. a seismic crew member who lays cable and plants geophones.

junk. a tool or broken pipe that has fallen to the bottom of a well. (fish) (jnk)

junk basket. a fishing tool. Drilling mud is circulated down along the outside of the tool to pick up pieces of junk on the bottom of the well. The mud then circulates up along the inside of the tool where the junk is caught in a basket.

junk mill. a fishing tool that is rotated a) to dress a fish in preparation for another fishing tool or b) to reduce the fish to metal flakes.

Jurassic. a period of geological time from 206 to 144 million years ago. It is part of the Mesozoic Era.

k. permeability.

K. coefficient.

K-Monel. a nonmagnetic metal used in some drill collars. It is used on well surveys made with a magnetic compass.

karst. a highly dissolved limestone.

KB. kelly bushing.

kelly. a strong, four- or six-sided steel pipe that is located at the top of the drillstring. It runs through the kelly bushing.

kelly bushing. a device that is fitted on the master bushing and rotating table. The kelly runs through the kelly bushing. (KB)

kelly cock. a valve run on a drillstring just above or below the kelly. A wrench is used to open and close it. It is used to stop fluids from flowing up the drillstring.

kerogen. insoluble organic matter in sedimentary rocks.

keyseat. a section in a well being drilled that has a cross section similar to a key hole. The smaller diameter portion was abraded by the drillpipe and the larger portion by the bit. Drill collars can become stuck in the smaller portion of the keyseat.

kick. 1) the flow of subsurface fluids into a well or 2) a distinctive deflection on a well-log curve.

kick off point. the location in a well where it begins to deviate. (KOP)

kick off. to start to drill a well out at an angle.

kill. to stop and halt.

kill fluid. a liquid used in a well to stop reservoir fluids from flowing into the well in preparation for a workover.

kill mud. heavy drilling mud used to stop a kick and control a well.

kilogram. a unit of weight in the metric system. It is each to 1000 grams or 2.2 pounds.

kilojoule. the metric system unit of heat used to measure the heat content of natural gas. It is 1000 joules and is equal to about one British thermal unit. (kJ)

kitchen. the deep part of a basin where gas and oil are formed.

KOP. kick off point.

l. lower.

L. 1) liter or 2) length.

LACT. lease automatic custody transfer.

LACT unit. *see* lease automatic custody transfer unit.

lag time. the time that it takes the well cuttings to circulate from the bottom of the well, up the well to the screens on the shale shaker.

land. to transfer the weight of a casing string to the casing hangers.

landed at. depth to which a casing string was set in a well.

landman. an oil company employee or independent who identifies mineral rights owners and negotiates leases.

land farming. spreading used, fresh water-based drilling mud out on agricultural land to improve the crop.

Landsat. one of six, unmanned, remote-sensing satellites operated by the United States. Landsat pictures of the earth are made in visible light and infrared and are transmitted back to earth.

Laterolog. a type of wireline electrical log used in conductive muds to measure the true resistivity of the rocks. (LL)

lava. molten rock on the surface of the earth.

lbm/cu ft. pounds per cubic foot.

lbm/gal. pounds per gallon.

LC. lost circulation.

LCM. lost circulation material.

lean gas. 1) natural gas containing a minor amount of liquid condensate in contrast to rich gas or dry gas or 2) natural gas with less than 2.5 gallons of natural gas liquids per thousand standard cubic feet of natural gas.

lease. a legal document between an oil company (lessee) and a mineral rights owner (lessor) for the purpose of obtaining drilling and production rights on the land under lease. (lse)

lease automatic custody transfer unit. a system that uses automatic equipment to measure, sample, test and transfer oil in the field and to record that transaction. (LACT unit)

left-lateral strike-slip fault. a fault that moves horizontally, with the opposite side of the fault moving toward the left as you face the fault.

Len. lens.

lent. lenticular.

lessee. the recipient of a lease.

lessor. the mineral rights owner who grants a lease.

lge. large.

lift gas. inert gas, usually natural gas, that is used for gas lift.

Lig or **lig.** lignite.

light oil. fluid, low-density oil with a °API above 35.

lightweight additive. an additive used to decrease the density of a fluid.

lignite. soft coal. (Lig or lig)

LIH. left in hole.

limb. one side of a fold in sedimentary rocks.

lime. a driller's term for limestone.

lime mudstone. a type of limestone with a very small percentage of large, sand-sized grains and a considerable amount of fine-grained material.

limestone. a common sedimentary rock composed of $CaCO_3$. It can range from fine to coarse grained and can be a reservoir rock. (Ls or ls)

line shooting. a method used to acquire data for a 3-D seismic survey at sea. The seismic is acquired by running the seismic sources and hydrophone streamers in closely spaced, parallel lines. The ship tows at least two arrays of air guns or twin streamers.

linear spread. a geometric pattern of geophone groups that are arranged in a line.

line pressure. the pressure on a fluid in a flowline or pipeline.

liner string. a string of tubulars similar to casing in a well but the liner does not run all the way up to the surface as a casing string does. It may or may not be cemented into the well. (lnr)

liquefied natural gas. methane gas (CH_4) that has been compressed and super cooled into a liquid. (LNG)

liquefied petroleum gas. propane gas in liquid form in the United States. It can be a propane/butane mixture in Europe. (LPG or LP-Gas)

liter. the unit of volume in the metric system. It is equal to 1000 cm^3 or 0.264 U.S. gallons in the English system. (L)

Lith or **lith.** lithology.

lithofacies. one particular rock type such as a sandstone in a rock layer.

lithofacies map. a subsurface map showing changes in the physical properties of a particular rock layer such as an isolith map.

lithologic log. a record of the physical properties of rocks in a well. It includes composition, texture, color, presence of pore spaces and oil staining. (sample log)

lithology. the composition of a rock such as sandstone. (Lith or lith)

lithostatic pressure. the pressure on rocks at a specific depth. It is caused by the weight of the overlying rocks. (earth pressure)

lmy. limy.

LNG. liquefied natural gas.

lnr. liner.

LO. load oil.

load water or **load oil.** water or oil filling a well to maintain pressure on the bottom of the well. (LW or LO)

loc abnd. location abandoned.

log. a record of rock properties in a well.

logged depth. *see* total depth.

logging-while-drilling. a real time log made by sensors in the drillstring above the bit. Measurements include gamma ray, resistivity, neutron porosity and formation density. The measurements are digitized and transmitted to the surface by pressure pulses in the drilling mud. It is used on directional wells and with measurements-while-drilling. (LWD)

long normal resistivity. a wireline resistivity measurement made with electrodes spaced far apart (64 in.).

loose. an emulsion that readily separates.

LORAN. a ocean navigational system that uses angles and distances from fixed radio transmitters.

lost circulation. a drilling problem in which large quantities of drilling mud flow into a permeable rock layer in the well. Very little, if any, drilling mud circulates back up the well. (LC)

lost circulation additive, control agent or **material.** an additive to drilling mud or cement slurry that clogs the pores of a lost circulation zone. (bridging material) (LCM)

lost circulation zone. a very permeable rock layer in a well. It takes large amounts of drilling mud during drilling. (thief zone)

lost section. the section of rock that is missing when drilling through a normal fault.

low-temperature separator. an installation that passes natural gas through a expansion choke to cool the gas and separate the natural gas liquids. (LTX)

low-velocity zone. the layer of loose sediments that occurs near the surface of the earth and has a relatively low seismic velocity. Statics corrects seismic data for the low-velocity zone. (weathering layer)

LP. low pressure.

LPG or **LP-Gas.** liquefied petroleum gas.

LPG Drive. an enhanced oil recovery method in which liquefied petroleum gas is injected into a depleted oil reservoir. The LPG is miscible with the oil and makes it more fluid.

Ls or **ls.** limestone.

lse. lease.

lt. light.

Ltl. little.

LTX. low-temperature separator.

LW. load water.

LWD. logging-while-drilling.

m. 1) slope or 2) medium

µ. viscosity.

m³. cubic meter.

magnetometer. an instrument that measures the earth's magnetic field intensity. It is able to detect variations in the magnetite content of the rocks. The units of measurement are gauss or nanoteslas.

maintain angle. to drill a deviated well out straight.

make up. to screw together pipe.

making a connection. to add another joint of drill pipe to a drillstring.

making a trip. to pull the drillstring from the well and put it back in.

making hole. drilling a well.

manometer. an instrument that measures fluid pressure in a well.

marble. metamorphosed limestone. (Mbl)

marine riser. a long length of flexible, steel tubular used to connect the blowout preventer stack on the bottom of the ocean to a floating drilling rig. The drillstring is run down the marine riser.

Mark II. an oil well beam pumping unit that uses levers to balance the weight of the sucker rod string.

marker bed. a thin, distinctive sedimentary rock layer such as volcanic ash used in correlation.

marn. marine.

Marsh funnel. a funnel used on a drilling rig to measure the viscosity of drilling mud. The time in seconds that the mud takes to drain through the funnel is related to the mud viscosity.

mass. massive.

mast. 1) the steel tower that sits on the drill floor of a drilling or workover rig or 2) the steel tower on the bed of a service unit. A mast is raised as a single unit in contrast to a derrick.

master bushing. a device that attaches to the rotary table. The kelly bushing fits on the master bushing.

materials balance equation. an equation that relates the volume of produced fluids from a reservoir to the change of reservoir pressure to calculate the remaining oil and gas.

materials man. an employee of the operator who is responsible for calculating the amount and ordering the supplies and supervising their timely delivery to a drilling rig.

matrix. the fine-grained particles that bind a poorly sorted sedimentary rock. (Mtrx)

mature area. an area in which many wells have been drilled.

mature oil. light oil generated at deep depths in the oil window.

max. maximum.

maximum efficient rate. a production rate for a field that balances the economics of rapid production against the waste caused by bypassing of subsurface oil during rapid production. (MER)

maximum potential flow. the maximum rate a well can produce with zero bottomhole pressure. (absolute open flow)

Mbl. marble.

Mbr. member.

Mcf. 1000 cubic feet.

md. millidarcy.

Mdst. mudstone.

meander. a river channel bend.

measured depth. *see* total depth.

measurements-while-drilling. a real time log made by sensors in the drillstring above the bit. It measures bit orientation (azimuth and inclination) and tool face direction. The measurements are digitized and transmitted to the surface by pressure pulses in the drilling mud. It is used on all directional wells and is often used with logging-while-drilling. (MWD)

mechanical integrity test. a test used to determine if casing in a well is leaking. A liquid is pumped down the well under pressure. The pumping is stopped and the liquid pressure is monitored for a period of time. If the pressure drops, the casing is leaking. (MIT)

med. medium.

member. a distinctive but local bed that occurs in a formation. It is given a formal, two-part name, similar to a formation name. (Mbr)

MER. maximum efficient rate.

Mesozoic. an era of geological time from 248 to 65 million years ago.

metamorphic rock. a rock that has been altered by heat and/or pressure. Gneiss and marble are examples. (Meta)

meteoric water. fresh, subsurface water.

meter prover. a device that calibrates a meter. It compares the amount of gas or liquid flowing through the meter prover to the meter reading on a meter as the same or equal amount of fluid flows through it. (prover)

methane. a hydrocarbon composed of CH_4. It is a gas under surface conditions and is a major component of natural gas. (C_1)

metric ton. the metric system unit for measurement of crude oil weight. A metric ton weighs 2240 pounds and is the equivalent of 7.5 barrels of average weight oil.

mica. a common mineral that occurs as thin, elastic flakes. Two types are white mica and black mica. (Mic or mic)

micellar-polymer flood. an enhanced oil recovery method in which a surfactant is injected into a depleted oil reservoir to form a microemulsion of the remaining oil. Polymer-thickened water is then injected to drive the oil to producing wells.

micrite. a very fine-grained limestone. (Micr or micr)

microbial gas. *see* biogenic gas.

microemulsion. an emulsion in which oil occurs as very small droplets suspended in water.

microfossil. the preserved remains of a plant or animal that are so small that a microscope is needed for identification. (Microfos or microfos)

micropaleontologist. a person who studies and identifies microfossils. (bug picker)

microresistivity. resistivity measured over a very short distance.

MICU. moving in completion unit.

Mid. middle.

migration. 1) the vertical and horizontal flow of oil and gas from the source rock to the trap or its ultimate destination or 2) a computer process that moves dipping seismic reflections into more accurate positions on a seismic record.

mill. 1) to grind up or pulverize or 2) a fishing tool with diamond or tungsten-carbide cutting edges used to grind a fish.

milled-teeth tricone bit. a tricone drill bit in which the teeth have been machined out of the steel cones. (steel-tooth tricone bit)

millidarcy. one thousandth of a darcy, a permeability unit. (md)

milligals. the units of gravity measurement.

MIM. moving in materials.

mineral. a naturally occurring, relatively pure chemical compound. It can occur as either a crystal or an amorphous grain. Rocks are composed of minerals. Quartz and calcite are examples. (Min or min)

mineral rights. the legal ownership of oil and gas below fee land. The mineral rights owner can explore and drill for gas and oil on that land. The mineral rights owner also owns and can produce the gas and oil. Mineral rights can be transferred by a lease.

Miocene. an epoch of time from 24 to 5.3 million years ago. It is part of the Tertiary Period.

MIPU. moving in pulling unit.

MIR. moving in rig.

MIRT. moving in rotary tools.

MIRU. moving in and rigging up.

mis-tie. a problem in correlating seismic horizons between intersecting seismic lines.

miscible gas drive. an enhanced oil recovery method in which gasses that mix with oil in reservoir, such as carbon dioxide or liquefied petroleum gas, are injected into a depleted reservoir to produce more oil.

MISR. moving in service rig.

Mississippian. a period of geological time from 365 to 320 million years ago. It is part of the Paleozoic Era.

mist extractor. wire mesh or vanes that are used to separate liquid droplets from gas.

MIT. mechanical integrity test.

mixed-base crude oil. a refiner's term for crude oil that contains both paraffin and asphalt.

ML. mud logger.

MLU. mud logging unit.

MMcf. 1,000,000 cubic feet.

MMS. Minerals Management Service.

mnr. minor.

MO. moving out.

MOCU. moving out completion unit.

mod. moderate.

MODU. mobile offshore drilling unit.

molecular sieve. a substance, such as the mineral zeolite, that can filter molecules based on size or structure.

Moll. mollusk.

monkeyboard. a small platform located near the top of a derrick or mast on a drilling rig. The derrickman stands on the monkeyboard when making a trip.

monobore. a well with only one inner diameter from the top to the bottom.

moon pool. the hole in a drillship through which the drillstring runs.

MOR. moving out rig.

morning report. a report made by the tool pusher or company man on the last 24 hours of activity on a drilling rig such as footage drilled and supplies used. (daily drilling report)

MORT. moving out rotary tools.

motorman. the person in charge of maintaining the prime movers on a drilling rig.

mott. mottled.

mouse hole. a hole in the drill floor used to hold the next joint of drillpipe to make a connection.

Mtrx. matrix.

mud. *see* drilling mud.

mud acid. a mixture of hydrochloric and hydrofluoric acids.

mud cake. the solid ring made of clay particles that were plastered against the sides of the well by drilling mud. (filter cake)

mud filtrate. the liquid and fines from drilling mud that are forced into the pores of rocks adjacent to the wellbore as a well is drilled.

mud hogs. mud pumps on a drilling rig. (slush pumps)

mud hose. the rubber hose that connects the mud pumps to the swivel on a drilling rig. (rotary hose)

mud log. a record of hydrocarbons in the drilling mud and well cuttings made when a well is being drilled by a service company.

mud man. *see* drilling fluids engineer.

mud motor. *see* downhole mud motor.

mud pit. an earthen excavation near the drilling rig where drilling mud is temporarily stored.

mud tanks. several steel tanks in a line that hold drilling mud. They are located on the ground next to a drilling rig.

mud up. to increase the density of drilling mud.

mud-gas separator. a vessel mounted on the mud tanks that is used to separate any gas out of the drilling mud coming from the well.

mudstone. a sedimentary rock composed of silt- and clay-sized particles. (mdst)

multi-component seismic. *see* nine-component seismic.

multilateral well. a well with several, smaller branches (laterals) drilled out from the main well.

multinational. an oil company that operates in several countries.

multiple stage compressors. several, in line compressors that increase gas pressure in increments.

multistage cementing. a cement job in which several sections of a casing string are cemented in successive stages.

muscovite. white mica. (Musc)

MWD. measurements-while-drilling.

N. 1) dimensionless number or 2) neutron log.

n.s. no sample.

n.v.p. no visible porosity.

n/s. no show.

nanoteslas. the units of magnetic measurement.

national company. *see* host company.

natural gamma ray log. a wireline well log that measures the natural radioactivity of rocks. (GR Log) (gamma ray log)

natural gas. a gas composed of a mixture of hydrocarbon molecules that have one, two, three and four carbon atoms.

natural gas liquids. condensate, butane, propane and ethane that have been removed from natural gas in a natural gas processing plant. (NGL)

natural gas processing plant. an installation that removes natural gas liquids from natural gas by cooling or absorption. (gas plant)

natural gasoline. *see* condensate.

NB. new bit.

net revenue interest. 100% minus all royalties on a well or property. (NRI)

neutron log or **neutron porosity log.** a radioactive wireline well log that is used to measure porosity. (N or NL)

NGL. natural gas liquids.

nine-component seismic. seismic exploration that uses three vibrator sources (vertical, inline horizontal and cross-line horizontal) and three detectors (vertical, inline horizontal and cross-line horizontal) at each station. It emits and detects both P and S waves. (vector component or multi-component seismic exploration)

nipple up. to connect equipment such as blowout preventers. (NU)

NL. neutron log.

NMR. nuclear magnetic resonance.

no returns. no well cuttings were obtained for that interval in the well. (NR)

noise. unwanted seismic energy recorded with the signal. It is everything except direct (primary) reflections that represent the subsurface geology.

noise log. a production log that records sounds with depth in a well.

nominal weight. calculated weight.

nonassociated gas. natural gas that is not in contact with crude oil in the reservoir.

nonexclusive. data shared by several parties.

normal fault. a fault with predominantly vertical movement (dip-slip), in which the hanging wall has been lowered in relation to the footwall. It creates a lost section.

normal pressure. fluid pressure in subsurface rocks due to the weight of the overlying fluids. (hydrostatic pressure)

nose. the lobate surface pattern of an eroded, plunging anticline.

nozzle. an orifice in a tricone drill bit between two cones. Drilling mud jets out the nozzle.

NR. 1) no report, 2) no recovery or 3) no returns.

NRI. net revenue interest.

NS. no show.

NU. nippling up.

nuclear magnetic resonance log. a wireline well log that uses magnetism to measure porosity and pore sizes. It can be used to calculate permeability and determine types of fluids in the reservoir. (NMR log)

num. numerous.

O. oil.

O & G. oil and gas.

OC. oil cut.

occ. occasional.

OCM. oil-cut mud.

OCS. outer continental shelf.

OD. outer diameter.

OEG. oil equivalent gas.

OF. open flow.

off structure. located off the top of a trap.

offset. the horizontal distance from the seismic source to the receiver.

offshore production platform. *see* production platform.

oil. *see* crude oil. (O)

oil cut. diluted with oil.

oil equivalent gas. the number of barrels of oil that is equal in heat content (Btus) to a volume of natural gas. The ratio is about 1 barrel of oil to 6000 cubic feet of natural gas. (barrels of oil equivalent or energy equivalent barrel) (OEG)

oil in place. the total amount of oil located in the pores of a subsurface reservoir in an oil field. (OIP)

oil sand. a common term for sandstone containing oil.

oil shale. a fine-grained sedimentary rock containing organic matter called kerogen which, when heated, forms oil.

oil string. the smallest diameter and longest casing string in a well. (production casing)

oil wet. a reservoir rock in which water occurs in the center of the pores and oil coats the rock surfaces.

oil window. the zone in the earth where crude oil is generated from organic matter in source rocks.

oilfield brine. very saline water that is produced with oil.

oil-in-water emulsion. droplets of oil suspended in water.

oil-water contact. the boundary between oil and water in a reservoir. (OWC)

OIP. oil in place.

Oligocene. an epoch of time from 34 to 24 million years ago. It is part of the Tertiary Period.

on structure. located on top of a trap.

OOIP. original oil in place.

oolite. a sand- or silt-sized sphere of calcium carbonate precipitated from water. (Ool or ool)

oolitic limestone. a limestone composed predominately of oolites.

op. opaque.

open-hole completion. a well with casing run and cemented down to the top of the reservoir rock, which is left uncased. (barefoot or top set completion)

open-hole log. a wireline well log such as an electric log that can only be run in a well without casing.

operator. the company who a) contracts to drill a well, b) is responsible for maintaining a producing lease or c) is in charge of operations in a working interest area.

Ordovician. a period of geological time from 500 to 425 million years ago. It is part of the Paleozoic Era.

org. organic.

orifice. a hole in a plate through which fluids can flow. An orifice is described by its diameter.

orifice gas meter. a meter that measures the volume of natural gas flowing through a line by passing the gas through a specific size orifice. The drop in pressure through the orifice is related to the velocity of the gas.

original pressure. *see* initial pressure.

ORR. overriding royalty.

ORRI. overriding royalty interest.

orthoquartzite. a sandstone composed of well-sorted quartz sand grains. It can be an excellent reservoir rock.

OS. oil show.

OSR. oil source rock.

outer continental shelf. the portion of the sea bottom where the federal government owns the mineral rights. It is from the state limit (3 nautical miles from the shoreline) to 200 nautical miles from the shoreline. (OCS)

outpost well. a well drilled to significantly increase the area of a producing field. If the outpost well is successful, it is called an extension well.

overbalance. the condition in a well in which the pressure of the drilling mud is more than the pressure of the fluids in the surrounding rocks.

override or **overriding royalty interest.** an interest in production that is free and clear of any costs. (ORRI)

overshoot. a fishing tool that is run around down and around a pipe (fish) on the bottom of the well. It grips the outside of the pipe to pull the fish out of the well.

overthrust belt. a zone of thrust faults that moved during the formation of a mountain range. (disturbed belt)

overturned fold. a fold in sedimentary rocks in which the axis is not vertical and the limbs are not symmetrical.

ox. oxidized.

p. pressure.

Ø. porosity.

P & A. plugged and abandoned.

P & P. porosity and permeability.

packed hole assembly. a downhole assembly that uses several stabilizers to make the well be drilled out straight.

packer. a cylinder of rubberlike material which is run on a tubular string and compressed to expand and seal the well at that level. (pkr)

packer flowmeter. an instrument used to force fluid to flow up the well through an orifice in a packer to measure the flow.

packstone. a type of limestone with large, sand-sized grains touching each other and having fine-grained material in between. (Pkst)

paid up lease. a type of lease that does not require delay rental payments to maintain the lease during the primary term.

paleo pick. a horizon in sedimentary rocks defined by fossils.

Paleocene. an epoch of geological time from 65 to 55 million years ago. It is part of the Tertiary Period.

paleogeographic map. an interpretation of the land surface during a certain time of earth history.

paleontologist. a geologist who studies fossils.

paleontology. a branch of geology that studies fossils.

Paleozoic. an era of geological time from 570 to 248 million years ago.

palynologist. a person who studies fossil spores and pollen. (weed and seed person)

Par or **par.** particle.

paraffin. a member of the hydrocarbon series of molecules. They are straight chains with single bonds. All hydrocarbon molecules in natural gas and some in crude oil are paraffins. Long paraffins are waxes that are solids at low temperatures.

paraffin inhibitor. an additive to crude oil that prevents formation of waxes during production.

paraffin knife and **paraffin scratcher.** tools that use knives to scrape wax out of a tubing string.

paraffin-base crude oil. a refiner's term for crude oil with little or no asphalt. It will yield a relatively high percentage of paraffin wax, high quality lubricating oil and kerosene when refined.

parallel-free pump. an oil well hydraulic pump system that uses two tubing strings. One is for the power oil that drives the pump and the other is for the produced fluids.

patch reef. a small detached reef.

pay. 1) the zone producing gas and/or oil in a well or 2) the vertical thickness of the producing zone. Pay can be measured as either gross pay, including nonproductive zones or net pay, including only productive zones.

pay sand. a sandstone that produces gas and/or oil.

pay zone. the vertical portion of a reservoir in a well that produces gas and/or oil.

payout. a criteria used to evaluate an investment in an oil or gas well. It is the time necessary for the net production revenue (minus royalties) to equal the costs of drilling, completing and operating the well up to that time.

PB. plugged back.

Pbl or **pbl.** pebble.

Pc. capillary pressure.

PDCB. polycrystalline diamond compact bit.

pendulum assembly. a downhole assembly that uses the weight of a drill collar below a stabilizer to cause the bit to drop and decrease the angle (drop angle) of a deviated well.

Pennsylvanian. a period of geological time from 320 to 290 million years ago. It is part of the Paleozoic Era.

percentage map. a map that uses contours to show the percentage of a specific rock type such as sandstone in a formation.

PERF. perforated.

perf. perforate.

perf csg. perforated casing.

perforating gun. a tool run on a wireline or tubing string that shoots perforations (holes) in the casing or liner. It uses either steel bullets or shaped-explosive charges. The gun is either expendable or retrievable. (gun)

perforation. a hole shot in casing or liner and cement to allow oil and/or gas to flow into the well.

period. a subdivision of an era of geological time.

permeability. a measure of the ease with which a fluid flows through a rock. The units are millidarcies or darcies. (Perm, perm or k)

permeameter. an instrument used to measure the permeability of a rock sample.

Permian. a period of geological time from 290 and 250 million years ago.

permit man. a seismic contractor employee who obtains permission from landowners to run seismic exploration across their land.

Pet or **pet.** petroleum.

petrochemicals. products made from petroleum feedstocks.

petroleum. crude oil and natural gas. (Pet or pet)

petroleum engineer. an engineer who is trained to drill and complete wells and produce petroleum.

petroleum geologist. a geologist who specializes in the search for and exploitation of petroleum.

petrophysics. the study of the physical properties of rocks and their fluids.

phase. the angular separation of perforations.

Phos or **phos.** 1) phosphate or 2) phosphatic.

PI. production index.

pick. an interpretation of where the top or bottom of a subsurface rock layer occurs on a well log.

piercement salt dome. a salt dome that has risen to breakthrough overlying sedimentary rocks.

pill. a batch of a substance or additive such as lost circulation material.

pilot hole. a small diameter wellbore drilled out from a well to kick off the well at an angle.

pin. a male-threaded connection that mates with a box.

pinch out. a rock such as sandstone which thins to zero thickness in another rock such as shale. (wedge out)

pinnacle reef. a small, cone-shaped reef.

pipe elevators. clamp-like devices that are attached to the traveling block. They are designed to attach onto the drillpipe.

pipe rack. a steel framework on the ground next to a drilling rig. It is used to store horizontal joints of drillpipe.

pipe ramp. a flat, steel incline in the front of a drilling rig. It is used to drag drillpipe and casing up through the V-door and onto the drill floor.

pipe rams. two, large blocks of metal with inserts cut into then. They are designed to close around drillpipe in a well to close the well. Pipe rams are used in a blowout preventer stack.

pipeline oil. crude oil that does not exceed a maximum basic sediment and water content and meets pipeline transportation specifications. (sales-quality oil)

pipeline-quality gas. natural gas that has been processed to remove impurities and meets pipeline contract specifications including chemical composition and pressure (sales-quality gas)

pit volume totalizer. a series of floats in the mud tanks of a drilling rig. They record the volume of mud in the tanks and send an alarm when the volume is decreasing or increasing.

pitman. the steel beam that connects the rotary counterbalance with the walking beam on a beam pumping unit.

pkr. packer.

Pkst. packstone.

plant. to position a geophone for a seismic survey.

plate tectonics. a theory in which the crust of the earth is divided into large plates which are moving.

play. a proven combination of reservoir rock, caprock and trap type that contains commercial amounts of petroleum in an area.

pld. pulled.

Pleistocene. an epoch of time, from about 1.8 million years ago to 10,000 years ago, during which glaciers occupied much of the land area. It is part of the Quaternary Period. (Ice Ages)

plg. pulling.

plgd. plugged.

Pliocene. an epoch of geological time from 5.3 to 1.8 million years ago. It is part of the Tertiary Period.

plug. 1) a small cylinder of rock drilled from a core. It is used to measure porosity and permeability or 2) to place cement in a well to abandon the well sealer a depleted zone in the well.

plug back. to plug and abandon one zone and complete another zone higher in the well. (PB)

plugged and abandoned. a well that has had a cement plug(s) placed in it to close the well. The depleted zone and the upper part of the well where fresh water reservoirs occur are sealed by cement. A steel plate is welded to the top of the well. Dry holes and depleted wells are plugged and abandoned. (P & A)

plunging anticline. an anticline oriented at an angle to horizontal.

plutonic rock. an igneous rock which crystallized from a hot melt below the surface of the ground such as granite.

pneumatic drilling. drilling with either a) air or b) air and water (mist) as the circulating fluid.

point bar. a sand bar deposited on the inside of a river meander.

polished rod. the polished, brass or steel rod that oscillates up and down through the stuffing box of an oil well rod pumping unit. It is located at the top of the sucker-rod string.

polycrystalline diamond compact bit. *see* fixed cutter bit. (PDCB)

polymer. a long-chain, high-weight molecule. When mixed with water, polymers form a thick, viscous fluid.

pony rod. a short sucker rod.

pool. a separate reservoir of gas and/or oil.

poorly sorted. a rock or sediments with clastic grains having a large range of sizes.

POP. put on pump.

Por or **por.** porosity.

pore. the space between solid particles in a rock.

porosimeter. an instrument used to measure porosity in a rock.

porosity. the percent volume of a rock that is pore space. (Por, por or Ø)

porosity cutoff. a minimum porosity value such as 8% for reservoir rock that is used as a guideline in deciding whether or not to complete a well or for reserve computations.

positive-displacement meter. a meter that measures the volume of a fluid in specific increments of a volume, one at a time.

poss. possible.

potential test. a test that measures the maximum amount of fluids that a well can produce in 24 hours. (PT)

pour point. the lowest temperature at which a particular crude oil will still flow. It is an indication of the wax content of the oil.

PP. pulled pipe.

ppg. pounds per gallon.

PPM or **ppm.** parts per million.

PR. poor returns.

PR&T. pulled rods and tubing.

Precambrian. an era of geological time from the beginning of the earth (4.5 billion years ago) until 570 million years ago.

precipitated. crystallized from dissolved salts in water.

pred. predominant.

prepacked liner. a double wall liner with permeable, synthetic sandstone filling the space between the walls.

pres. preserved.

pressure bomb. an instrument run on a wireline in a well to record pressures. It consists of a pressure sensor, recorder and clock drive.

pressure buildup curve. a plot of pressure increase after a gas well has been shut in.

pressure maintenance. a system in which produced gas is injected into the free gas cap and produced water is injected into the reservoir below the oil-water contact. It is used during primary production to maintain pressure on the remaining oil and increase ultimate production.

pressure transient test. a test that measures pressures and flow rates in a well.

prestack migration. the migration of seismic data before the data is stacked.

prim. primary.

primacord. an explosive cord used as a seismic source on land and for a back off operation on stuck pipe in a well.

primary cementing. a cement job done as the casing is being run.

primary drive. the original force which causes oil to flow through the reservoir rock and into a well.

primary production. the oil that naturally flows into the well due to the reservoir drive. It does not include oil produced during waterflood or enhanced oil recovery.

primary stratigraphic trap. a petroleum trap formed by the deposition of a reservoir rock such as a reef that is encased in shale.

primary term. the time granted in a lease for exploration and drilling.

prime movers. the diesel engines that supply the power to a drilling rig.

prob. probable.

Prod. production.

producer. a well that can produce commercial amounts of petroleum.

producing gas-oil ratio. the number of standard cubic feet of natural gas that a well produces per barrel of oil.

production casing. the smallest diameter and longest casing string in a well. (oil string)

production foreman. an employee of the operator of a field who receives orders from the field superintendent and gives orders to the pumpers and work crews.

production index. the downhole pressure drawdown in psi divided by the production in bbls/D from a well. (PI)

production log. a log run in a producing well to evaluate a problem.

production platform. an offshore platform that treats and separates produced fluids from offshore wells. It can have the wellheads on the platform or receive the produced fluids through flowlines from satellite wells or a wellhead platform in deeper water. The oil or gas goes ashore through a submarine pipeline. Two types are steel jacket and gravity based.

production rig. a portable well service or workover hoisting unit.

production sharing contract. a contract between a foreign government and a multinational company. The multinational company bears the entire cost of exploration, drilling and production. The multinational company is reimbursed for expenditures from the oil that is produced. After reimbursement, the oil is split by an agreed formula.

production test. a test that measures the amount of gas, oil and water that a well contributes to a central processing unit.

production tree. the fittings, values and gauges that are bolted to the wellhead of a flowing well to control the flow from the well. (Christmas tree)

productivity index. the flow rate that a well can produce per psi difference between reservoir and bottomhole pressures. It is an indicator of that well's ability to produce oil. (J)

productivity test. a well test that determines the effect of different flow rates on the reservoir. Fluid pressure is measured with the well shut-in and at different, stabilized rates.

profit oil. produced oil that is split between a host company and a multinational company by an agreed formula after the multinational company has been reimbursed for expenditures.

prograde. to deposit sediments out into a basin.

prom. prominent.

propane. a hydrocarbon composed of C_3H_8. It is a gas under surface conditions and is found in natural gas. (C_3)

proppants or **propping agents.** small spheres suspended in the frac fluid that is pumped down a well during a frac job. They hold the fractures open.

proprietary. secret.

prospect. a location where both geological and economic conditions favor drilling a well.

protection casing. a casing string with an intermediate length and diameter. It is used to isolate a problem zone in the well as the well is being drilled. (intermediate casing)

prover. *see* meter prover.

ps. pseudo-.

PSA. packer set at.

psi. pounds per square inch.

psia. pounds per square inch absolute.

Pt or **pt.** part.

PT. potential test.

PTR. pulled tubing and rods

pull rods. to pull the sucker rods out of a well during a workover.

pulling unit. a winch and mast on a truck that is used to work over a well. The crew usually consists of an operator, derrickman and floorman. (service unit)

pulsed neutron log. a type of neutron log that can be used to distinguish gas and oil from water behind casing in a well.

pumpability time. the time a cement slurry remains fluid enough to be pumped. (thickening time)

pumpdown. to pump equipment down a producing well to service the well.

pumper. 1) a well that requires a pump to bring the oil to the surface or 2) a mechanic who is responsible for maintaining producing equipment in the field. The pumper receives orders from a production foreman.

putting the well on pump. to replace the production tree on a well that has lost pressure with a sucker-rod pumping unit.

pyrobitumen. a naturally-occurring, hard, dark hydrocarbon such as tar. (Pybit)

q. rate.

Qtz or qtz. quartz.

Qtzt or qtzt. quartzite.

quartz. a common mineral composed of SiO_2. Sandstones are usually composed of quartz sand grains. (Qtz or qtz)

quartzarenite. a sandstone composed of more than 95% quartz sand grains. It can be an excellent reservoir rock.

quartzite. a very hard sandstone composed primarily of quartz sand grains. (Qtzt or qtzt)

Quaternary. a period of geological time from 1.8 million years ago to the present. It is part of the Cenozoic Era.

quick-look log. a wellsite computer generated log that uses two or more logging measurements. It can show water saturation, porosity, percentages of sandstone, limestone and shale and the location of fractures.

r. 1) radius or 2) rare.

ρ. density.

R. resistivity.

R/H. ran in hole.

radioactivity log. a wireline log that uses a radioactive source to bombard the rock with either atomic particles or energy. Examples are the neutron porosity and formation density logs.

radiolaria. a single-cell animal that floats in the ocean and has a silicon dioxide shell. It is a type of microfossil.

range. 1) a system of north-south strips six miles wide that are used in land subdivision or 2) the geological time extent that a fossil species existed. (rng)

rank wildcat. an exploratory well drilled at least two miles away from the nearest production.

raster. scanned. Raster well logs have been scanned into a computer data base.

rat hole. a hole in the drill floor used to hold the swivel and kelly when tripping out.

RB/D. reservoir barrels per day.

rd. red.

RD. 1) rigged down or 2) rigging down.

reamer. a sub used to enlarge the wellbore or casing.

reboiler. a distillation vessel that heats wet glycol to separate glycol and water.

Rec or **rec.** recovered.

Recent. an epoch of geological time from 10,000 years ago until the present. It is part of the Quaternary Period. (Holocene)

recharge area. an aquifer outcrop where fresh water enters.

recomp. recomplete.

recomplete. to plug and abandon one zone in a well and complete in another. This can be done during a workover by either plugging back or drilling deeper. (recomp)

recoverable oil. the amount of oil that can be produced from a reservoir under current economic conditions. It is a fraction of the oil in place.

recovery factor. the percentage of oil and/or gas in place that will be produced from a reservoir.

red bed. red-colored sedimentary rocks with an iron oxide coating thought to be originally deposited in a desert environment.

reef. a ridge or mound-like structure with wave-resistant, framework building organisms such as corals. (Rf or rf)

reflection coefficient. the percentage of seismic energy reflected off a surface.

regression. a retreat of the sea from the land.

regular acid. hydrochloric acid.

relative permeability. the ratio between effective permeability of a fluid at partial saturation to the permeability of that fluid had it been at 100% saturation.

relief well. a well drilled close to a blowout well in order to decrease the pressure on the abnormal high pressure zone that is causing the blowout. Heavy drilling mud (kill mud) is then pumped into the uncontrolled well.

Rem. remains.

repeat formation tester. a wireline tool that samples reservoir fluids and measures reservoir pressures at several levels in a well. (RFT)

repeated section. *see* double section.

Repl or **rep.** replaced.

reprocess. the application of new computer processing methods to older seismic data that was recorded digitally.

Res or **res.** residue.

reserve pit. an earthen pit, often lined with plastic, located next to a drilling rig. It holds drilling mud that is not being used and well cuttings.

reserves. the calculated amount of gas and/or oil that is expected to be produced from a well or a field. Proven reserves are calculated with reasonable certainty. Developed reserves can be produced from existing wells whereas undeveloped reserves cannot. Unproven reserves are not as certain due to technical and economic reasons as proven reserves. Probable and possible reserves are even less certain.

reservoir. the subsurface deposit of oil and/or gas located in the pores of a reservoir rock. Fluids cannot flow from one reservoir to another.

reservoir barrel. one liquid barrel of oil in the subsurface reservoir. When brought to the surface and gas bubbles out, the volume of oil will shrink. (*see* formation volume factor and stock barrel of oil) (res bbl)

reservoir drive. the source of pressure on subsurface fluids that forces them through the reservoir rock and into the well. It comes from fluid expansion, rock expansion and gravity. Some types are solution gas, free gas cap, water, gravity and expansion gas.

reservoir pressure. the pressure on fluids in the pores of rock at a specific depth. Normal reservoir pressure is due to the weight of the overlying waters. (fluid or formation pressure)

reservoir rock. a rock that has porosity and permeability. It can hold and transmit fluids.

residual gas. the gas that exits a natural gas processing plant after the natural gas liquids have been separated. (tail gas)

residual water. pore water that will not move. (irreducible water)

resistivity. the opposition of a substance to the flow of an electrical current. It is a measurement made on an electric and induction wireline log.

restricted basin. a body of water which is separated from the ocean by a shallow sill or bar at the entrance and has limited water circulation.

retarder. an additive that slows a process such as cement setting.

retention time. the time produced fluids spend in a separator.

retrograde gas. A hydrocarbon gas under initial, subsurface reservoir conditions. A liquid condensate forms in the reservoir as production decreases the reservoir pressure.

return on investment. a criteria used to evaluate an investment in an oil or gas well. It is the estimated net production revenue during the life of the well divided by the drilling and completion costs.

returns. drilling mud and well cuttings that flows up a well as it is being drilled.

reverse fault. a fault with predominantly vertical movement (dip-slip) in which the hanging wall has moved up in relation to the footwall. It creates a double or repeated section.

rexlzd. recrystallized.

Rf or rf. reef.

RFT. repeat formation tester.

RI. royalty interest.

rich gas. natural gas that contains a significant amount of condensate in contrast to lean gas or dry gas or 2) moderately rich gas that contains between 2.5 and 5 GPM and very rich gas that contains more than 5 GPM.

rift. a large fault with predominantly horizontal movement.

rift valley. a deep, wide fracture.

rig down. to disassemble a drilling rig. (RD)

rig up. to assemble a drilling rig. (RU)

right-lateral strike-slip fault. a fault that moves horizontally, with the opposite side of the fault moving toward the right as you face the fault.

risk. the number of wells completed as producers divided by the number of wells drilled. It is expressed as a decimal or percent. (success rate)

rng. range.

rock. a naturally occurring aggregate of mineral grains and crystals such as sandstone or granite.

rod basket. a steel platform with sides that is located near the top of the mast on a well service unit. The derrickman stands in the rod basket to place the sucker rods in the rod fingers as the rods are pulled from the well.

rod pump. a common type of oil well downhole pump driven by a sucker-rod string. It is run as a complete unit on the sucker-rod string through the tubing string. (insert pump)

rod pumping system. a common artificial lift system for an oil well. A surface pumping unit drives a sucker-rod pump on the bottom of the tubing string. A sucker-rod string that runs down the tubing connects the surface pumping unit with the sucker-rod pump.

ROL. rig on location.

roller cone bit. a rotary drilling bit that has rotating cones mounted on bearings. A tricone bit is very common.

rollover anticline. a large fold formed in sedimentary rocks on the basin side of a growth fault. It can be a petroleum trap.

ROP. rate of penetration.

rotary drilling rig. a drilling rig that rotates a long length of steel pipe with a bit on the bottom to cut the well. Four major systems on the rig are power, hoisting, rotating and circulating.

rotary helper. *see* roughneck.

rotary hose. *see* mud hose.

rotary table. a revolving plate on the drill floor that is driven by the prime movers. The master and kelly bushings are attached to it. It turns the drill-string which runs down through the center of it. (RT)

roughneck. a drilling crew member who operates and maintains the equipment under orders from the driller. (rotary helper)

roustabout. 1) a general helper on producing wells and well service units or 2) a member of the offshore drilling crew who helps bring supplies and equipment aboard under orders from the head roustabout.

royalty. a percentage of the revenue from oil and gas production that is paid to the mineral rights owner and any other royalty owner. It is free and clear of the costs of production.

royalty interest. an ownership in production that bears no cost of production. The royalty interest owners receive their share of production revenue before the working interest owners. (RI)

RT. rotary table.

RU. 1) rigged up or 2) rigging up.

run. 1) to raise and lower equipment in a well or 2) to just lower equipment into a well.

run ticket. a form filled out when oil is transferred from stock tanks to a tank truck or pipeline. It lists the quality and quantity of the oil and is used to pay the operator of the wells.

RUR. rigging up rotary.

RURT. rigging up rotary tools.

S. saturation.

S/. swabbed.

Sa or **sa.** salt.

sack. a container for dry cement (94 lbs), bentonite clay (100 lbs), barite (100 lbs) and other dry supplies. (sk or sx)

sales-quality gas. *see* pipeline-quality gas.

sales-quality oil. *see* pipeline-quality oil.

salt dome. a large mass of salt (salt plug) which is or has been flowing upward through the overlying sedimentary rocks. The salt dome also includes the surrounding and overlying sedimentary rocks that have been deformed.

sample log. a record of the physical properties of rocks in a well. It includes composition, texture, color, presence of pore spaces and oil staining. (lithologic log)

Samson post. the steel beam assembly on which the walking beam pivots on an oil well beam pumping unit.

sand. a clastic particle between 2 and $\frac{1}{16}$ mm in diameter. (sd)

sand cleanout. a workover in which saltwater or drilling mud is circulated to remove loose sand from the bottom of a well.

sand control problem. loose sand clogging the bottom of a well.

sand frac. *see* hydraulic fracturing.

sand/shale ratio map. a map that uses contours to show the ratio of sandstone to shale in a formation.

sandstone. a common sedimentary rock composed primarily of sand grains. It can be a reservoir rock. (Sst or ss)

satellite well. a subsea well in a remote part of an offshore field or marginal field with a flowline that conducts produced fluids to a production platform for treating. The well was not drilled from the platform. It was drilled from a jackup rig or floater.

saturated. the condition in which a liquid has dissolved all the gas or salt it can hold. (Sat or sat)

saturated pool. an oil reservoir with a free gas cap.

saturation. the percentage of different fluids such as gas, oil and water in the pore space of a rock. (S)

scale. salts that have precipitated out of water. Calcium carbonate, barium sulfate and calcium sulfate are common in oil fields.

scale inhibitors. a chemical used to prevent salt formation in a well.

scalped anticline. *see* bald-headed anticline.

scf. standard cubic foot.

scf/D. standard cubic feet per day.

scout card or **scout ticket.** a form completed by an oil company scout on engineering and geological information gathered on a specific well being drilled.

scout. an oil company or commercial scouting company employee who gathers information on petroleum related activities of other companies in a regional area.

SCR. silicon controlled rectifier.

scratchers. wires on a collar that are attacked to the lower part of a casing string being run into a well to remove the mud cake from the well walls.

screen liner. a short section of pipe similar to casing that is set in the bottom of a well. Screens wrapped around the liner prevent sediments from flowing through holes in the liner during production.

scrubber. equipment used to remove liquid from gas.

scs. scarce.

sd. sand.

SD. shut down.

sdtrk. sidetrack.

sdy. sandy.

seafloor seismic method. seismic exploration at sea with hydrophone streamers positioned on the seabed.

seafloor spreading. a theory in which the earth's crust (seafloor) is formed by basalt volcanoes along the mid-ocean ridge. It is then split and spreads to either side of the ridge because of convection currents in the molten interior of the earth. The crust is destroyed in subduction zones.

seal. an impermeable rock layer that forms the cap on top of an oil or gas reservoir. (caprock)

sealing fault. a fault that does not allow fluid flow along or across the fault.

seating nipple. a small fitting that is run on a tubing string. It has a constricted inner diameter that stops any tool that has fallen down the tubing string.

sec. secondary.

secondary cementing. a cement job done as a workover on a producing well.

secondary faults. a relatively minor fault oriented parallel to a major fault.

secondary gas cap. a free gas cap that forms from the solution gas that bubbles out of the oil as reservoir pressure drops during production.

secondary recovery. a process of injecting gas or water into a reservoir to restore production when the primary drive has been depleted.

secondary stratigraphic trap. a petroleum trap formed by an angular unconformity

secondary term. the time granted in a lease for production. It occurs after the primary term and continues as long as commercial amounts of petroleum are being produced.

section. a surveyed square of land one mile on a side. Thirty six sections make a township.

sediment. loose particles or salts. Sediments are deposited out of water, air, or ice. (Sed or sed)

sedimentary rock. a rock composed of sediments deposited on the surface of the ground or bottom of the ocean. Sandstone, limestone and shale are examples. Salts precipitated out of water also form sedimentary rocks.

seep or **seepage.** a natural occurrence of oil and/or gas that has leaked onto the surface.

seismic contractor. a company that maintains and operates seismic equipment.

seismic horizon. a reflection that can be traced on a seismic record.

seismic method. a petroleum exploration method in which sound energy put into the earth with a source. The sound energy reflects off subsurface sedimentary rock layers and is recorded by detectors on the surface. An image of the sub-surface rock layers is made with seismic data to find petroleum traps.

seismic option. a type of mineral rights acquisition in which the lessee pays the lessor a bonus for the right to run seismic exploration on the land and to have the option of leasing the land after reviewing the seismic data.

seismic record or **section.** a record of seismic reflections recorded off sub-surface rock layers displayed similar to a vertical cross section of the earth. Shot points are located along the top of the section. The original vertical scale is in seconds. Zero seconds is always at or near the surface of the earth or at the surface of the ocean. A header with seismic information is located on the record.

seismic stratigraphy. the recognition and use of unconformities on seismic records to correlate and map sedimentary rock packets called sequences. Each sequence was deposited during a major cycle of sea level fall, rise and fall. Seismic facies (seismic reflection characteristics) identify the depositional environments in each sequence. (*see* sequence stratigraphy)

selenite. *see* gypsum.

self potential. *see* spontaneous potential

semisubmersible (semi). a type of floating, offshore, exploratory drilling rig system anchored above the drill site. It has large, submerged flotation chambers (pontoons) located on short columns below the drilling platform.

sep. separator.

separator. a long, steel tank used to separate produced fluids from oil wells. Separators use gravity, impingement, centrifugal force, filters and other methods. They can be either horizontal or vertical. (sep)

sequence stratigraphy. the use of timelines such as unconformities to map and correlate packets of sedimentary rocks called sequences. A sequence was deposited during an interval of geologic time and can be subdivided in parasequence sets and further into parasequences. (*see* seismic stratigraphy)

sequestering agent. an additive used during acidizing a well to prevent the formation of an iron gel or precipitate.

series. a time-rock division of rocks deposited during an epoch.

service company. a company that supplies services such as logging or mud engineering.

service unit. a winch and mast on a truck that is used to work over a well. The crew usually consists of an operator, derrickman and floorman. (pulling unit)

set. 1) to position such as set pipe in a well or 2) to harden such as set cement.

set in the dark. to run and cement a string of casing to the top of the reservoir rock in a well without first drilling and testing the reservoir rock.

set pipe. to case and complete a well.

set through completion. a well completion in which casing or liner has been cemented into the reservoir. The casing or liner is then perforated.

sft. soft.

Sg. gas saturation.

SG&O. show of gas and oil.

Sh or sh. shale.

shake-out test. a method used to determine the basic sediment and water content of oil by centrifuging a sample.

shaker. *see* shale shaker.

shale. a very common sedimentary rock composed of clay-sized particles. Black shales are source rocks for petroleum. (sh)

shale oil. an organic-rich shale that is old enough but has never been buried deep enough for heat to generate oil.

shale shaker. a set of vibrating screens in a steel frame on the mud tanks of a drilling rig. It is used to separate well cuttings from drilling mud coming from the well.

shear rams. two, large blocks of metal with chisel edges. They are designed to shear across any drillpipe in the well and close the well. Shear rams are used in a blowout preventer stack.

shield. a low-lying, stable area of basement rocks on the surface of the earth.

shock sub. *see* vibration dampener.

shoestring sandstone. a long, narrow, lens-shaped sandstone usually encased in shale and originally deposited as a barrier island, river channel, bar finger or valley fill.

short normal resistivity. a wireline resistivity measurement made with electrodes spaced close together.

shot. 1) an explosion used to artificially fracture reservoir rocks in a well to stimulate production or 2) an explosion used as a seismic exploration source to put sound energy into the ground.

shot hole. a shallow hole drilled through loose surface sediments into hard sedimentary rocks. An explosive used for a seismic source is planted on or near the bottom of the hole.

shot point array. the pattern of several seismic sources used simultaneously at a shot point to reduce source noise.

shot point. the location at which a seismic source such as dynamite, Vibroseis or air gun was activated.

show. hydrocarbons in an amount above background. (Shw)

show evaluation. a detailed analysis of the composition of hydrocarbons in a show.

shrinkage factor. the decimal amount to which a barrel of reservoir oil shrinks to on the surface of the ground after the pressure has dropped and the gas has bubbled out of the oil. (*see* reservoir factor and stock tank barrels of oil)

shut in. to cease production from a well. (SI)

shut-in pressure. pressure on a fluid that is not moving. (static pressure)

Shw. show.

SI. shut in.

SIBHP. shut-in bottomhole pressure.

SICP. shut-in casing pressure.

sidetrack. 1) a new section of wellbore drilled from an existing well. It is often the deviated portion of a well drilled around a fish or 2) to drill a sidetrack (sdtrk or ST).

sidewall core. a 1 in. diameter core from the sides of a well. It is obtained by either by an explosive-propelled tube or by drilling. (SWC or S.W.C.)

sieve. a screen used to size particles.

signal. the desired seismic energy (direct or primary reflections) received from the subsurface in contrast to noise.

Sil or **sil.** siliceous.

silicon controlled rectifier. a device that converts alternating current to direct current. It is used on a diesel-electrical drilling rig. (SCR)

sill. an igneous rock that was injected as a molten liquid between sedimentary rock layers.

silt. a clastic particle between $\frac{1}{16}$ and $\frac{1}{256}$ mm in diameter. (Slt)

siltstone. a sedimentary rock composed primarily of silt-sized particles. (Sltst)

Silurian. a period of geological time that occurred from 425 to 405 million years ago. It is part of the Paleozoic Era.

sim. similar.

SIP. shut-in pressure.

sk. sacks.

skeletal sands. sands formed by fragments of shells.

skin damage. *see* formation damage.

SL. sea level.

SLAR. side-looking airborne radar.

sli. slight

slick assembly. a downhole assembly that has no stabilizers. It is used to drill a straight hole.

slick line. a single strand of wire that is used to raise and lower equipment in a well in contrast to a wireline.

slim hole. a well with a small diameter wellbore ($6\frac{3}{4}$ to $4\frac{3}{4}$ in). Slim holes are less expensive to drill than normal diameter wells and are often used for exploration.

slips. a steel wedge with teeth used in the bowl of a rotary table to grip and prevent the drillstring from falling down the well.

slotted liner. a short section of pipe similar to casing that is set in the bottom of a well. Long, narrow slots in the liner prevent sediments from flowing into the liner during production.

Slt. silt.

Sltst. siltstone.

slty. silty.

sluff. the collapse of well walls into the hole. (cave)

sloughing shale. shale along the walls of a well that absorbs water and expands.

slug. a batch of chemicals.

slurry. wet cement.

slush pumps. *see* mud hogs.

sml. small.

smwt. somewhat.

sniffers. a chemical devise towed behind a ship to detect hydrocarbons in ocean water.

snub. to run tools or pipe into a high pressure well that is still flowing.

So. oil saturation.

SO. show of oil.

SO&G. show of oil and gas.

soil investigation. *see* subsea site investigation.

soil. a surface layer of weathered rock particles.

SOL. % solids.

solution gas. the dissolved natural gas that bubbles out of crude oil on the surface when the pressure drops during production.

solution gas drive. a reservoir drive in which the drop in reservoir pressure during production causes dissolved gas to bubble out of the oil and force the oil through the rock. It has a very low oil recovery efficiency. (dissolved gas or depletion drive)

solution gas-oil ratio. the scf of natural gas dissolved in one barrel of oil in the reservoir. (formation or dissolved gas-oil ratio)

sonde. a cylinder filled with instruments that is run in a well on a wireline to make a well log. It senses the electrical, radioactive and sonic properties of the rocks and their fluids and the diameter of the wellbore.

sonic amplitude log. a wireline well log that measures the attenuation of sound through rocks to detect fractures.

sonic log. a wireline well log that measures the sound velocity through rocks in a well. It can be used to calculate the porosities of the rocks. (acoustic velocity log) (SL or SONL)

sorting. a measure of the range of different size particles in a clastic rock.

sour. gas or oil with a high sulfur content.

source rock. a sedimentary rock rich in organic matter which can or has been transformed under certain geological conditions into gas and/or oil.

SP. spontaneous potential.

spacing. *see* drilling and spacing unit.

spar. a long, vertical, buoy-like vessel that is anchored above or near an offshore oil or gas field. Flowlines from subsea completion wells bring produced fluids to the spar where they are separated and treated.

spear. a fishing tool that is run into pipe (fish) on the bottom of the well. It grips the inside of the pipe as the fish is being pulled out of the well.

spec survey. a seismic survey paid for and run by a seismic contractor. Various exploration companies can pay to view the nonexclusive data.

specific gravity. the ratio between the weight of a solid or liquid and the weight of an equal volume of water.

spiking. to add condensate to crude oil to lighten the density of the oil.

spill point. the lowest elevation down to which a trap can be filled with gas and oil. If the structure is filled down to the spill point, the addition of more gas or oil will cause the oil to flow out at this point.

spinning chain. a chain used on the floor of a drilling rig to wrap around drillpipe to start screwing together or finish unscrewing the pipe.

spinning wrench. a pneumatic or hydraulic operated wrench that is suspended above the drill floor by a cable. It is used to grip and turn the drillpipe when screwing together and unscrewing the pipe.

spiral-grooved drill collar. a drill collar with three, spiraling grooves cut into the outer wall. It is used to reduce the drill collar surface area in contact with the well walls to prevent stuck pipe.

Spl. sample.

spontaneous potential. a wireline measurement of the electrical current caused by the contact of mud filtrate in the pores of a reservoir rock with the natural waters in the rock. (self potential) (SP)

spot. to place.

SPOT. one of two unmanned, remote-sensing satellites operated by France. They take pictures of the earth in visible light and infrared.

spot a well. to locate and put a well on a base map.

spotting fluid. a lubricant put in a well to loosen struck pipe.

spread. the geometric pattern of geophone groups in relation to the seismic source. They are described by names such as split spread, cross, end-on and in-line offset.

sps. sparse.

spsly. sparsely.

spud or **spudding in.** starting to drill a well.

spud date. the day a well is started.

squeeze cementing. to pump cement under pressure down a cased well to force the cement through casing perforations.

srt. sorted.

srtg. sorting.

Sst or **ss.** sandstone.

ST. sidetrack.

stab. a) stabilized, or b) to guide the end of a pipe such as casing into a coupling or tool joint to make a connection.

stabilized. steady. (stab)

stabilizer. a sub with blades running along the length of it. It is designed to keep the downhole assembly in the center of the well.

stack. the number of seismic reflections used in stacking to make a common mid-point stack. It can be expressed either as a number or a percentage with 100% equal to 1.

stacking. the combining of several different seismic reflections off the same point in the subsurface.

stage. a time-rock subdivision of rocks deposited during an age.

stage separation. the use of two or more, decreasing-pressure separators in line to treat oil and retain more of the lighter fractions in the liquid.

stake a well. to survey the exact location and elevation of a proposed well and make a map of the site.

stand. two, three, or four joints of drillpipe.

standard cubic foot. the English system unit of natural gas volume measurement under standard temperature and pressure (STP) which is defined by law. It is often a surface temperature of 60°F and a surface pressure of 14.65 psia. (scf)

standing valve. one of two valves in a downhole pump driven by a sucker-rod string. The standing valve does not move up and down.

stands. lengths of drillpipe racked in the derrick of a drilling rig. A stand can be composed of two, three, or four joints of drillpipe.

static pressure. pressure on a fluid that is not moving. (shut-in pressure)

statics. corrections applied to seismic data for elevation and the thickness and velocity of the loose sediments near the surface in the low-velocity zone.

STB. stock tank barrel.

STB/D. stock tank barrels per day.

steam flood. an enhanced oil recovery method used on heavy oil reservoirs. Very hot steam is pumped down injection wells to heat the heavy oil and make it more fluid. The steam condenses into hot water that drives the heated, heavy oil to producing wells.

steam injection. *see* cyclic steam injection.

steel jacket. the legs on an offshore, fixed production platform.

steel jacket production platform. an offshore production platform that is held in place by piles driven into the ocean bottom. They are bolted, welded or cemented to the legs (steel jacket).

steel-tooth tricone bit. a tricone drill bit in which the teeth have been machined out of the steel cones. (milled-teeth tricone bit)

steerable downhole assembly. a downhole assembly that includes a bent sub, stabilizers, a downhole turbine motor and diamond bit. It is used in the rotating mode to maintain angle and in the sliding mode to drop or build angle in deviated wells.

step out well. a well drilled to the side of a discovery well to determine the extent of the new field. (appraisal or delineation well)

stepping out. drilling to the sides of a discovery well to determine the limits of the reservoir.

Stn or **stn.** stain.

stock tank. a large, bolted or welded, steel tank that holds oil in the field. It has a thief hatch on the top for sampling and an oil sales outlet near the bottom for transferring the oil.

stock tank barrel. one stabilized barrel of oil on the surface after the gas has bubbled out. (*see* reservoir barrel) (STB)

STP. standard temperature and pressure.

straddle packers. two packers on a drillstem. They are expanded above and below a zone to be tested to isolate that zone.

straddle plant. an installation on a pipeline that removes condensate from natural gas.

straight hole. a vertical well that nowhere deviates more than 3°/100 ft and stays within a vertical cone with a 5° angle.

strapping. to measure the height and volume of oil in a specific tank to prepare a tank table.

strat test. *see* stratigraphic test well.

strata. layers of rocks. (Strat or strat)

stratigraphic column. a column showing the vertical succession of rock layers in an area.

stratigraphic cross section. a cross section made by correlating well logs that have been hung from a common marker bed or horizon in each well.

stratigraphic test well. a well drilled primarily to determine the characteristics of the subsurface rocks. (strat test)

stratigraphic trap. a petroleum trap formed during the deposition of the reservoir rock such as a limestone reef (primary stratigraphic trap) or by erosion of the reservoir rock such as an angular unconformity (secondary stratigraphic trap).

streamer. a long, plastic tube containing hydrophones and a cable connecting them. It is towed behind a boat or left on the bottom for seismic exploration at sea.

strike. the horizontal, compass direction of a plane such as a sedimentary rock layer or fault.

strike-slip fault. a break in the rocks with horizontal movement of one side with respect to the other. It can be either a right or left lateral strike slip fault.

string. a long length of tubulars such as casing, tubing or drillpipe made by screwing together joints.

strip. to remove a liquid from a gas.

stripper well. a well that is barely profitable.

Strk or strk. 1) streak or 2) streaking.

structural casing. *see* conductor casing.

structural cross section. a cross section made by correlating well logs that have been hung by modern sea level in each well.

structural map. a map that uses contours to show the elevation of the top of a subsurface rock layer. It is made from well data. (structure-contour map)

structural trap. a petroleum trap formed by the deformation of the reservoir rock such as a fold or fault.

structure-contour map. a map that uses contours to show the elevation of the top of a subsurface rock layer. It is made from well data (structured map).

stuck pipe. a drillstring stuck along the sides of a well.

stuffing box. the steel container on the wellhead of a rod pumping unit on an oil well. It contains packing that seals around the polished rod that oscillates up and down through it.

Su or su. sulfur.

sub. a short section of pipe run on the drillstring between or below the drill collars. A stabilizer is a common sub.

subcrop map. a geologic map of rock layers cropping out under an angular unconformity.

subduction zone. an area described in the seafloor spreading theory as the place two opposite-moving seafloors collide.

submarine fan. a large wedge of sediments deposited in deep water at the base of a submarine canyon.

submersible electrical pump. *see* electric submersible pump.

subsalt. sedimentary rock structures located below a layer of salt.

subsea completion or **well.** a well with the wellhead equipment such as the production tree located on the bottom of the ocean. It can be either wet or dry. It is drilled from a jackup or floater rig and is tied to a production platform, semi or fpso vessel by flowline.

subsea site investigation. a survey of the ocean bottom to determine slope, composition and load-bearing capacity for a drilling rig or platform. (soil investigation)

substructure. the steel framework used to elevate the drill floor on a drilling rig above the ground.

subsurface safety valve. a valve run in a tubing string for a well located in the ocean. The valve closes when pressure drops below a specific level.

subsurface trespass. to illegally drill a well under land without permission from the mineral rights owner.

success rate. the number of wells completed as producers divided by the number of wells drilled. It is expressed as a decimal or percent. (risk)

sucker rod. a narrow diameter, solid metal rod with threaded ends. A sucker-rod string is run in a well down the tubing to connect a rod pumping unit on the surface with a downhole pump.

sucker-rod pumping system. *see* rod pumping system.

supply company. a company that provides materials such as casing.

surface casing. the largest diameter and shortest string of casing in a well. It is used to protect freshwater aquifers and prevent the sides of the well from collapsing.

surface rights. the legal ownership of the surface of fee land. The surface right owner can build, ranch or farm on that land.

surfactant. a detergent-like chemical used in enhanced oil recovery to reduce the surface tension of oil and wash it from the rock surfaces and out of small pores.

susp. suspended.

suspended well. a well that has been producing but is shut-in. It eventually will have to be put on production again or plugged and abandoned.

Sw. water saturation. Gas or oil saturation is equal to 100% minus Sw.

SW or **S.W.** salt water.

swab. to remove liquids from a well with a swabbing tool. (swb)

swage. a tapered tool that is run on a workstring to reopen collapsed casing in a well

swath shooting. a method used to acquire data for 3-D seismic exploration on land. The receiver cables are laid out in parallel lines. The shot points are run perpendicular to the receiver lines.

swbd. swabbed.

swbg. swabbing.

SWC or **S.W.C.** sidewall core.

sweep. the frequency range that is injected into the subsurface by a vibrator truck at a shot point for seismic exploration.

sweep efficiency. the ratio of pore volume contacted by an injected fluid to the total pore volume in a reservoir during waterflood or enhanced oil recovery.

sweep length. the time during which a vibrator truck shakes the ground at a shot point for seismic exploration.

sweet. gas or oil with a low sulfur content.

sweetening. removal of acid gasses such as hydrogen sulfide and carbon dioxide from natural gas.

swivel. a device on a drilling rig which allows the drillstring to rotate while being suspended from the derrick. It is located at the top of the kelly and hangs from the hook on the traveling block.

SWU. swabbing unit.

sx. sacks.

syncline. a large, long fold of sedimentary rocks that are bent downward.

syncrude. a crude oil made from processing tar sands.

synthetic seismogram. an artificial, computer-generated seismic record made from the acoustical impedances of subsurface rock layer contacts.

system. a time-rock division of rocks deposited during a period.

t. time.

T. temperature.

t.b. thin-bedded.

T/. top of.

T/pay. top of pay.

tadpole plot. a diagram which shows the dip of subsurface rock layers in a well as determined by a dipmeter.

tail gas. the gas that exits a natural gas processing plant after the natural gas liquids have been separated. (residual gas)

tank battery. two or more stock tanks connected in line.

tank table. a table that relates the height of oil in a stock tank to the volume of the oil. (gauge table)

tar. a viscous material composed of very heavy, high-molecular weight hydrocarbons.

tar sands. a natural deposit of very heavy oil mixed with sand.

target. 1) the potential reservoir rock to which a well is drilled, or 2) the proposed bottomhole location for a deviated well.

tax royalty participation contract. a contract between a foreign government and a multinational company. The multinational company receives an exclusive concession and bears the entire cost of exploration, drilling and production. The host government is paid bonuses, taxes and royalties from production. (concession agreement)

tbg. tubing.

Tcf. trillion cubic feet.

TD. total depth.

temperature log. a production log that records fluid temperatures at various levels in a well. (TL)

tendon. one of many long, steel tubes about 2 ft in diameter that connects a tension leg platform to the anchor weights on the bottom of the ocean.

tension leg platform. a floating, wellhead and production platform held in place by large weights on the bottom of the ocean. (TLP)

tension leg well platform. a floating, wellhead platform. It is similar to a tension leg platform except production is sent by submarine pipeline to a processing platform in shallow water for treating. (TLWP)

ter. terrigenous.

Tertiary. a period of geological time from 65 to 1.8 million years ago. It is part of the Cenozoic Era.

tertiary recovery. a process used after the primary drive mechanism has been depleted and secondary recovery has been completed on an oil reservoir. Either a) chemicals or steam is injected into a reservoir or b) the subsurface oil is set afire.

Tex or **tex.** texture.

tgh. tough.

TH. tight hole.

thermogenic gas. natural gas formed by subsurface heat on organic matter or by the thermal breakdown of oil. It can be either dry or wet gas.

thickening time. the time a cement slurry remains fluid enough to be pumped. (pumpability time)

thief. 1) a brass or glass container that is used to obtain an oil sample from a stock tank or 2) to obtain an oil sample.

thief hatch. the hatch on the roof of a stock tank. It is used to gain access to the tank to measure the height of the oil and obtain a sample.

thief zone. 1) a highly permeable zone in reservoir rock through which water-flood or enhanced oil recovery fluids flow, bypassing oil in other parts of the reservoir or 2) a very permeable rock layer in a well that takes large amounts of drilling mud during drilling. (lost circulation zone)

thin section. a paper-thin slice of rock mounted on a glass slide.

thk. thick.

thn. thin.

thread protector. a plastic or metal cap screwed to the ends of tubulars such casing and drillpipe to prevent damage to the threads.

3-D seismic method. a petroleum exploration method that shows the seismic reflectors in three dimensions. It is usually displayed on a computer monitor. The record can be rotated and slices (time or horizontal slices) taken out at various levels.

throw. the vertical displacement on a fault.

thru. throughout.

thrust fault. a reverse fault with a dip of less than 45° from horizontal. The hanging wall has been thrust over the footwall.

ti. tight.

tie back. to connect something such as a subsea well by flowline to a production platform.

tie in. 1) to run seismic lines together or 2) to run a seismic line through a well.

tight. 1) an emulsion that resists separation or 2) an impermeable rock.

tight hole. a well being drilled in which the results are being kept secret. (TH)

tight sands. 1) a sandstone with little or no permeability or 2) a general term for a reservoir rock of any composition with very low permeability.

tightness. the degree to which an emulsion resists separation.

TIH. trip in hole.

time interval map. a map that uses contours to show the span in time (milliseconds) between two seismic horizons. (isotime or isochron map)

time slice. a flat, horizontal section made at a specific depth in time from 3-D seismic data. It shows where each seismic reflector intersects the slice. (horizontal slice)

time-lapse seismic. The seismic differences between several 3-D seismic surveys run at different times over the same reservoir during production from that oil field. Changes in seismic responses from the reservoir such as amplitude can show the flow of fluids through the reservoir. (4-D seismic)

time-structure map. a map that uses contours to show the depth in time (milliseconds) to a seismic horizon.

time-to-depth conversion. a seismic process in which the vertical scale on a seismic record is converted from time in milliseconds to depth in feet or meters.

title opinion. a legal history of mineral rights ownerships on a parcel of land.

tn. tan.

to the right. clockwise.

TOC. top of cement.

TOF. top of fish.

TOH. trip out of hole.

TOL. top of liner.

tongs. a wrench-like device that is suspended above the drill floor by a cable. It is used to grip and hold the drillpipe as it is being screwed together and unscrewed by the spinning wrench.

tool face. the direction the drill bit is facing.

tool joint. a short, steel cylinder with female-threads. It is used to connect joints of pipe. (collar or coupling)

tool pusher. a drilling company employee at the drill site who is in charge of the drilling crews and the rig.

top drive. 1) a hydraulic or electrical motor that drives the drillstring from the swivel. It replaces the rotary table and kelly bushing or 2) a drilling rig with a top drive motor.

top set completion. *see* open-hole completion.

topographic map. a map that uses contours to show the elevation of the surface of the ground.

total acid number. a measure of the acidity and corrosiveness of a crude oil. It is reported in units of mg KOH/g. Higher numbers are more corrosive.

total depth. the depth of a well measured along the wellbore. (measured, logged or driller's depth) (TD)

tour. a crew shift on a drilling rig. There are usually three 8-hour tours on a land rig and two 12-hour tours on an offshore rig.

township. a surveyed square of land 6 miles on a side. Townships are divided into 36 sections.

Tp. top.

tr. trace.

trace. the response of the seismic detector due to seismic energy such as a reflection. It is recorded as a vertical line with peaks and troughs to the right and left sides that represent recorded seismic energy. (wiggle trace)

trace fossil. indirect remains of a plant or animal in a sedimentary rock. Tracks, burrows, root casts and trails are examples.

tracer log. a log that uses a radioactive tracer and detector to measure fluid flow characteristics in a well.

transgression. the advance of seas onto the land.

transportable gas. natural gas that has had minimal field processing so that it can be transported to a final processing plant.

trap. a high area on the reservoir rock such as a dome or reef that is overlain by caprock. Oil and gas can accumulate in a trap. Structural, stratigraphic and combination are types of traps.

traveling block. a steel frame with steel wheels on a horizontal shaft. It is suspended in the derrick or mast by the hoisting line.

traveling valve. one of two valves in a downhole pump driven by a sucker-rod string. The traveling valve moves up and down with the sucker-rod string.

treater. a vessel used to separate an emulsion. A heater-treater uses heat and an electrostatic treater uses high-voltage electric grids. Chemicals, called emulsion breakers, can also be used.

trend. the area along which a petroleum play occurs. (fairway)

Triassic. a period of geological time from 248 to 206 million years ago. It is part of the Mesozoic Era.

tribble. three joints of drillpipe.

tricone bit. a common drill bit with three rotating cones on the bottom. Two types are milled teeth and insert.

trip. a complete cycle of pulling the drillstring out of the well (trip out) and putting it back in (trip in).

trip tank. a small tank, usually on the floor of a drilling rig, that holds drilling mud that is added to a well during tripping out.

triplex pump. a mud pump with three single-acting pistons in cylinders. The mud is pumped only on the forward stroke of the pistons.

tripping in. putting the drillstring into the hole.

tripping out. pulling the drillstring out of the hole.

trnsl. translucent.

trnsp. transparent.

trt. treat.

trtd. treated.

trtg. treating.

true vertical depth. the depth of a well measured straight down. (TVD)

truncated. the lateral termination of rocks, usually either by erosion or faulting.

tst. test.

tstd. tested.

tstg. testing.

tubing. a small diameter ($\frac{3}{4}$ to $4\frac{1}{2}$ in.), steel tubular that is used in a well to conduct the produced fluids up the well. (tbg)

tubing anchor. a device that grips the casing to secure the bottom of the tubing string.

tubing packer. a packer run on the bottom of the tubing to seal the space between the tubing and casing. (completion packer)

tubing pressure. pressure on the fluid in the tubing string. It can be either flowing or static.

tubing pump. a sucker-rod pump that is run on the tubing string.

tubing swage. a tool with a cylindrical body that tapers toward the bottom. It is run on a wireline to open collapsed tubing.

tubinghead. the forged or cast steel fitting on the top part of the wellhead. It contains the tubing hangers that suspend the tubing string in the well.

tubingless completion. an oil well completion in which the production is brought up the casing without using a tubing string.

tubular. a long metal cylinder such as a joint of drillpipe, casing, tubing or drill collar.

tungsten carbide. an extremely hard alloy (W_2C) used in granular form to hardface drilling tools.

turbidite. a layer of sedimentary rocks deposited by a turbidity current coming to rest. It can be graded with the coarsest grains such as sand on the bottom and the finest on the top.

turbidity current. a dense mixture of water and sediment flowing down a submarine slope.

turbine. a motor driven by fluid flowing through revolving vanes on a shaft.

turbine meter. a gas or liquid meter that measures the volume by the turns per unit time on the turbine shaft.

turbine motor. *see* downhole turbine motor.

turn to the right. to drill a well.

turnkey contract. a drilling contract based on a fixed fee to drill to contract depth.

TVD. true vertical depth.

twin. to drill a well adjacent to an existing well.

two-way travel time. the recorded time from the seismic source to the reflector and back to the surface detector.

u. upper.

ultimate oil recovery or **production.** the oil that can be economically recovered from a reservoir by primary production, waterflood and enhanced oil recovery.

unconformity. an ancient erosional surface. (Unconf)

unconsolidated sediments. loose sediments. (uncons)

underbalance. the condition in a well in which the pressure on the drilling mud is less than the pressure on fluids in the surrounding rocks.

undercompacted. sediments that have not compressed as much as would be expected at that depth.

underreamer. a tool run on a drillstring to enlarge the bottom of a well using cones on arms that expand. The drillstring is rotated to ream out the cavity.

undersaturated. a liquid which can dissolve more of a gas or a salt.

undershoot. to make a 3-D seismic image of the subsurface of an area without the seismic equipment ever being on that land. The geophones are positioned on one side and the source(s) on the other side of the land.

undly. underlying.

uni. uniform.

unitize. the coordination of all operators in a unit or field to increase ultimate oil production with a common pressure maintenance, waterflood or enhanced oil recovery project.

unload. to remove liquid from a well.

unsaturated pool. an oil reservoir without a free gas cap.

updip. in a direction located up the angle of a plane such as a sedimentary rock layer.

uplift. an area which has been forced upward.

upset. a thicker section on a tubular. It is used to strengthen the pipe where it has been threaded.

upstream. petroleum exploration, drilling and production. Downstream involves transportation, refining and marketing.

upthrown. the side of a dip-slip fault that moved up.

v. 1) velocity or 2) very.

V. 1) volt, 2) volume or 3) mud viscosity in API seconds.

V.P.S. very poor sample.

valve. a gate used to regulate fluid flow.

variable area wiggle trace. a method of displaying seismic data. It uses vertical lines with wiggles to the right and left. Wiggles to the right are reflections and are often colored black. (*see* variable density display)

variable bore rams. two, large blocks of metal with inserts cut into them. They are designed to close around a range of different size drillpipes and collars to close a well. Variable bore rams are used in a blowout preventer stack.

variable density display. a method of displaying seismic data using shades of gray to show reflections. The darker the color, the stronger the reflection. (*see* variable area wiggle trace)

V-door. the inverted, V-shaped opening in the front of a mast or derrick. It allows the drillpipe and casing to be brought onto the drill floor.

vector component seismic. *see* nine-component seismic.

vert. vertical.

vertical seismic profiling. a method used to measure seismic velocities of rock layers in a well. The seismic source is located on the surface next to the well. A borehole geophone is raised in the well to measure seismic velocities at various depths. It is similar to a check shot but the geophone stations are closer together. (VSP)

vgt. variegated.

vibration dampener. a sub used in a downhole assembly to reduce vibrations. (shock sub)

vibrator. a truck with hydraulic motors on the bed of the truck and a steel pad below the motors. It is used as a seismic source to shake the ground by Vibroseis.

Vibroseis. a seismic method in which the energy is put into the ground with a vibrator truck or trucks that shake the ground.

virgin pressure. the original reservoir pressure before any production. (initial or original pressure)

vis. visible.

viscometer. a laboratory instrument used to measure the viscosity and gel strength of drilling mud.

viscosity. the resistance of a fluid to flow. The units of dynamic viscosity are centipoises (cp). Kinematic viscosity is viscosity divided by fluid density and is measured in centistokes (cs). (μ)

visualization center. a room used to project and display 3-D seismic data. There are several other names such as visionarium, decisionarium and visualization station for these rooms.

vitrenite reflectance. a method that uses a microscope to measure the amount of light reflected off a type of organic matter (vitrenite) in shale to determine if the shale has generated oil or gas.

vol %. volume percent.

volat. volatile.

volatile oil. an oil that contains a relatively high percentage of intermediate and short hydrocarbon molecules.

Volc or volc. volcanic rock.

volumetric drive. a gas field reservoir drive in which the expanding gas produces the energy to force the gas through the rocks. (gas-expansion drive)

VSP. vertical seismic profiling.

vug. a roughly-spherical solution pore in limestone. (Vug or vug)

W. 1) watt or 2) mud weight in ppg.

w/. with.

w/o. without.

wackestone. a type of limestone with significant amounts of large, sand-sized particles supported in fine-grained material. (Wkst)

WAG. water-alternating-gas.

wait-and-weigh method. a method used to control a well with a kick. After the blowout preventers have been thrown, kill mud is prepared and the kick-diluted mud is replaced by kill mud during one circulation.

waiting on cement. the time that operations on a well are shut down as the cement sets behind a casing string. (WOC)

walking beam. the steel beam that pivots up and down on the Samson post of an oil well beam pumping unit.

wall scratcher or **scraper.** a ring of metal wires that is attached to the outside of a casing string. It is rotated or reciprocated to scratch the mud cake off the well walls.

wash job. acidizing a well to remedy skin damage on the wellbore.

wash tank. a settling tank that uses gravity to separate a loose emulsion. (gun barrel separator)

washout. 1) excessive erosion and enlargement of the wellbore by drilling mud or 2) damage on a drillstring caused by fluids flowing through the walls of a tubular.

washover pipe or **washpipe.** a fishing tool that consists of a section of casing. It is run onto the fish. Drilling mud is pumped out the bottom of the washover pipe to clear debris from around the fish. Another tool is then used to retrieve the fish.

water back. to dilute drilling mud with water.

water cut. 1) diluted with water or 2) the percentage of water than an oil well produces.

water drive. a reservoir drive in which the expansion of water beneath or besides an oil or gas reservoir forces the oil or gas through the rocks. It has a high oil recovery efficiency but moderate gas recovery efficiency.

water encroachment. water flowing into the oil-producing part of a reservoir.

water hauler. a service company that uses a tank truck to pick up and take oilfield brine to a disposal well.

water table. the subsurface level below which the pores in the soil or rock are filled with water.

water washing. a process in which water flowing by crude oil removes the lighter fractions by solution.

water wet. a reservoir rock in which oil or gas occupies the center of the pores and water coats the rock surfaces.

water-alternating-gas. an enhanced oil recovery method in which slugs of inert gas and water are alternately injected into a depleted oil reservoir. The water helps prevent fingering of the gas and early breakthrough. (WAG)

watercutmeter. an instrument used to measure the water content of fluids in a well.

waterflood. a method used to produce more oil from a depleted reservoir. Water is pumped down injection wells into the reservoir in order to force oil through the reservoir to producing wells.

water-in-oil emulsion. droplets of water suspended in oil. It is the most common emulsion produced from an oil well.

waxy crude. crude oil with a high wax (paraffin) content. It has a high pour point.

weathering. the physical and chemical breakdown of rock.

weathering layer. *see* low-velocity zone.

wedge out. *see* pinch out.

weed and seed person. *see* palynologist.

weighting material. a drilling mud additive such as barite used to increase the density of the mud.

well cuttings. rock flakes made by the drill bit.

well intervention. a general term for work on a producing well. It includes repairing, replacing, and installing equipment and well stimulation.

well log. a continuous record of rock properties measured in a well. Some types are sample, mud and wireline.

well shooting. to explode nitroglycerin in a torpedo at reservoir depth in a well to fracture the reservoir and stimulate production. (explosive fracturing)

well site geologist or **well sitter.** a geologist at the drill site who is responsible for sampling and testing.

well spotting. to locate wells on a base map.

well stimulation. an engineering method used to increase the permeability of a reservoir around the wellbore to increase production. It includes acidizing and hydraulic fracturing.

wellbore. the hole made by a drilling rig.

wellhead. the forged or cast steel fitting on the top of a well. It consists of casingheads located on the bottom and a tubinghead on the top. It is bolted or welded to the top of the surface casing.

wellhead equipment. equipment attached to the top of the tubing and casing strings in a well. It includes the casingheads, tubinghead, Christmas tree, stuffing box and pressure gauges.

well-log library. a place where copies of well logs from a region are on file.

well-sorted. a rock or sediments composed of clastic grains that are all about the same size.

West Texas Intermediate. the benchmark crude oil for the United States. It has 38–40 °API and 0.3% S. (WTI)

wet. 1) a reservoir that produces only water or 2) a well that encountered no commercial hydrocarbons.

wet combustion. a fireflood in which air and water are alternately injected into the reservoir. The steam generated from the water helps drive the oil to producing wells. (combination of forward combustion and water-flooding)

wet gas. a hydrocarbon gas under both initial reservoir conditions and during production as the pressure decreases in the reservoir. A liquid condensate separates from the gas after production under surface conditions but not in the reservoir.

wetting fluid. the fluid such as oil or water that coats the rock surfaces of pores in a reservoir. A water wet oil reservoir has water located on the outside of the pores and oil in the center.

wh. white.

whip. whipstock.

whipstock. a long, metal wedge used in a well to bend the drillstring and kick off a deviated well. (whip or WS)

white oil. *see* condensate.

WI. 1) working interest or 2) water injection.

wiggle trace. *see* trace.

wild well. a well that is blowing out of control.

wildcat well. a well drilled to find new gas or oil reserves. It can be drilled to 1) test a trap that has never produced (new-field wildcat), 2) test a reservoir that has never produced in a field and is shallower or deeper than the producing reservoir(s) or 3) extend the known limits of a producing reservoir in a field. (controlled exploratory well)

wind gas. nitrogen.

window. a hole cut in casing to kick off a deviated well.

wing. the fittings on the side of a production tree that directs the produced fluids to a flowline. It can be either a single or double wing tree.

wiper plug. a cylinder of aluminum and rubber used during a cement job. It is pumped down the casing to remove the cement slurry from the casing and to separate fluids in the casing. A top and bottom plug is used.

wireline. wire rope, usually between $\frac{5}{16}$ to $\frac{7}{16}$ inches in diameter, made of numerous twisted-strands of steel wires. It is used to raise and lower tools in a well. Less commonly used is a slick line.

wireline log. a record of rock properties and their fluids that are measured by an instrument (sonde) raised up the well on a wireline. The diameter of the wellbore can also be measured. The properties are recorded as curves on a long strip of paper called a well log. Examples are electrical, gamma ray, and neutron porosity.

wireline spear. a fishing tool that uses barbs to engage and remove a wireline fish from a well.

wk. weak.

wko. workover.

wkor. workover rig.

Wkst. wackestone.

Wl or **wl.** well.

WO. 1) waiting on or 2) workover.

WOB. weight on bit.

WOC. waiting on cement.

WOCR. waiting on completion rig.

WOO. waiting on orders.

WOPL. waiting on pipeline.

WOR. waiting on rig.

work over. to have a service company do work (a workover) such as pull rods or sand cleanout on a producing well. A production rig, either a workover rig or a smaller service or pulling unit is used. (wko or WO)

working interest. an ownership in a well that bears 100% of the cost of production. The working interest owners receive their share of the production revenue after the royalty owners have taken their share and after expenses have been deducted. (billing interest) (WI)

working pressure. the maximum pressure that equipment is designed to operate under. (WP)

workover. the methods used to maintain, restore or improve production from a well. (wko or WO)

workover rig. A portable rig with a mast and hoisting system used in the workover of a well. A workover rig can drill and circulate. (wkor)

workstring. a length of tubulars run in a well during a workover.

WORT. waiting on rotary tools.

WOT. waiting on tools.

WOW. waiting on weather.

WP. working pressure.

WS. whipstock.

WSRT. well sorted.

wt%. weight percent.

wthd. weathered.

WTI. West Texas Intermediate.

wtr. water.

WW. 1) wash water or 2) water well.

wxy. waxy.

X-hvy. extra heavy.

X-stg. extra strong.

XBD, X-bd, X-bdd or **x-bdd.** crossbedded.

Xln or **xln.** crystalline.

XTAL or **xtal.** crystal.

Xtree. Christmas tree.

yel. yellow.

YTD. year to date.

Z. 1) elevation or 2) compressibility or Z factor.

Z or Zn. zone.

Z factor. a number, usually between 1.2 and 0.7, that compensates for natural gas not being an ideal gas under high pressure. It is used in gas reserve equations and can is found in tables of gas composition, temperature and pressure. (compressibility factor) (Z)

zeolyte. a naturally occurring aluminosilic mineral that can also be manufactured. It is used in treating crude oil and natural gas.

zone. 1) a rock layer identified by a characteristic microfossil species such as the Siphonia davisi zone (Z or Zn) or 2) *see* pay zone.

figure references

All line drawings in this book are original. Many have been modified, especially simplified, from the sources listed below:

Arabian American Oil Company staff, 1959 Ghawar Oil Field, Saudi Arabia, *AAPG Bull.* v. 43, 434–454.

Baker, R., 1979, *A Primer of Oil-Well Drilling*, Petroleum Extension Service, University of Texas at Austin, 94 p.

Baker, R., 1983, *The Production Story*, Petroleum Extension Service, University of Texas at Austin, 81p.

Barlow, J.A.Jr., and J.D. Haun, 1970, Salt Creek Field, Wyoming, in M.T. Halbouty, ed., *Geology of Giant Petroleum Fields*, AAPG Memoir 14, 147-157.

Baugh, J.E., 1951, Leduc D-3 zone pool, *World Oil*, v. 132, p. 210–214.

Beebe, B.W., 1961, Drilling the exploratory well, in G.B. Moody, ed., *Petroleum Exploration Handbook*, McGraw-Hill Book Co., New York, 17-1-17-36.

Berg, R.R., 1968, Point-bar origin of Fall River Sandstone reservoirs, northeastern Wyoming, *AAPG Bull.*, v. 52, 2116–2122.

Beydown, Z.R., 1991, Arabian Plate Hydrocarbons Geology and Potential – A Plate Tectonics Approach, *AAPG Studies in Geology*, v. 33, 77 p.

Bishop, W.F., 1968, Petrology of the Upper Smackover Limestone in northern Haynesville Field, Clairborne Parish, Louisiana, *AAPG Bull.*, v. 52, 92–128.

Bruce, L.G. and G.W. Schmidt, 1994, Hydrocarbon fingerprinting for application in forensic geology; review with case histories, *AAPG Bull.*, v. 78, 1692–1710

Burke, D.C.B., 1972, Longshore drift, submarine canyons and submarine fans, *AAPG Bull.*, v. 56, 1975–1983.

Busch, D.A., 1974, Stratigraphic traps in sandstones–exploration techniques, *AAPG Memoir 21*, 174 p.

Charles, H.H., 1941, Bush City oil field, Anderson County, Kansas, in A.I. Levorsen ed., *Stratigraphic Type Oil Fields*, A Symposium, AAPG, Tulsa, 43–56.

Collingwood, D.M. and R.E. Retter, 1926, Lytton Springs Oil Field, Texas, *AAPG Bull.*, v. 10, 953–975.

Colter, V.S. and D.J. Harvard, 1981, *Petroleum Geology of the Continental Shelf of North-West Europe*, Institute of Petroleum, London 521 p.

Craig, D.H.,1988, Caves and other features of Permian karst in San Andres Dolomite, Yates field reservoir, west Texas, in N.P. James and P.W. Choquette eds., *Paleokarst*, Springer Verlag, New York, 342–365.

Exploration Logging Inc., 1979, *Field Geologist's Training Guide: An Introduction to Oil Field Geology, Mud Logging and Formation Evaluation*, Sacramento

Frey, M.G. and W.H. Grimes, 1970, Bay Marchand-Timbalier Bay-Caillou Island salt complex, Louisiana, in M.T. Halbouty, ed., *Geology of Giant Petroleum Fields*, AAPG Memoir 14, 277-291.

Galloway, W.E., T.E. Ewing, C.M. Garrett, N. Tyler and Bebout, 1983, Austin/Buda fractured chalk, *Atlas of Major Texas Oil Reserves*, Bureau of Econ. Geol., Austin, 41–42.

Gallup, W.B., 1982, A brief history of the Turner Valley oil and gas field, *Canadian Soc. of Petrol. Geol. Guidebook*, N. 8, 81–86.

Gatewood, L.E., 1970, Oklahoma City Field-anatomy of a giant, in M.T. Halbouty ed., *Geology of Giant Petroleum Fields*, AAPG Memoir 14, 223-254.

Gearhart-Owen, *Chart Book*, Fort Worth

Gerding, M., 1986, *Fundamentals of Petroleum*, Petroleum Extension Service, University of Texas at Austin, 452 p.

Gill, D., 1985, Depositional facies of middle Silurian (Niagaran) pinnacle reefs, Belle River Mills Gas Field, Michigan basin, southeastern Michigan, in P.D. Roehl and P.W. Choquette eds., *Carbonate Petroleum Reservoirs*, Springer Verlag, New York, 121–140.

Halbouty, M.T., 1979, *Salt Domes, Gulf Region, United States and Mexico*, Gulf Publishing Co., Houston, 61 p.

Halbouty, M.T., 1991, East Texas Field-USA., *in* E.A. Beaumont and N.H. Foster, eds., *Stratigraphic Traps II, Treatise of Petroleum Geology Atlas*, AAPG, Tulsa, 189-206.

Harris, P.M. and S.D. Walker, McElroy Field-USA, in E.A. Beaumont and N.H. Foster eds., *Stratigraphic Traps I, Treatise of Petroleum Geology Atlas of Oil and Gas Fields*, AAPG, Tulsa

Helander, D.P., 1983, *Fundamentals of Formation Evaluation*, Oil & Gas Consultants International Inc., Tulsa, 332 p.

Heritier, F.E., P. Lossel and E. Wathne, 1980, Frigg Field: large submarine-fan trap in lower Eocene rocks of the Viking graben, North Sea, in M.T. Halbouty, ed., *Giant Oil and Gas Petroleum Fields of the Decade 1968–1978*, AAPG Memoir 30 59-80.

Hull, C.E. and H.R. Warman, 1970, Asmari Oil Field of Iran, *in* M.T. Halbouty ed., *Geology of Giant Petroleum Fields*, AAPG Memoir 14 428-437.

Hurley, N.F. and R. Budros, 1990, Albion-Scipio and Stoney Point fields-USA, in E.A.Beaumont and N.H. Foster, eds., *Stratigraphic Traps I, Treatise of Petroleum Geology Atlas of Oil and Gas Fields*, AAPG, Tulsa 1-38.

Hyne, N.J., 1991, *Dictionary of Petroleum Exploration, Drilling and Production*, PennWell Books, Tulsa, 624 p.

Hyne, N.J., 1995, Sequence stratigraphy: a new look at old rocks, in N.J. ed., *Sequence Stratigraphy of the Mid-Continent*, Tulsa Geol. Soc. Sp. Pub. N. 4, 5–17.

Jackson, M.L., 1991, Port Acres Field-USA, *in* E.A. Beaumont and N.H. Foster eds., *Stratigraphic Traps II, Treatise of Petroleum Geology Atlas of Oil and Gas Fields*, AAPG, Tulsa, 329-348.

Jamison, H.C., L.D. Brockett and R.A. McIntosh, 1980, Prudhoe Bay at 10 year perspective, in M.T. Halbouty, ed., *Giant Oil and Gas Fields of the Decade 1968–1978*, AAPG Memoir 30 289-314.

Jardine, D., 1974, Cretaceous oil sands of western Canada, Oil Sands: Fuel of the Future, *Canadian Soc. Petrol. Geol. Memoir 3*, 50 –67.

Jardine, D., D.P. Andrews, J.W. Wishart and J.W. Young, 1977, Distribution and continuity of carbonate reservoirs, *Jour. Petroleum Technology*, 873–885.

Kirk, R.H., 1980, Statfjord Field: a North Sea giant, in M.T. Halbouty, ed., *Giant Oil and Gas Fields of the Decade 1968–1979*, AAPG Memoir 30 95-116.

Kolb, C.R. and J.R. Van Lopik Jr., 1966, Depositional environments of the Mississippi River Delta plain-southeastern Louisiana, *in* M.L. Shirley, ed., *Deltas in their Geological Framework*, Houston Geological Society, Houston, 17-62.

Lamb, C.F., 1980, Painter Reservoir Field, *in* M.T. Halbouty, ed., *Giant Oil and Gas Fields of the Decade 1968–1978*, AAPG Memoir 30 281-288.

Landes, K.K., 1970, *Petroleum Geology of the United States*, John Wiley and Sons, New York, 571 p.

LeMay, W.J., 1972, Empire Abo Field, southeast New Mexico, in R.E. King, ed., *Stratigraphic Oil and Gas Fields-Classification of Exploration Methods and Case Histories*, AAPG Memoir 16, 472-480.

Levorsen, A.I., 1967, *Geology of Petroleum*, W.H. Freeman and Company, New York, 724 p.

Lyday, J.R., 1990, Berlin Field-USA, in E.A. Beaumont and N.H. Foster, eds., *Stratigraphic Traps I, Treatise of Petroleum Geology Atlas of Oil and Gas Fields*, AAPG, Tulsa, 39-68.

MacKay, A.H. and A.J. Tankard, 1990, Hibernia Oil Field-Canada, in E.A. Beaumont and N.H. Foster, eds., *Structural Traps III, Treatise of Petroleum Geology Atlas of Oil and Gas Fields*, AAPG, Tulsa, 145-176.

Martinez, A.R., 1970, Giant fields of Venezuela, in M.T. Halbouty ed., *Geology of Giant Petroleum Fields,* AAPG Memoir 14, 326-336.

Mayuga, M.N., 1970, Geology and development of California's giant Wilmington Field,in M.T. Halbouty, ed., *Geology of Giant Petroleum Fields*, AAPG, Tulsa, 158-184.

McGreger, A.A. and C.A. Biggs, 1970, Bell Creek Oil Field, Montana, a rich stratigraphic trap, *in* M.T. Halbouty, ed., *Geology of Giant Petroleum Fields*, AAPG Memoir 14, 128-146.

McGreger, A.A. and C.A. Biggs, Bell Creek Oil Field, Montana, *in* R.E. King ed., *Stratigraphic Oil and Gas Fields*, SEG Spec. Pub. n. 10, 367-375.

Mero, W.E., 1991, Point Arguello Field-USA, *in* E.A. Beaumont and N.H. Foster eds., *Structural Traps V, Treatise of Petroleum Geology Atlas of Oil and Gas Fields*, AAPG, Tulsa, 27-58.

Mills, H.G., 1970, Geology of Tom O'Conner Field, Refugio Co., Texas, in M.T. Halbouty ed., *Geology of Giant Petroleum Fields*, AAPG, Tulsa, 292-300.

Morgridge, D.L. and W.B. Smith Jr., 1972, Geology and discovery of Prudhoe Bay Field, eastern Arctic Slope, Alaska, in R.E. King, ed., *Stratigraphic Oil and Gas Fields*, SEG Spec. Pub. n. 10, 489-501.

Murphy, J.A., 1952, Good field data are necessary for good reservoir engineering, *The Petroleum Engineer*, v. 24, B-91 – B-99.

North, F.K., 1980, *Petroleum Geology*, Allen & Unwin, Boston, 607 p.

Ottman, R.D., P.L. Keyes and M.A. Ziegler, 1976, Jay Field, Florida-a Jurassic stratigraphic trap, *in* J. Braunstein ed., *North American Oil and Gas Fields*, AAPG Memoir 24, 276-286.

Pippen, L., 1970, Panhandle-Hugoton Field, Texas-Oklahoma-Kansas-the first fifty years, in M.T. Halbouty ed., *Geology of Giant Petroleum Fields*, AAPG Memoir 14, 204-222.

Proctor, R.M., G.C. Taylor and J.A. Wade, 1983, Oil and natural gas resources of Canada, *Canadian Geological Survey Paper*, 81–83.

Schlumberger, 1987, *Log Interpretation Principles/Applications*, Schlumberger Educational Services, Houston, 198 p.

Schumard, C.B., 1991, Stockholm Southwest Field-USA, in E.A. Beaumont and N.H. Foster eds., *Stratigraphic Traps II, Treatise of Petroleum Geology Atlas of Oil and Gas Fields*, AAPG, Tulsa, 269-304.

Sieverding, J.L. and F. Royse Jr., 1990, Whitney Canyon-Carter Creek fields-USA, in E.A. Beaumont and N.H. Foster eds., *Structural Traps III, Treatise of Petroleum Geology Atlas of Oil and Gas Fields*, AAPG, Tulsa 1-30.

Sills, S.R., 1992, Drive mechanisms and recovery, in D. Morton-Thompson and A.M. Woods eds., *Development Geology Reference Manual*, AAPG Methods in Exploration Series, n. 10, 518-522.

Stafford, P.T., 1957, Scurry Field, Scurry, Kent and Borden counties, Texas, in Herald ed., *Occurrence of Oil and Gas in West Texas*, 295–302.

Stafford, P.T., 1959, Geology of part of the Horseshoe atoll in Scurry and Kent counties, Texas, *U.S. Geol. Survey Bull.* 315-A.

Stanley, T.B. Jr., 1970, Vicksburg fault zone, in M.T. Halbouty ed., *Geology of Giant Petroleum Fields*, AAPG Memoir 14, 301-308.

State of Ohio, 1966, *Oil & Gas Fields Map of Ohio*, Department of Natural Resources, Division of Geological Survey.

Viniegra, F. and C. Castillo-Tejero, 1970, Golden Lane fields, Veracrus, Mexico, *in* M.T. Halbouty ed., *Geology of Giant Petroleum Fields*, AAPG Memoir 14, 307-325.

unit conversions

English to metric

one inch = 2.54 centimeters
one foot = 0.3048 meters
one mile = 1.609 kilometers
one pound = 0.4536 kilograms
one gallon(U.S.) = 3.7854 liters
°F - 32 = 5/9°C

metric to English

one centimeter = 0.394 inches
one meter = 3.2808 feet
one kilometer = 0.6214 miles
one kilogram = 2.2046 pounds
one liter = 0.2642 gallons (U.S.)
9/5°C + 32 = °F

oil and gas records

First oil wells

1858 – Oilsprings, Ontario, Canada. The well was dug to 60 ft by James M. Williams.

1859 – Titusville, Pennsylvania, USA. The Drake well was drilled to 69.5 ft by William "Uncle Billy" Smith.

First dry hole

1859 – Grandin well, Warren County, Pennsylvania. 134 ft TD

First recorded lease

July 4, 1853 – Cherrytree township, Venango Co., Pennsylvania with a 5-year term

First use of a rotary drilling rig for oil

Between 1895 and 1900 by M.C. and C.E. Baker in Corsicana, Texas

First casing

1808 – Great Buffalo Lick, West Virginia using wood strips in a saltwater well

First use of drilling mud

1900 – Spindletop, Texas

First wireline well log
September 5, 1927 in the Pechelbronn oil field, France by Marcel and
Conrad Schlumberger

First well stimulation
1865 – Ladies well, Titusville, Pennsylvania using a tin torpedo with a gun-
powder primer and a fuse

First hydraulic frac job
1947 – Hugoton field, Oklahoma

First offshore well
1897 – Summerland, California on a pier

First offshore well drilled out of sight of land
1947 – Gulf of Mexico southeast of Morgan City, Louisiana in 18 ft of water
by Kerr-McGee Co.

First subsea well
1943 – Lake Erie, United States

First refraction seismic discovery
1924 – Orchard salt dome, Texas

First reflection seismic discovery
1928 – Maud pool (Seminole field), Oklahoma

Deepest well drilled for gas or oil
31,441 ft. – Bertha Rogers No. 1. Washita Co., Oklahoma in 1974 – it was
a dry hole

Most oil from a single well
57,000,000 bbls from Cerro Azul No. 4, Mexico in the early 1900s

Most oil from a well in a single day
260,858 bbls from Cerro Azul No. 4, Mexico on February 19, 1916

Largest conventional oil field
Ghawar, Saudi Arabia, 82 billion bbls of recoverable oil

Largest gas field
Urengoy, Siberia, 210 trillion cu ft of recoverable natural gas

First reported waterflood
1880 – Pithole City, Pennsylvania

Deepest producing gas well in United States
26,536 ft (TVD) in Wheeler Co., Texas

Deepest producing oil well in United States
18,876 ft (TVD) – on land, in Kern Co., California
22,936 ft (TVD) – offshore, Gulf of Mexico (includes water depth)

index

H

M

Stock tank barrel of oil, 432, 434

Stockholm Southwest oil field, 82

Storage and measurement (petroleum), 370–374

Storm choke, 389

Straddle packer, 330–331

Straddle plant, 370

Straight hole, 285–287:
  drilling, 285–286;
  straight-in well, 286–287

Straight hole drilling, 285–286

Straight-in well, 286–287

Stranded gas, 430

Stratification, 20, 22

Stratigraphic column, 123–124

Stratigraphic cross section, 203

Stratigraphic test well/strat test, 228

Stratigraphic trap, 73, 168, 181–187, 201:
  secondary trap, 181–183;
  angular unconformity, 181–183;
  primary trap, 183–187

Stratigraphy, 200

Streamer, 218–219

Strike (orientation), 122–123

Strike-slip fault, 62–63, 70–71:
  right lateral, 70–71;
  left lateral, 70

String shot, 278

Stripper well, 418

Stripping, 368

Structural cross section, 201–203

Structural map, 124–126, 168

Structural mast, 394

Structural trap, 49, 169–181, 186, 200:
  anticline, 169–172;
  dome, 169–172;

growth fault, 172–177;
  rollover anticline, 172–177;
  drag fold, 178–181

Structure contour map, 170–171, 223

Structure of Earth crust, 22–26

Stuck pipe, 277–279:
  log, 278

Stuck-pipe log, 278

Stuck-point indicator tool, 278

Stuffing box, 355

Sub (pipe), 258

Subduction, 138–140, 143–144:
  zone, 138–140

Subduction zone, 138–140

Submarine canyon, 130–132

Submarine fan, 131–133

Submarine mudflow, 391–392

Subsea completion/well, 390–391

Subsea manifold, 390

Subsea site, 376

Subsea work, 389

Subsidence, 107, 427–428:
  surface, 427–428

Substructure (framework), 251, 270

Subsurface conditions, 273–275

Subsurface map, 124–128, 201:
  structural map, 124–126;
  isopach map, 127–128;
  percentage map, 128

Subsurface safety valve, 347

Subsurface trespass, 289

Success rate/ratio, 273

Sucker rod coupling, 352–353

Sucker rod pump, 350–354

Sucker rod string, 352–353

Sucker-rod guide, 354

T